Principles and Advances in Human Genetics: A Primer on Genetic Disorders

Principles and Advances in Human Genetics: A Primer on Genetic Disorders

Cheryl Natividad

Delve
PUBLISHING

www.delvepublishing.com

Principles and Advances in Human Genetics: A Primer on Genetic Disorders

Cheryl Natividad

Delve Publishing
224 Shoreacres Road
Burlington, ON L7L 2H2
Canada
www.delvepublishing.com
Email: orders@arclereducation.com

ISBN: 978-1-77407-760-3 (Hardcover)

ABOUT THE AUTHOR

Cheryl Agdaca- Natividad graduated with a BS Biology degree (magna cum laude) at the Saint Louis University, Baguio City in the Philippines. She took the licensure examination for secondary teachers in 2003 where she ranked no. 10 among the examinees. She earned her masters degree in Genetics at the premier state university, University of the Philippines Los Baños. She taught in the same university from 2006-2017 handling lecture and laboratory classes in general biology, cell biology, genetics, and molecular genetics. Apart from teaching, her work involved proposing and revising laboratory manuals and mentoring students in their research work.

TABLE Of CONTENTS

LIST OF FIGURES

Figure 1.8. In a pedigree, the Roman numerals represent the generation number. Circles represent the female members of the family while squares are the males. Those who are unshaded do not show the trait while those who are shaded do. A horizontal midline connecting two individuals denotes marriage or mating between the two. A vertical line connects the parents to their children and these line branches out to connect to each child. (Source: Modified from National Institutes of Health, Public Domain)

Figure 1.9. These two pedigrees show the classic skipping of generations which usually indicates that the trait is due to an autosomal recessive gene. Specifically, affected individuals were only observed in the third (for the upper pedigree) and fourth (for the lower pedigree) generations. Thus, affected individuals have a homozygous recessive genotype. The arrow indicates that the individual is the **proband**, or the starting point for the genetic study. It is interesting to note that the parents of the affected female in the lower pedigree are first cousins. Such consanguineous mating is denoted by a double horizontal midline. Mating between relatives increases the likelihood that two parents are carriers since they come from the same family, which has a history of the trait. (Source: AlĕsPanczak, Creative Commons License)

Figure 1.10. In this pedigree of a Chinese family with members suffering from autosomal dominant late- onset nonsyndromic hearing loss, there are six affected females and six affected males which are observed in all four generations. These patterns are characteristic of an autosomal dominant trait. Gray squares and circles represent members with unavailable status. All unaffected individuals are assumed to have homozygous recessive genotypes. Affected individuals may either have homozygous dominant or heterozygous genotypes. For the I2 female, she is likely to be a heterozygote since after marrying a normal male they had three affected children but one son (individual II1) was unaffected. Given these genotypes of the parents, all the affected children (generation II) are inferred to be heterozygotes. All the three affected siblings married normal mates and so it is expected to observe affected heterozygotes (+/-) as well as normal individuals (-/-) among their offspring (generation III). (+ represents the dominant mutant allele) (Source: Zhang, C.; Wang, M.; Xiao, Y.; Zhang, F.; Zhou, Y.; Li, J.; Zheng, Q.; Bai, X.; Wang, H. Neural Plasticity 1512831, 10 pp., 2016. Creative Commons License)

Figure 1.11. Thomas Hunt Morgan studied the inheritance of traits in *Drosophila* including those located in the sex chromosomes. The pattern of inheritance of these traits deviated from that of autosomal traits. (Source: National Institutes of Health, Public Domain)

Figure 1.12. The allele for eye color in *Drosophila* is located in the X chromosome with the allele for red eye color (w+) being dominant and that for white color (w) being recessive. A red-eyed male is expected to produce X-bearing and Y-bearing gametes. All the X-bearing sperms have the allele for red eye color. A white-eyed female is homozygous recessive and so all its gametes bear the X chromosome with the recessive allele for white eye color. A cross between these two is expected to produce red-eyed females and white-eyed males. (Source: Charles Molnar and Jane Gair, Creative Commons License)

Figure 1.13. In the first cross where a true-breeding red-eyed female is crossed with a white-eyed male, the F_1 progeny is composed of heterozygote red-eyed females and red-eyed males. When a reciprocal cross is performed involving a white-eyed female and a red-eyed male, the F_1 progeny consists of heterozygote red-eyed females and white-eyed males. (Source: Todd Nickle and Isabelle Barrette-Ng, Creative Commons License)

Figure 1.14. All the individuals affected with retinal degenerative disease in this pedigree of a Chinese family are males which is one indication for an X-linked recessive inheritance. It is possible that individual I1, who is deceased, might have been affected or at least a carrier since both her sons are affected. The mothers of the affected individuals IV2 and IV5 are normal but they must be carriers of the mutant allele, which they inherited from their affected father. This was confirmed by DNA sequencing in the case of individual III2. The dotted circles denote female carriers and the diagonal shape represents an unborn infant. (Source: Guo, H.; Li, J.; Gao, F.; Li, J.; Wu, X.; Liu, Q. BMC Ophthalmology 15, 85, 2015, Creative Commons License)

Figure 1.15. A family's pedigree for Charcot-Marie-Tooth disease type 1 (CMTX1) shows more affected females rather than males which is one indication for X-linked dominant inheritance. Dot symbols are used to represent affected individuals whose phenotype was deduced with a medical history. Most of the living family members were genetically tested and the symbol "+" means the individual has the dominant allele while "–" denotes the absence of the dominant allele. Individuals with no symbol were not genetically tested. It can be inferred that all affected females are heterozygous since they only had one affected parent while the other parent is normal. This was confirmed for most of the affected females because they were able to have normal children. (Source: Luo, S.; Jin, H.; Chen, J.; Zhang, L. Frontiers in Neurology 10, 413, 2019, Creative Commons License)

Figure 1.16. A Y-linked inheritance can be deduced from a pedigree where all affected members are males and all the sons of the affected father are affected while the daughters are normal. (Source: Madibc68, Creative Commons License)

Figure 1.17. (A)Walther Flemming was a German anatomist who observed, illustrated, and correctly deduced the sequence of chromosome behavior during mitosis. (B) He illustrated how the chromosomes appeared under the microscope at various stages of mitosis. (Source: A: Public Domain; B: Walther Flemming, Public Domain)

Figure 1.18. A replicated chromosome is made up of two sister chromatids joined by a centromere (2). The ends of the chromosome arms are the telomeres (1). The upper arm of the chromosome is called the p arm (3) while the lower one is the q arm (4). Source: Magnus Manske, Creative Commons License)

Figure 1.19. Chromosomes may be classified based on the placement of their centromeres. A telocentric chromosome (I) has its centromere located at the tip of the chromosome. An acrocentric chromosome's (II) centromere is positioned near the end of the arms while that of a submetacentric chromosome (III) is positioned just off-center. Finally, a metacentric chromosome (IV) has its centromere in the middle.

(Source: Fockey003, Creative Commons License)

Figure 1.20. (A) The proteins of the inner kinetochore (purple) assemble on the chromosome centromere (gray). The outer kinetochore proteins (SPC105:light blue, NDC80: yellow, and DASH: blue) is where the microtubules bind while the proteins making up the central kinetochore (red) links the inner and outer kinetochore subcomplexes. (B) Kinetochores are involved in the attachment and polymerization of microtubules during cell division. (Source: Duro, E.; Marston, A.L. Genes and Development 29(2), 109-122, 2015, Creative Commons License)

Figure 1.21. (A) Theodor Boveri inferred that Mendel's hereditary units are located in the chromosomes. (B) Boveri made drawings illustrating the various stages of cell division that occur during the embryonic development of *Parascaris equorum*. (Source: A: Hugo Freund and Alexander Berg (ed.), Public Domain; B: Theodor Boveri, Public Domain)

Figure 1.22. The cell cycle is made up of the interphase, which prepares the cell prior to cell division, and the M or division phase. Interphase is subdivided into three stages, i.e. G_1, S, and G_2. (Source: Simon Caulton, Creative Commons License)

Figure 1.23. Mitosis distributes the four replicated chromosomes into the two daughter cells so that the daughter cells also have four chromosomes. Apart from having the same number of chromosomes, the resulting cells are genetically identical to each other and to the parental cell. (Source: JPablo Cad, Juliana Osorio, Matthias M, and M3.dahl, Creative Commons License)

Figure 1.24. From a diploid parent cell with four chromosomes, the stages of meiosis I produce haploid daughter cells (n=2) that are genetically different from each other. (Source: Ali Zifan, Creative Commons License)

Figure 1.25. At the zygotene stage of prophase I, homologous chromosomes pair together along their length with the aid of proteins in the synaptonemal complex. (Source: OpenStax Biology 2nd Edition, Creative Commons License)

Figure 1.26. Crossing over is the exchanging of segments between the chromatids in the bivalent. If this occurs between non-sister chromatids, then there is genetic recombination of maternal and paternal genes. (OpenStax Biology 2nd Edition, Creative Commons License)

Figure 1.27. There are various ways that bivalents may align themselves along the equator at metaphase I. This results in gametes with assorted combinations of paternal and maternal chromosomes. This is also the explanation underlying Mendel's law of independent assortment. (Source: OpenStax Biology 2nd Edition, Creative Commons License)

Figure 1.28. The events of meiosis II are very similar to mitosis which distributes the genetic material of the haploid parent cell (n=2) equally to the haploid daughter cells (n=2). (Source Ali Zifan, Creative Commons License)

Figure 1.29. Cytokinesis during egg cell formation is characterized by the unequal distribution of cytoplasmic content so that after meiosis I, one large oocyte and a much smaller polar body are produced. Both cells undergo meiosis II but the large oocyte exhibits unequal cytokinesis while the first polar body's cytoplasmic content is divided equally. In the end, only the single large cell can function as an egg cell while the three polar bodies are non-functional. (Source Henry Vandyke Carter and Mysid, Public Domain)

Figure 1.30. Crossing-over during meiosis can separate linked genes producing recombinant or new combination of genes in the gamete. On the other hand, if chromosome regions did not undergo crossing-over, the original or parental combination of genes is retained. (Source: Abby Provenzano, Creative Commons License)

Figure 1.31. Alfred Sturtevant made the first genetic or linkage map showing the arrangement and relative distances of six genes in the X chromosome of *Drosophila* (fruit fly). The distances indicated are equivalent to the percent recombination or crossing-over between two adjacent genes. (Source: Adapted from Sturtevant, A.H., 1913. Journal of Experimental Zoology, 14, 43-59)

Figure 1.32. Friedrich Miescher was a young Swiss physician when he first isolated DNA from the nuclei leucocytes in pus, which he called nuclein. (Source: PBS, Public Domain)

Figure 1.33. The series of experiments performed by Frederick Griffith involved injecting the nonvirulent rough (R) strain, the virulent smooth (S) strain, or a combination of either dead or living R and S strains of *Streptococcus pneumoniae* in mice. His results showed that the mouse died when a mixture of heat-killed S strain and viable R strain were injected, and viable S strain was recovered from the dead mouse. This indicates that the R strain was transformed to become a virulent S strain by a certain substance from the dead S strain. (Source: OpenStax: Microbiology, Creative Commons License)

Figure 1.34. Oswald Avery, Colin Macleod, and Maclyn McCarty provided experimental proof that the transforming substance in Frederick Griffith's experiment is DNA. They isolated the transforming principle from S strain of *Streptococcus pneumoniae* and observed that it converted R cells to become the S strain. The transforming activity was not affected by the destruction of proteins and RNA, but it stopped when DNA was digested. (Source: OpenStax: Microbiology, Creative Commons License)

Figure 1.35. Alfred Hershey and Martha Chase performed an experiment where they separately labeled the DNA and protein components of T2 bacteriophage with radioactive isotopes of phosphorus (^{32}P) and sulfur (^{35}S), respectively. Their results proved that it is mostly the viral DNA that entered the host *Escherichia coli* during infection and that this is responsible for producing new bacteriophages. (Source: OpenStax: Microbiology, Creative Commons License)

Figure 1.36. A nucleotide is formed when a nitrogenous base is joined to the pentose via the 1' carbon and a phosphate group is linked to the 5' carbon. The pentose in the nucleotide of DNA s deoxyribose while in RNA, it is ribose. The nitrogenous bases are of two types, pyrimidines and purines. Pyrimidine bases include cytosine, thymine (found only in DNA), and uracil (found only in RNA). Purine bases are adenine and guanine. (Source: OpenStax Biology 2e, Creative Commons License)

Figure 1.37. DNA is made up of two polynucleotide strands with the phosphate-deoxyribose outer backbone forming two antiparallel strands. The nitrogenous bases exhibit specific pairing (i.e., adenine is always paired with thymine while guanine pairs with cytosine) in the inner portion and are linked by hydrogen bonds. (Source: Madeleine Price Ball, Public Domain)

Figure 1.38. (A) Rosalind Franklin was a British chemist and X-ray crystallographer who took X-ray diffraction photographs of DNA in an effort to determine its structure. (B) Franklins' famous photo 51 clearly showed that the DNA is a helix. (Source: Wellcome Library, Creative Commons License)

Figure 1.39. The DNA model proposed by James Watson and Francis Crick shows a double-stranded helix that winds clockwise and forms a minor and major groove per complete turn. The sugar and phosphate components of the nucleotide form the outer backbones while the nitrogenous bases are stacked inside the molecule, exhibiting specific base pairing. (Source: Zephyris, Creative Commons License)

Figure 1.40. The linear eukaryotic chromosome is made up of the loosely packed euchromatin, and thus easily accessed by the DNA binding proteins for gene expression, and the condensed heterochromatin, and thus less transcriptionally active. (Source: Modified from Wickramasekara, R.N.; Stessman, H.A.F. Biology, 8(1), 11, 2019; Creative Commons License)

Figure 1.41. The DNA is condensed several thousand times in order to fit inside the microscopic nucleus. The DNA double helix is coiled into nucleosomes to form the 10-nm chromatin primary structure. This is further coiled into the secondary structure, the 30-nm chromatin, and then finally the tertiary structure of the chromosome. (Source: Iyer, B.V.S.; Kenward, M.; Arya, G. BMC Biophysics 4:8, 2011, Creative Commons License)

Figure 1.42. There are three proposed modes of DNA replication. In the conservative model, the parental strands remain together after replication and two new strands make up the daughter DNA molecule. In the semiconservative model, the parental strands separate from each other and serve as templates for the synthesis of new strands so that after replication, there are two DNA molecules made up of an old and new strand joined together. In the dispersive model, the products of replication have regions of double-stranded parental DNA and regions of double-stranded daughter DNA. (Source: OpenStax: Microbiology, Creative Commons License)

Figure 1.43. Human chromosomes, or eukaryotic chromosomes in general, are made up of linear and very long DNA so that there is a need for multiple origins of replication and replication bubbles in order for DNA synthesis to be completed quickly. (Source: OpenStax: Microbiology, Creative Commons License)

Figure 1.44. At the replication fork, the helicase enzyme unwinds the double-stranded parental DNA to generate single -stranded templates for replication. The supercoils produced by this localized unwinding of the DNA are untangled by topoisomerase. Single- stranded binding proteins associate with the template strands to stabilize them. A primase-DNA polymerase α complex synthesizes a short RNA primer followed by a short DNA segment. This primer is elongated by the DNA polymerase δ. The leading daughter strand is extended continuously. On the other hand, the lagging strand has several primers that are extended by DNA pol δ producing Okazaki fragments. (Source: Frank Boumphrey, Creative Commons License)

Figure 1.45. The length of telomeres of eukaryotic chromosomes is maintained by the activity of the telomerase enzyme, which uses its built-in RNA as a template in order to extend the DNA sequence of the telomere in one strand. It then repeats until the template of the lagging strand has been lengthened considerably. The newly synthesized segment can then be used as a template by primase to make a new primer in the lagging strand. The primer is then extended by DNA polymerase so that the full telomere sequence is replicated. (Source: OpenStax: Microbiology, Creative Commons License)

Figure 1.46. The main enzyme of the transcription process is RNA polymerase which opens the double- stranded DNA and copies the gene sequence from the DNA template to produce RNA in a 5'→3' direction. (Source: Genomics Education Programme, Creative Commons License)

Figure 1.47. The preinitiation complex for transcription is formed by the sequential assembly of transcription factors (TFs) and RNA polymerase II at the TATA box. The first to bind the TATA promoter is TFIID via its TATA-binding protein (TBP) subunit. TFIID is assisted by TFIIA. The next to join the assembly is TFIIB. This is followed by the binding of the preformed complex of TFIIF and RNA polymerase II. Then TFIIE and TFIIH join, respectively, completing the complex. (Source: OpenStax Biology 2nd Edition, Creative Commons License)

Figure 1.48. Promoters and enhancers are both cis-acting elements that are distant from each other in the eukaryotic DNA. RNA polymerase and transcription factors bind the promoter in order to initiate transcription. On the other hand, activators bind the enhancer and interact with the transcription factors in order to increase the efficiency of transcription initiation. DNA bending protein facilitates looping of the DNA to enable the physical interaction of proteins in the promoter and enhancer regions. (From: OpenStax Biology 2nd Edition, Creative Commons License)

Figure 1.49. Eukaryotic transcription starts with the formation of the preinitiation complex, which includes various types of transcription factor II (TFII proteins) and RNA polymerase II. This complex is bound to the core promoter region, which consists of binding sites of the TFII proteins and RNA polymerase II as well as the transcription start site. In addition, distant cis-regulatory sequences modulate the transcription process. Enhancers increase transcription rate with the aid of a Mediator complex that transmits the signal from the activator bound to the enhancer to the preinitiation complex(green arrow) while silencers decrease or stop transcription (red T bar). Barrier-form insulators demarcate the boundary of active regulatory regions and enhancer-blocking (EB) insulators block regulatory sequences from influencing adjacent genes. (Source: Cowie, P., Ross, R., Mackenzie, A. Biology 2(1), 64-84, 2013, Creative Commons License)

Figure 1.50. TFIIS is an elongation factor for eukaryotic transcription. It recognizes backtracked or paused RNA polymerase II then stimulates cleavage of the transcript.

This allows RNA polymerase II to resume elongation of the transcript. Without transcript cleavage, transcript elongation is disturbed. (From Zatreanu, D., Han, Z., Mitter, R., Tumini, E., Williams, H., Gregersen, L., Dirac-Svejstrup, A.B., Roma, S., Stewart, A., Aguilera, A., Svejstrup, J. Molecular Cell 76(1), 57-69.e9, 2019, Creative Commons License)

Figure 1.51. The immediate product of transcription in eukaryotes is a precursor mRNA that undergoes modifications such as the addition of a cap at the 5'end and a poly-adenylate tail at the 3' end (step 3), the removal of non-coding introns (step 4), and joining together the exons (step 5). (Source: Kelvin Song, Creative Commons License)

Figure 1.52. The eukaryotic precursor mRNA needs to have its introns removed. This is accomplished by the spliceosome complex which cuts at the 5' GU end of the intron. The freed guanylate is linked to an adenylate residue within the intron to form a lariat. Another cut is made at the 3' AG end of the intron, which removes the intron and allows the exons to be joined together. (Source: BCSteve, Creative Commons License)

Figure 1.53. The three types of RNA produced during transcription are necessary for protein synthesis. The ribosomal RNA (rRNA) associates with ribosomal proteins to form the ribosome. The messenger RNA (mRNA) contains the codes (or the codons) for the amino acid sequence of the protein to be synthesized while the transfer RNA (tRNA) brings the amino acid to the site of protein synthesis. The anticodon in the tRNA is complementary to the codon in the mRNA, which specifies the amino acid carried by that tRNA. (Source: Modified from WebStockReview, Creative Commons License)

Figure 1.54. The genetic code chart lists the 64 possible codons in the messenger RNA. One of these is the start codon (AUG in red), three are stop codons (in bold fonts), and the rest correspond to any of the 20 amino acids. (Source: National Institutes of Health, Public Domain)

Figure 1.55. The translation process requires the activation of tRNAs (step 1) catalyzed by aminoacyl tRNA synthetases. At the initiation phase (steps 2 and 3), the tRNA bearing methionine is positioned opposite the start codon. The tRNA carrying the amino acid that corresponds to the next codon is then put in place. The previous tRNA leaves the translation complex and the ribosome moves every three nucleotides to expose the next codon. These steps occur repeatedly resulting in elongation of the amino acid chain (steps 4, 5, and 6). Finally, when the stop codon is encountered by the ribosome, termination of translation ensues (step 7). (From Kelvin Song, Creative Commons License)

Figure 1.56. Heat shock protein 70 (Hsp70), a class of molecular chaperones, recognizes the hydrophobic sequences of misfolded or damaged protein. This initiates a cascade of reactions in the ubiquitination pathway that eventually results in the degradation of the protein. This is one of the ways of protein quality control since the accumulation of misfolded and damaged proteins leads to a host of diseases. (Source: Ciechanover, A.; Kwon, Y.T. Frontiers in Neuroscience 11, 185, 2017, Creative Commons License)

Figure 1.57. The immediate product of translation is called pre-proinsulin. Its signal peptide (blue segment) is eventually cleaved forming proinsulin. Proinsulin is folded in the ER then transported to the Golgi apparatus where the C-peptide (light green segment) is cleaved to form a free C- peptide and mature insulin. (Source: Anoel8, Creative Commons License)

Figure 1.58. Acetylation at the N-terminus of the protein involves N-terminal acetyltransferase (NAT), which catalyzes the transfer of an acetyl group from acetyl-CoA (Ac-CoA) to the free α-amino group. An N-terminal deacetyltransferase (NDAC) is yet to be discovered. Another type is lysine acetylation where lysine acetyltransferase (KAT) catalyzes the transfer of acetyl group to the ε-amino group of the lysine (K) side chains. The attached acetyl group may be removed by the activity of lysine deacetyltransferase (KDAC). (Source: Ree, R.; Varland, S.; Arnesen, T. Experimental & Molecular Medicine 50:90, 2018, Creative Commons License)

Figure 1.59. (A) If the carbohydrate group is attached to the nitrogen atom in the side chain of asparagine, the protein modification is called N-linked glycosylation. (B) If it is attached to the oxygen atom of serine or threonine, then it is called O-linked glycosylation. (Source: Tpirojsi, Public Domain)

Figure 1.60. (A) Hydroxylation of proline residues of collagen is catalyzed by procollagen prolyl 4-hydroxylase (P4H) and prolyl 3-hydroxylase (P3H). This improves the structural stability of collagen. (B) Procollagen lysine 5-hydroxylase (PLOD) hydroxylates the lysine residues of collagen creating cross-linking sites, which provides further stability. (Source: van Staalduinen, L.M.; Jia, Z. Frontiers in Microbiology, https://doi.org/10.3389/fmicb.2014.00798, 2015, Creative Commons License)

Figure 1.61. Methylation occurs in specific arginine (R) and lysine (K) residues of histone 3 (H3) and histone 4 (H4) tails. Histone methylation, depending on its location, is associated with various functions such as heterochromatin formation, gene silencing, alternative splicing, DNA damage response, transcription regulation, cell-cycle checkpoint, and cellular development. (Source: Kim, J-H; Lee, J.H.; Lee, I-S.; Lee, S.B.; Cho, K.S. International Journal of Molecular Sciences, 18(7), 1404, 2017, Creative Commons License)

Chapter 2

Opening figure (Source: C. Burnett, Creative Commons License)

Figure 2.1. The karyogram of a normal human male shows 22 pairs of autosomes as well as the X and Y sex chromosomes. The autosomes are arranged in descending order according to size except for chromosome 21, which is actually smaller than chromosome 22. (Source: Wessex Reg. Genetics Centre, Creative Commons License)

Figure 2.2. Aneuploidy may result from nondisjunction or non-separation of homologous chromosomes during meiosis I (top left) or of sister chromatids during meiosis II (top right). This gives rise to gametes, mostly egg cells, with extra or missing chromosomes. (Bottom) When an aneuploid egg cell is fertilized by a normal sperm, the resulting zygote is an aneuploid. (Source: Ahmad, M.; Silvera-Redondo, C.; Rodriguez, M.H. SaludUninorte 26(1), 117-133, 2010, Creative Commons License)

Figure 2.3. The most common viable trisomy in humans is trisomy 21, i.e. having an extra copy of chromosome 21 due to nondisjunction. (Source: Wessex Reg. Genetics Centre, Creative Commons License)

Figure 2.4. Down syndrome is characterized by certain physical features like a flat, broad face and nose bridge, upwardly slanted eyes with epicanthal fold, white Brushfield spots on the iris, small ears, and a large protruding tongue. (Source: Centers for Disease Control and Prevention, Public Domain)

Figure 2.5. Random nondisjunction of chromosome 13 during meiosis may lead to gametes with extra chromosome 13, which when involved in fertilization will result in an infant with Patau's syndrome. (Source: Wessex Reg. Genetics Centre, Creative Commons License)

Figure 2.6. Cleft lip and palate are some of the many physical deformities associated with trisomy 13, which is the case for this male infant. (Source: Gashi, A.M. Journal of Pregnancy and Reproduction 1, 2017, Creative Commons License)

Figure 2.7. Nondisjunction of chromosome 18 during meiosis results in gametes having an extra copy of this chromosome. When these are involved in a fertilization event with a normal gamete, then the resulting individual exhibits trisomy 18. (Source: Wessex Reg. Genetics Centre, Creative Commons License)

Figure 2.8. (Left) One physical feature of trisomy 18 is the clenched fist with overriding fingers. (Right) While most infants with this condition do not live past their first month, this boy is over 2 years of age and apart from the tracheostomy to aid his breathing, is quite stable medically. (Source: Cereda, A.; Carey, J.C. Orphanet Journal of Rare Diseases 7 (81), 2012, Creative Commons License)

Figure 2.9. A case of mosaic trisomy 10 in a female child showed a karyotype of 47 chromosomes (extra chromosome 10), XX in 42% of metaphases that were observed in her blood cells. (Source: Gao, Y.; Ma, Y.-C.; Ju, Y.-H.; Li, Y.-N. BMC Pediatrics 18 (266), 2018, Creative Commons License)

Figure 2.10. A 14-month old girl with mosaic trisomy 10 showing morphological features such as (a) low-set ears, (b) small palpebral fissure, (c) retrognathia, (d) fused teeth, and (e) patchy and uneven skin pigmentation. (Source: Gao, Y.; Ma, Y.-C.; Ju, Y.-H.; Li, Y.-N. BMC Pediatrics 18 (266), 2018, Creative Commons License)

Figure 2.11. One of the two copies of X chromosomes in females is randomly inactivated. The inactivated X chromosome (Xi) condenses into a compact Barr body (indicated by the arrow), appearing as a densely staining structure at the periphery of the nucleus of ribosome fluid cell. (Top) The inactive and active X chromosome (Xa) were detected by fluorescence *in situ* hybridization (FISH) and visualized with a confocal microscope. (Bottom) The same nucleus was stained with DAPI and recorded with a charged coupled device (CCD) camera. (Source: Eils, R.; Dietzel, S.; Bertin, E.; Schrock, E.; Speicher, M.R.; Ried, T.; Robert-Nicoud, M.; Cremer, C.; Cremer, T. Journal of Cell Biology 135, 1427-1440, 1996, Creative Commons License)

Figure 2.12. A type of aneuploidy specific among females is having three copies of the X chromosome or the triple X syndrome. (Source: Wessex Reg. Genetics Centre, Creative Commons License)

together to allow the probe to hybridize with its complementary target sequence. The fluorescence emitted by the probe is detected by a fluorescence microscope and the fluorescence data is then analyzed. (Source: Ratan, Z.A.; Zaman, S.B.; Mehta, V.; Haidere, M.F.; Runa, N.J.;Akter, N. Cureus 9(6), e1325, 2017, Creative Commons License)

Figure 2.24. (A) Direct fluorescence *in situ* hybridization (FISH) utilizes a probe with fluorochrome that is incorporated directly into the nucleotides. (B) Indirect FISH, on the other hand, employs a probe with a hapten which is then detected by a fluorochrome-labeled antibody. (Source: Adapted from Escott16, Creative Commons License)

Figure 2.25. (A) The result of fluorescence *in situ* hybridization (FISH) shows a red signal, which is specific for the X chromosome, and a green one for the Y chromosome. (B) The probe used to detect the X chromosome is labeled with digoxigenin, which is detected by incubating it with rhodamine (i.e., red in color)-conjugated antibodies. The probe targets the centromeric region. The probe for the Y chromosome is directly labeled with a green fluorochrome and it targets a satellite DNA in the long arm of the chromosome. (Source: Hsu, J.H.; Zeng, H..; Lemke, K.H.; Polyzos, A.A.; Weier, J.F.; Wang, M.; Lawin-O'Brien, A.R.; Weier, H.-U.G.; O'Brien, B. International Journal of Molecular Science 14(1), 57-71, 2013, Creative Commons License)

Figure 2.26. Polymerase chain reaction (PCR) is an *in vitro* technique that consists of a series of temperature cycles composed of template DNA denaturation, annealing to the template of a pair of primers that flank the target region, and the elongation of the primers by a DNA polymerase using deoxyribonucleotide triphosphates (dNTPs) as substrates. This cycle is repeated 20-40 times in order to amplify the target DNA segment and produce millions of copies from a small quantity of DNA template. (Source: Quan, P.-L.; Sauzade, M.; Brouzes, E. Sensors 18(4), 1271, 2018, Creative Commons License)

Figure 2.27. Detection of sex chromosome aneuploidies via quantitative fluorescence-polymerase chain reaction (QF-PCR) requires primers that target STRs located in both X and Y chromosomes such as X22 as well as those exclusively located in the X chromosome like DXS1053, DXS981, DXS6809, DXS1187, and DXS8377. Two known sex-determining genes can also be included such as the *SRY* gene found only in the Y chromosome and the two alleles of the amelogenin gene, one found in the X chromosome (AMX) and the other in the Y chromosome (AMY). (Source: Xie, X. and Liang, Q. PLoS ONE 2014, 9(9), e106307, Creative Commons License)

Figure 2.28. After capillary electrophoresis, analyzing the resulting fluorescence peaks of the products of quantitative fluorescence-polymerase chain reaction (QF-PCR) reveals significant differences between (A) normal XX and (B) XY karyotypes and the sex chromosome aneuploidies such as (C) triple X syndrome, (D) Klinefelter syndrome, (E) Jacob syndrome, and (F) Turner syndrome. (Source: Xie, X. and Liang, Q. PLoS ONE 2014, 9(9), e106307, Creative Commons License)

Figure 2.29. Comparative genome hybridization (CGH) allows for the screening of the whole genome to determine changes in the amount of DNA in either the chromosome or region of interest. There are two probes used, the sample or test DNA and a known normal DNA that are labeled with different fluorochromes. Cot-1 DNA is added into the probe mixture to block non-specific hybridization with repetitive DNA. These are then incubated with the target, which consists of normal metaphase DNA. The resulting fluorescence intensity is analyzed to determine if there is no change in DNA amount at the region of interest and if there is, whether it is a net loss or net gain. (Source: Gbdivers, Creative Commons License)

Figure 2.30. Array comparative genomic hybridization (aCGH) employs two sets of DNA samples, one that is extracted from the patient (test DNA) and a normal control (reference DNA). These are labeled with different fluorochromes, respectively (e.g. red cyanine-5 for the test DNA and green cyanine-3 for the reference). These are then allowed to simultaneously hybridize with thousands of known DNA sequences, called probes, arranged on specific spots on an array chip. These probes represent the entire genome. The labeled DNA that successfully annealed with a probe emits a fluorescent signal that is detected by a microarray scanner. A certain software computes the $\log_2$ ratio of the Cy5/Cy3 fluorescence. A higher intensity of the test sample color in a specific region of the genome relative to the control indicates the gain of DNA (or excess DNA) in that region while a higher intensity of the control sample color indicates DNA loss in that region. When the resulting color is intermediate of the two fluorochromes used (e.g., yellow in the case of red and green), then that indicates normal DNA amount in that region. (Source: Colaianni, V.; Mazzei, R.; Cavallaro, S. Neurological Sciences 37(12), 2016, Creative Commons License)

Figure 2.31. The results of array comparative genome hybridization (aCGH) show the $\log_2$ ratio of the red-green fluorescence for each chromosome. In these examples, the test DNA is labeled green and the reference DNA is labeled red. Thus, green fluorescence indicates the gain of DNA while red indicates a loss. Examples are: (A) A sample from a female embryo showing loss of chromosome 22; (B) a cell from a male embryo with trisomy 21; (C) a sample from a male embryo showing trisomy 18 and loss of chromosome 14; (D) a cell taken from another male embryo exhibits multiple aneuploidies, an extra copy of chromosome 3 and 15 as well as a missing copy of chromosome 22; (E) a normal female embryo shows an intense fluorescence for the X chromosome; and (F) a cultured cell with a duplicated long arm of chromosome 22 that served as a positive control. (Source: Chung, M.K.; Jeong, H.J.; Lee, J.; Park, S.-J. Molecular Cytogenetics 6(1), 22, 2013, Creative Commons License)

Figure 2.32. If the lost segment is located within the chromosome, then it is classified as interstitial deletion. (Source: National Human Genome Research Institute, Public Domain)

Figure 2.33. A female with chromosome 4p deletion syndrome, specifically a microdeletion, at age (a, b) 3 months and (c, d) 13 years. She exhibits microcephaly, micrognathia, protruding eyeballs, broad nasal bridge, and a high forehead, which are characteristic facial features of the syndrome. (Source: Sogaard, M.; Tümer, Z.; Hjalgrim, H.; Hahnemann, J.; Friis, B.; Ledaal, P.; Pedersen, V.F.; Baekgaard, P.;

centromere. It contains a number of genes such as *TBX1* and *COMT*, which when lost are associated with the facial features as well as behavioral and mental abnormalities exhibited by those carrying a deletion in this segment. (Source: genetics4medics, Creative Commons License)

Figure 2.43. (A) An individual with Prader-Willi syndrome may become obese due to excessive eating, and also exhibits distinct facial features such as (B) narrow forehead, (C, D) small hands and feet. (Source: A. Cortés, F.M.; Alliende, M.A.R.; Barrios, A.R.; Curotto, B.L.; Lorena, S.M.V.; Barraza, X.O.; Troncoso, L.A.; Mellado, C.S.; Pardo, R.V. Chilean Medical Journal 133, 33-41, 2005, Creative Commons License. B,C,D. National Human Genome Research Institute, Public Domain)

Figure 2.44. Prader-Willi syndrome is due to the deletion of a portion of the long arm of chromosome 15, which may span 15q11.2 (either from breakpoint/BP 1 or 2) to 15q13.1 (BP3). The deleted segment may be divided into four segments: (1) a nonimprinted region containing four genes, with both paternal and maternal alleles expressed (biallelic expression), (2) a region with only the paternal copies of the genes expressed, which include protein-coding genes and short nucleolar RNA (snoRNA) genes, (3) a region with maternally-expressed genes, and (4) a distal nonimprinted region with genes that are biallelically expressed. (Source: Costa, R.A.; Ferreira, I.R.; Cintra, H.A.; Gomes, L.H.F.; Guida, L.d.C. Frontiers in Endocrinology 10, 864, 2019, Creative Commons License)

Figure 2.45. If the duplicated region is located within the chromosome, then it is classified as interstitial duplication. Moreover, if it is positioned next to the original segment, then it is considered a tandem duplication. (Source: National Human Genome Research Institute, Public Domain)

Figure 2.46. (A) A boy with chromosome 1q.21.1 duplication syndrome shows some facial dysmorphisms such as mild hypertelorism, small nose, and protruding ears. (B) Developmental profiles of the patient were determined according to Battelle Developmental Inventories. At 2 years and 8 months (blue), he showed mild developmental delay, mostly affecting his language, social and adaptive abilities. At 5 years and 4 months (orange), his cognitive skills seemed more impaired. (C) The Luria-Nebraska Neuropsychological Battery for Children (LNNB-C) suggests that at 5 years and 6 months, the boy has significant impairment of language and executive functions. (D) The Inventory for Client and Agency Planning (ICAP), which was administered in the same period, indicates an almost 2-year delay in the child's global development, especially his social and communication skills. (Source: Benítez-Burraco, A., Barcos-Martínez, M., Espejo-Portero, I., Fernández-Urquiza, M., Torres-Ruiz, R., Rodríguez-Perales, S. & Jiménez-Romero, M.S. *Frontiers in Pediatrics*, 6, 163, 2018. Creative Commons License)

Figure 2.47. Approximately half of those with chromosome 7q11.23 duplication syndrome exhibit aortic dilation. This may lead to the life-threatening aortic dissection, where a tear in the innermost layer (intima) of the aorta causes blood to flow between the layers of the aortic wall causing the layers to get separated. (Source: U.S. National Library of Medicine, Public Domain)

Figure 2.48. A nearly 15-year old boy with duplicated 12q13.13 region showing certain physical features such as (a) long and narrow face, high arched heavy eyebrows with medial flare, hypertelorism, broad nasal bridge and tip, and short philtrum; (b) receding anterior hairline with widow's peak; (c) bifid uvula and abnormal teeth; (d) low-set ear; and (e) abnormal toenails. (Source: Hu, J., Ou, Z., Infante, E., Kochmar, S.J., Madan-Khetarpal, S., Hoffner, L., Parsazad, S., Surti, U. Molecular Cytogenetics 10, 24, 2017, Public Domain)

Figure 2.49. One feature observed commonly among those with Charcot-Marie-Tooth type 1 is pes cavus (high-arched foot) with hammertoes. (Source: Itoh, N.; Hoshijima, H.; Takeuchi, R.; Nakamura, C.; Iwase, Y.; Nagasaka, H.; Sato, T.; Yoda, T. Stomatological Disease and Science 2,2, 2018, Creative Commons License)

Figure 2.50. Onion bulbs (black arrows), which form concentric layers around nerve cells, are usually observed in Charcot-Marie-Tooth type 1A disease. These are made up of accumulated Schwann cells, axon, and myelin sheath. (Source: Jensflorian, Creative Commons License)

Figure 2.51. Members of the same family (and b are first cousins, c and d are siblings) with *M*ECP2 duplication syndrome exhibiting typical facial features of the condition such as midface hypoplasia, low-set ears with large lobe and crease, small mouth, and protruding jaw. (Source: Li, X.; Xie, H.; Chen, Q.; Yu, X.; Yi, Z.; Li, E.; Zhang, T.; Wang, J.; Zhong, J.; Chen, X. BMC Medical Genetics 18(1), 131, 2017, Creative Commons License)

Figure 2.52. Pericentric inversion occurs if the inverted chromosome segment includes the centromere, but if it does not, it is called paracentric inversion. (Source: Boundless.com, Creative Commons License)

Figure 2.53. When one of the two homologues has an inverted segment, an inverted loop is formed by one chromosome in order to ensure that homologous regions interact during synapsis. (Source: Boundless.com, Creative Commons License)

Figure 2. 54. (A) The normal structure of the *F8* gene with 26 exons (light blue boxes). Intron 22 is its largest intron with *int22h1* repeat that is homologous with telomeric repeats, *int22h-2*and *int22h-3*. (B) Homologous recombination may occur between *int22h-1* and *int22h-3* (producing a type 1 inversion) or between *int22h-1* and *int22h-2* (type 2 inversion). (C) The recombination disrupts the F8 gene so that exons 1 to 22 are displaced toward the telomere and are inverted in orientation (Xq^{tel}: X chromosome q arm telomere. Xq^{cen}: X chromosome q arm centromere). (Source: Hwang, S.H.; Kim, H.-J.; Kim, H.S. Hemophilia, 2012, Creative Commons License)

Figure 2.55. (A) A repeat sequence within intron 1 (*int1h-1*) is homologous with an extragenic repeat (*int1h-2*) near the promoter region. (B) Homologous recombination may occur between the two repeats. (C) This inverts the direction of transcription of exon 1, opposite the location of the promoter, so that a complete Factor VIII protein cannot be synthesized (Xq^{tel}: X chromosome q arm telomere. Xq^{cen}: X chromosome q arm centromere). (Source: Hwang, S.H.; Kim, H.-J.; Kim, H.S. Hemophilia, 2012, Creative Commons License)

Figure 2.56. A reciprocal translocation occurs if two nonhomologous chromosomes exchange segments (in this case, chromosome 4 and chromosome 20). It is a balanced type of translocation if DNA is neither gained nor lost during the exchange. The rearranged chromosomes are called derivative (der) chromosomes and the particular translocation is designated as "t". After translocation, the chromosomes in this example are referred to as der(4) t(4; 20) for the derivative chromosome with a centromere derived from chromosome 4 and der(20); t(20;4) for the other derivative chromosome with its centromere from chromosome 20. (Source: National Human Genome Research Institute, Public Domain)

Figure 2.57. Robertsonian translocation happens when the long arms of two acrocentric chromosomes are joined at a single centromere and the two short arms are lost. (Source: yourgenome, Creative Commons License)

Figure 2.58. Translocation carriers are usually phenotypically normal but may produce abnormal gametes. (a) In reciprocal translocation, the derivative chromosomes are partially homologous with each other and with their original homologue. This results in the formation of a quadrivalent. This can segregate in an alternate manner producing either normal gametes or gametes with a balanced translocation. If segregation is in an adjacent manner, gametes with unbalanced translocations are formed. If there is non-disjunction, aneuploid gametes are formed. (b) In Robertsonian translocation, synapsis among the hybrid chromosome and the original homologues results to a trivalent. Normal gametes are produced only if two normal chromosomes segregate together. Otherwise, gametes with unbalanced translocations or aneuploid gametes are formed. (Source: Tulay, P. IntechOpen, 2017, Creative Commons License)

Figure 2.59. Balanced Robertsonian translocation between chromosomes 13 and 14 is the most common type of this chromosomal rearrangement, in which a novel chromosome made up of fused long arms of these two chromosomes are formed. Since there is no net gain or loss of genetic material, carriers of this aberration do not manifest clinically significant phenotypes. (Source: Wessex Reg. Genetics Centre, Creative Commons License)

Figure 2.60. Robertsonian translocation of chromosomes 13 and 14 causes irregular pairing and segregation of these chromosomes during gamete formation. As a result, the resulting offspring may inherit the unbalanced type of Robertsonian translocation. One possible outcome is translocation trisomy 14, in which there are three copies of chromosome 14 long arms. This does not produce a viable embryo. (Source: Wessex Reg. Genetics Centre, Creative Commons License)

Figure 2.61. Down syndrome may be a result of unbalanced translocation where a portion of chromosome 21 is fused to another chromosome, such as chromosome 14. Since there is already a pair of chromosome 21, the fused segment is an extra copy of genetic material causing the syndrome. (Source: Wessex Reg. Genetics Centre, Creative Commons License)

Figure 2.62. A reciprocal translocation may occur between the q34 region of chromosome 9, which contains the *ABL* gene, and the q11 region of chromosome 22. This results in a shorter chromosome 22 derivative called the Philadelphia chromosome. This chromosome contains the hybrid *BCR-ABL* gene. (Source: Aryn89, Creative Commons License)

Figure 2.63. A Nigerian boy with a large facial tumor due to Burkitt lymphoma. The tumor is made up predominantly of aberrant B lymphocyte cells. (Source: Robert S. Craig, Center for Disease Control and Prevention, Public Domain)

Figure 2.64. Burkitt lymphoma results from the translocation of *MYC* protooncogene at region q24 of chromosome 8 to breakpoints in chromosomes 2, 14, and 22 where the genes for *IGHK* (k), *IGH* (H), and *IGHL* (λ) are located, respectively. This places the *MYC* gene under the transcriptional control of the strong promoter of these immunoglobulin genes. (Source: National Cancer Institute, Public Domain)

Figure 2.65. (a) Chromosomes have their distinct banding patterns as shown by these illustrations of normal chromosomes. (b-e) Five patients with premature ovarian failure (POV) exhibit translocations involving chromosome X as shown in the partial Q-banding karyotype on the left and the illustrations of the derivative chromosomes on the right. Specifically, X chromosome was observed to have undergone translocation with (b) chromosome 2, (c) chromosome 19, and (d and e) Y chromosome. (Source: Baronchelli, S.; Conconi, D.; Panzeri, E.; Bentivegna, A.; Redaelli, S.; Lissoni, S.; Saccheri, F.; Villa, N.; Crosti, F.; Sala, E.; Martinoli, E.; Volontè, M.; Mrozzi, A.; Dalprà, L. Journal of Biomedicine and Biotechnology 370195, 2011, Creative Commons License)

Figure 2.66. A karyotype produced after performing the G-banding technique showing translocation between chromosomes 7 and 18 (red arrows) with potential microdeletion on chromosome 7 or 18 or both. (Source: Tang, Z.; Berlin, D.S.; Toji, L.; Toruner, G.A.; Beiswanger, C.; Kulkarni, S.; Martin, C.L.; Emanuel, B.S.; Christman, M.; Gerry, N.P. Genes, Genomes, Genetics 3 (7), 1143-1149, 2013, Creative Commons License)

Figure 2.67. Partial karyotype from the same individual was constructed using G-banding, R-banding, and C-banding methods. All techniques reveal that one copy of chromosome 9 has additional bands on its short arm compared to its homologue. The C-banding indicates that the added chromosome material is heterochromatic. It also produces dark centromeric bands. (Source: Kasakyan, S.; Lohmann, L.; Aboura, A.; Quimsiyeh, M.; Menezo, Y.; Tachdjian, G.; Benkhalifa, M. Molecular Cytogenetics 1, 27, 2008, Creative Commons License)

Figure 2.68. C-banding pattern exhibits dark bands for heterochromatic regions such as centromeres, pericentric regions (e.g., chromosomes 1, 9, and 16), and the long arm of the Y chromosome. (Source: Di Tomaso, M.V.; Liddle, P.; Lafon-Hughes, L.; Reyes-Ábalos, A.L.; Folle, G. IntechOpen, 2013, Creative Commons License)

Figure 2.69. (A) After two rounds of cell division with 5 bromo-2' deoxyuridine (BrdU) and then treating the chromosomes with Giemsa stain, the sister chromatids are seen as a pair of light and dark-staining structures. (B) Sister chromatid exchange occurred if the chromatids exhibit a mixture of light and dark segments along their length (arrows). (Source: Sharma, S.; Brosh, R.M. Jr. PloS ONE 2(12), e1297, 2007, Public Domain)

Figure 2.70. It is now possible to combine fluorescent dyes in order to produce 23 colors to label the 22 autosomes and the X chromosomes in a human female karyotype. In the case of a male karyotype, an additional label can be used for the Y chromosome. (Source: National Human Genome Research Institute, Public Domain)

Figure 2.71. A karyotype of a human neck and squamous cell carcinoma cell line that is labeled via multiplex fluorescence *in situ* hybridization (m-FISH) shows several numerical and structural aberrations of the chromosomes. Aneuploidies are observed for chromosomes 1, 5, 7, 11, 19, 20 and 21. Translocation occurred between chromosomes 1 and 15, chromosomes 1 and 18. Deletion is also observed in chromosome 13. (Source: Singchat, W.; Hitakomate, E.; Rerkarmnuaychoke, B.; Suntronpong, A.; Fu, B.; Bodhisuwan, W.; Peyachoknagul, S.; Yang, F.; Koontongkaew, S.; Kornsorn, S. PLoS ONE 11(8), 2016, Creative Commons License)

Figure 2.72. Multiplex FISH (m-FISH) was used to determine the chromosomal origin of supernumerary marker chromosomes (sSMCs) in two cases where the sSMC is identified to originate from (a) chromosome 16 and (b) chromosome 19. (Source: Jang, W.; Chae, H.; Jiyeon, K.; Son, J.-O.; Kim, S.K.; Koo, B.K.; Kim, M.; Kim, Y.; Park, I.Y.; Sung, I.K. Molecular Cytogenetics 9(1), 2016, Creative Commons License)

Figure 2.73. Novel chromosomal aberrations in *de novo* chronic myelogenous leukemia (CML) patients in their chronic phase (CP) are detected by G-banding and then confirmed by spectral karyotype. (A) (i) G-banded karyotype shows translocation between chromosomes 9 and 22 and between chromosomes 11 and 15. (ii) Partial spectral karyotype confirms the translocation between chromosomes 11 and 15. (B) (i) G-banded karyotype shows translocation between chromosomes 9 and 22 as well as Robertsonian translocation between the two chromosomes 13. This is confirmed by the (ii) spectral karyotype and (iii) partial spectral karyotype. (Source: Chandran, R.K.; Narayanan, G.; Sakthivel, K.M.; Kumar, R.S.; Krishna, K.M.N.J.; Sreedharan, H. Frontiers in Oncology 9,88, 2019, Creative Commons License)

Figure 2.74. Restriction enzyme digestion, at certain restriction sites (RS), is performed on the genomic region prone to rearrangement to produce fragments that cover the breakpoint (BP) and that are composed of known (orange) and unknown (blue) sequences. The generated fragments are re-ligated to form circular templates for PCR with a reverse (RPrimer) and forward primer (Fprimer) relative to the BP. The amplicons are separated via gel electrophoresis and those containing a translocated segment form a higher molecular weight band than those without translocation. (Source: Adapted from Schneider, B.; Drexler, H.G.; MacLeod, R.A.F. Atlas of Genetics and Cytogenetics in Oncology and Haematology 15(1), 106-114, 2011.)

Figure 2.75. In the multiplex delta-PCR, three primers (arrows) are designed and used simultaneously to detect a specific translocation breakpoint or insertion/deletion locus. Two primers are oriented in the same direction with the external primer (red arrow) at a defined distance Δ (e.g. 30 bp) from the internal one (blue arrow). The third primer is oriented in the opposite direction of the other two. (Source: Adapted from Lin, M.-T.; Tseng, L.-H.; Rich, R.G.; Hafez, M.J.; Harada, S.; Murphy, K.M.; Eshleman, J.R.; Gocke, C.D. The Journal of Molecular Diagnostics 13(1), 85-92, 2011.)

Figure 2.76. In samples with supernumerary marker chromosomes (sSMCs), array CGH detected a (a) gain of 12.6 Mb derived from chromosome 2q11.1q12.3. (b) An identical gain was observed from the mother of the sample source, indicating that the sSMC was maternally inherited. (c) Another sample showed a gain in chromosome 2, but this time at a different region, i.e. 2q11.1q12.1. (Source: Jang, W.; Chae, H.; Jiyeon, K.; Son, J.-O.; Kim, S.K.; Koo, B.K.; Kim, M.; Kim, Y.; Park, I.Y.; Sung, I.K. Molecular Cytogenetics 9(1), 2016, Creative Commons License)

Figure 2.77. Molecular analysis was performed on genetic samples from a male infant with moderate developmental delay and minor dysmorphic features. (a) Array CGH using two alternative arrays, i.e. Cy3/Cy5 (red line) and Cy5/Cy3 (blue line), showed an interstitial deletion at the subtelomeric region of chromosome 10. (b) The deletion was confirmed by fluorescence *in situ* hybridization where one copy of chromosome 10 does not show a hybridization signal with the subtelomeric probe used. (c) The deleted region is 10q26.2q26.3 which contains a number of genes. (Source: Iourov, I.Y.; Vorsanova, S.G.; Kurinnaia, O.S.; Yurov, Y.B. Case Reports in Genetics 505832, 2014, Creative Commons License)

Figure 2.78. When performing single nucleotide polymorphism (SNP) microarray analysis, the patient's sample DNA is fluorescently labeled and then allowed to hybridize with the probes on the array slide. After the DNA hybridizes with probes having its complementary sequence, the fluorescence is measured and the intensity for each SNP allele is calculated. A heterozygous locus (AB) shows equal intensities for both alleles while a homozygous locus (AA or BB) shows higher intensity for the allele present (A or B). Copy number variation can also be extrapolated by comparing the data with a reference data file. Then each probe is plotted along the length of the chromosome depending on its location, fluorescence intensity, and B allele frequency. (Source: Karampetsou, E.; Morrogh, D.; Chitty, L. Journal of Clinical Medicine 3(2), 663-678, 2014, Creative Commons License)

Figure 2.79. Although the banding technique shows normal chromosome 4 in this sample, SNP microarray reveals a loss of a small segment spanning 1.48 Kb (red bar) in region 4q24. (Source: Song, Q.; Chu, Y.; Yao, Y.; Peng, M.; Yang, W.; Li, X.; Huang, S. Scientific Reports 7,10354, 2017, Creative Commons License)

Figure 2.80. SNP array may show the presence of mosaicism, in this case using placental tissue DNA. (A) The upper axis indicates the value of B allele frequency (BAF, blue dots). For diploid individuals, a normal chromosome (chromosome 18 used as a control in this example) may have three BAF values, i.e. 1 for BB, 0.5 for AB, and 0 for AA. The lower axis indicates the logR ratio (LRR, red line), where R is the ratio of each SNP compared to chromosome 18. The LRR of 0 ,in this case, means an actual ratio of 1. (B) The SNP array of placental tissue with mosaic trisomy 18 shows altered BAF values, i.e. 1 for BBB and BB, 0.6 for ABB and AB, 0.4 for AAB and AB, and 0 for AAA and AA. The increased LRR also indicates mosaicism. (Source: Pan, Q.; Sun, B.; Huang, X.; Jing, X.; Liu, H.; Jiang, F.; Zhou, J.; Lin, M.; Yue, H.; Hu, P.; Ning, Y. Molecular Cytogenetics 7(1), 48, 2014, Creative Commons License)

Figure 2.81. (A) The blue line in the SNP microarray karyogram represents the average fluorescent intensity of each SNP while the green tics represent heterozygous calls for each SNP. (B) The absence of heterozygous calls represents areas of uniparental disomy or loss of heterozygosity (LOH) in the chromosome (blue and pink bars). (Source: Gondek, L.P.; Szpurka, H.; McDevitt, M.A.; Maciejewski, J.P. PLoS ONE 2(11), e1225, 2007, Creative Commons License)

Figure 2.82. Mutation may be a result of base substitution. If a purine is substituted with another purine during replication or if a pyrimidine is replaced with another pyrimidine, then such substitution is called transition. If the substitution is between a purine and a pyrimidine, or vice versa, it is called transversion. (Source: Adapted by Emmanuel J.P. Douzery from Petulda, Creative Commons License)

Figure 2.83. Base substitution may change the codon in the mRNA. In turn, this may or may not change the amino acid in the protein product. If the changed codon specifies the same amino acid, then it is a silent mutation. If the amino acid is altered, it is a missense mutation. On the other hand, if the codon becomes a stop codon, it is a nonsense mutation. (Source: OpenStax: Microbiology, Creative Commons License)

Figure 2.84. Missense mutation in exon 1 of the *LMNA* gene, leads to the substitution of the 39th amino acid from asparagine to histidine (black arrow). This affects the electrostatic charge of the coil1A segment and weakens its electrostatic interaction with the tail domain. (Source: Ishiyama, A.; Iida, A.; Hayashi, S.; Komaki, H.; Sasaki, M.; Nonaka, I.; Noguchi, S.; Nishino, I. Human Genome Variation 5, 19, 2018, Creative Commons License)

Figure 2.85. The 42-amino acid peptide of amyloid β-protein is neurotoxic and aggregates to form amyloid plaques in the brain of Alzheimer's disease patients. The 23rd amino acid is aspartic acid that forms a network of hydrogen bonds with neighboring residues. (Source: Lermyte, F.; Everett, J.; Lam, Y.P.Y.; Wootton, C.A.; Brooks, J.; Barrow, M.P.; Talling, N.D.; Sadler, P.J.; O'Connor, P.B.; Collingwood, J.F. Journal of The American Society for Mass Spectrometry 30, 2123-2134, 2019, Creative Commons License)

Figure 2.86. A missense mutation changing the 342nd amino acid of α_1-antitrypsin from glutamic acid to lysine (encircled area) disrupts a "clasp" that regulates conformational change. As a result, the mutant protein adopts an intermediate conformation that rapidly polymerizes. (Source: Nyon, M.P.; Segu, L.; Cabrita, L.D.; Lévy, G.R.; Kirkpatrick, J.; Roussel, B.D.; Patschull, A.O.M.; Barrett, T.E.; Ekeowa, U.I.; Kerr, R.; Waudby, C.A.; Kalsheker, N.; Hill, M.; Thalassinos, K.; Lomas, D.A.; Christodoulou, J.; Gooptu, B. Structure 20-135(3-3), 504-512, 2012, Creative Commons License)

Figure 2.87. Normal hemoglobin molecules occur in red blood cells as solitary units. A missense mutation in the β-globin gene produces a mutant protein product that causes hemoglobin to form long strands when deoxygenated. This distorts the red blood cell so that it may assume a sickle shape that prevents it from flowing freely along small blood vessels. (Source: Diana Grib, Creative Commons License)

Figure 2.88. (Step 1) ATP-binding cassette transporter 1 (ABCA1) monomers translocate phospholipids and cholesterol across the membrane and then out of the cell. This process is ATP-dependent. (Step 2) The lipids are accumulated in the extracellular domain (ECD) of ABCA1. (Step 3) When enough lipids have been sequestered, ABCA1 forms a dimer and lipid-free apolipoprotein A-I (apoA-I) binds directly to the ECD. (Step 4) The apoA-I picks up the lipids and separates itself from the ECD to form nascent high- density lipoprotein (HDL). Then the ABCA1 dimer dissociates into monomers ready to perform step 1 once again. (Source: Ishigami, M.; Ogasawara, F.; Nagao, K.; Hashimoto, H.; Kimura, Y.; Kioka, N.; Ueda, K. Scientific Reports 8, 6160, 2018, Creative Commons License)

Figure 2.89. An 18-month old boy with hypophosphatasia exhibits (A) high-arched palate, which is an indication of skeletal problems in the maxilla, (B) open bite, and (C) crowding of hypomineralised teeth. (Source: Okawa, R.; Kokomoto, K.; Kitaoka, T.; Kubota, T.; Watanabe, A.; Taketani, T.; Michigami, T.; Ozono, K.; Nakano, K. PLoS ONE 14(10), e0222931, 2019, Creative Commons License)

Figure 2.90. After being synthesized, proteins are targeted to their destination, whether to be in the cytosol or to be in a specific part of a certain organelle. (Source: Clip Art, Public Domain)

Figure 2.91. Certain proteins like ABCA1, are synthesized by ribosomes on the endoplasmic reticulum and then transported through the Golgi apparatus by vesicles. These vesicles will then bring the proteins to their final location, such as the plasma membrane. (Source: Molnar C.; Gair, J. Concepts of Biology- 1st Canadian Edition, Creative Commons License)

Figure 2.92. After being synthesized by cytosolic ribosomes, the alanine-glyoxylate aminotransferase (AGT) (orange) monomer is partially folded when it interacts with Hsp40 and Hsp70 chaperones. The correct folding of the monomer occurs as it is transferred to the Hsp90 and Hsp60 machineries so that it dimerizes. The folded AGT dimer is then imported via the Pex5p import machinery into the peroxisome, where it functions for glyoxylate detoxification. Certain missense mutations generate a targeting sequence that is recognized by the TOM20 or TOM70 receptors leading to the importation of AGT into the mitochondrion. (Source: Pey, A.L.; Albert, A.; Salido, E. BioMed Research International 12, 687658, 2013, Creative Commons License)

Figure 2.93. Hemoglobin is a protein component of red blood cells made up of two α-globin chains and two β-globin chains. (Source: OpenStax: Anatomy and Physiology, Creative Commons License)

Figure 2.94. One of the common features observed among individuals with Marfan syndrome is having a displaced lens in one or both eyes. (Source: Farat, J.G.; Quieroz, M.F.N.; Lottelli, A.C. Arquivos Brasileiros de Oftalmologia 82(1), 2019, Creative Commons License)

Figure 2.95. Insertion or deletion of nucleotides in the coding sequence of the gene may result in a shift in the reading frame of the gene. This produces a mutant protein with an amino acid sequence that is different from that of the wild type. (Source: OpenStax: Microbiology, Creative Commons License)

Figure 2.96. The cystic fibrosis transmembrane conductance regulator (CFTR) protein made up of 1, 480 amino acids with five domains. It has two transmembrane domains (i.e., TMD1 and TMD2) that traverse the plasma membrane. It also has two nucleotide-binding domains (i.e., NBD1 and NBD2) which associate with and hydrolyzes ATP. Finally, it has one regulatory domain (RD) which is the site of phosphorylation. CFTR functions to transport chloride (Cl^-) across the plasma membrane. (Source: Lopes-Pacheco, M. Frontiers in Pharmacology 7(e98502), 275, 2016, Creative Commons License)

Figure 2.97. (A) Cystic fibrosis affects various organs. (B) A normal airway has a thin layer of mucus, but (C) those with cystic fibrosis have airways that are blocked by thick, sticky mucus containing blood and bacteria. (Source: U.S. National Heart, Lung, and Blood Institute, Public Domain)

Figure 2.98. β-hexosaminidase A is an enzyme in the lysosome that hydrolyzes GM2 gangliosides. Therefore, a mutation in this enzyme leads to the accumulation of GM2 gangliosides in the lysosome. This leads to progressive destruction of neurons causing reduced motor and mental functions. (Source: Solovyeva, V.V.; Shaimardanova, A.A.; Chulpanova, D.S.; Kitaeva, K.V.; Chakrabarti, L.; Rizvanov, A.A. Frontiers in Physiology 9, 1663, 2018, Creative Commons License)

Figure 2.99. The deletion of a 3-nucleotide codon (e.g. codon GTT) will not change the way that the rest of the mRNA is translated but it will produce a protein with a missing amino acid. (Source: Adapted from Sumukal, Creative Commons License)

Figure 2.100. Regions of substantia nigra in a patient in Parkinson's disease that were magnified 60 times. It shows the α-synuclein inclusions in the neurons that aggregated to form Lewy bodies. (Source: Suraj Rajan, Creative Commons License)

Figure 2.101. Psoriasis is a skin inflammatory disorder that is characterized by red elevated plaques on the arms and torso. (Source: James Heilman and The Wednesday Island, Creative Commons License)

Figure 2.102. A DNA library is prepared from the extracted genomic DNA that is eventually digested by restriction enzymes. The resulting fragments are inserted into vectors which are eventually taken up by bacterial cells. (Source: Aluquette, Creative Commons License)

Figure 2.103. A dideoxynucleotide (ddNTP) is similar in structure to a deoxynucleotide (dNTP) except for its lack of 3' hydroxyl (OH) group (shaded light blue). The 3' OH group is needed in forming a bond with the next nucleotide during DNA replication. For this reason, elongation of DNA stops when a ddNTP is incorporated in the DNA strand being synthesized. (Source: OpenStax CNX, Creative Commons License)

Figure 2.104. Capillary electrophoresis separates the DNA fragments from the Sanger reactions based on molecular weight. As each fragment moves past a laser, the fluorescently labeled dideoxynucleotide (ddNTP) at the end of the fragment gets activated. Each type of ddNTP emits fluorescence at a different wavelength. The resulting chromatogram is translated by computer software into a DNA sequence. (Sources: (Left image) Estevesj, Creative Commons License; (Right image) Modified from Abizar, Creative Commons License; (Arrow) SimpleIcon, Creative Commons License)

Figure 2.105. The commonly used second-generation sequencing platforms share certain similarities. They prepare their DNA library by making their starting DNA material double-stranded and ligating it to adapters. These are then amplified into thousands of copies by PCR prior to sequencing. The sequencing method varies among the platforms in terms of the data that is collected, whether it is fluorescence, light, or pH. (Source: Knief, C., Frontiers in Plant Science 5, 216, 2014, Creative Commons License)

LIST OF TABLES

PREFACE

Humans are innately interested about life. More so if it is our own life that is involved. Still, ethical considerations have limited the extent at to which humans can perform research or studies on fellow humans. For this reason, many of the principles in genetics were elucidated using microbes, plants, or simple forms of animals. At the DNA level, we share many similarities with these organisms. Therefore, insights derived from studying them proved very valuable in understanding the genetic mechanisms that occur in human cells. The advancement in terms of research techniques also allowed for studying the actual human DNA. This contributed further to the body of knowledge regarding human genetics. This book provides both basic concepts and the applications of such concepts, particularly in understanding diseases with genetic causes.

The first chapter of this book establishes the genetic foundation necessary to understand the concept of heredity and the relationship between genes and the observed biological characteristic. It presents how the concept of a gene was developed historically, from a cytological entity that is inherited from a parent to as a molecule that stores biological information. Therefore, this chapter encompasses cytogenetics, molecular genetics, and even biochemistry.

While some genetic disorders have a long history of being a subject of research, many have been relegated into the background. This is because many of these diseases are rare and complex. As the techniques for studying the DNA improved and has become less expensive, these so-called "orphan" diseases started to gain the attention of researchers. The second chapter, although by no means a complete compilation of genetic diseases, provides the various types of mutations that cause diseases or that increases the risk of having one. Some disorders are described in detail as a case study. It also presents the various methods by which these conditions are diagnosed.

This book uses simple and clear language in order to appeal even to the uninitiated reader. Medical terms are usually given additional description or definition. It also cites the references used and provides a complete reference list. This will allow the reader to find the original resource material if he/she wishes to learn more details about the topic. Photographs, images, and diagrams are also provided in order to complement the textual explanation with a visual representation.

Genetics enthusiasts, at various levels of expertise, as well as students or practitioners in the medical field will surely find this book as a useful reference to gain information or to aid in one's own research.

1

Fundamentals of Genetics

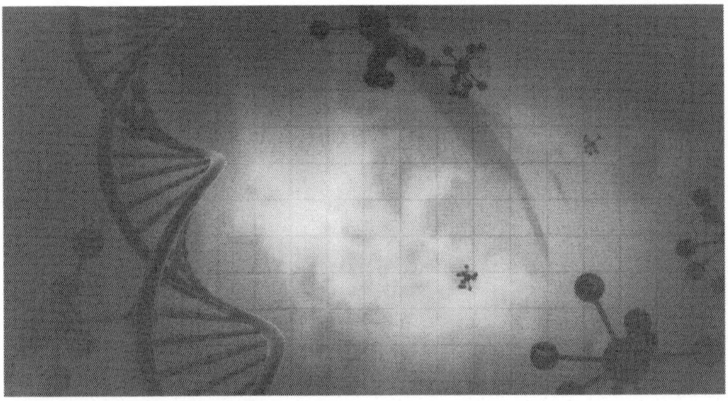

(Source: Microbiology, Pixabay, Public Domain)

Genetics is a field of study that focuses on the structure, function, and regulation of the genetic material. It peers into these aspects at various levels, i.e. molecular, cytological, organismic, and the population level. The expansion in this field gave rise to areas of specialization based on biological hierarchy or on the organism. Nevertheless, genetic principles are universal and are proven to be true regardless of the hierarchy or organism. In fact, these have contributed to a clearer understanding of other fields of biology like anatomy, physiology, taxonomy, or ecology as well as those delineated by their study organisms such as botany, zoology, microbiology, and the like. This chapter provides a concise discussion of these fundamental concepts, highlighting human examples if applicable.

1.1. THE LAWS OF HEREDITY BY MENDEL

Genetics started out as the study of **heredity**, which is the passing on of traits from parents to offspring. From the time that humans started to cultivate plants and domesticate animals, they undoubtedly observed that most offspring tend to resemble their parents. This observation of heredity led them to select the best parents to mate or cross in order to breed horses that were stronger or faster, cows that produced more milk, or crops with bigger or better-tasting fruits. In the human front, ancient Egyptians and Hebrews noticed that some diseases were passed down in families. Indeed, ancient people observed heredity and used it to their advantage, but they did not understand how it worked (Hasan, 2005).

Gregor Mendel, an Austrian monk, is considered the father of genetics and is popularly credited for coming up with the laws that govern heredity. However, there were others who attempted to explain heredity apart from Mendel. Notable among these pre-Mendelian expositions is that of **Francis Galton**, an English explorer, anthropologist, and eugenicist (Encyclopaedia Britannica, Inc., 2019a) (see Figure 1.1). He spoke about the regularity observed in heredity and attempted to determine the principle behind it. In 1897, he published a paper in which he proposed his law of ancestral heredity, stating that the parents contribute, on average between them, ½ to the offspring's heritable traits, the grandparents contribute ¼, the great-grandparents contribute 1/8, and so on. He came up with this conclusion by noting the pedigrees of Basset Hound, specifically noting the presence or absence of tri-colored fur (i.e., black, yellow, and white). Various matings between Basset Hound dogs of known ancestry were performed and the average numbers of tri-colored and non-tri-colored offspring corresponded with what was expected based on Galton's law (Bateson and Mendel, 2013; Bulmer, 1998). Galton's law was eventually surpassed by Mendel's law as Galton treated individual variations as errors while Mendel considered these as alternative expressions of discrete heritable factors, which we now call genes (Genome News Network, 2019).

Mendel proposed his three laws or principles of inheritance in 1865 (see Figure 1.2). These greatly contributed to a clearer understanding of how traits are inherited across generations. Prior to the proposal of these laws, inheritance was thought to be the result of the blending of the parental traits (Miko, 2008a). For instance, if a white flower-bearing plant is crossed with a red flower-bearing one, the offspring were expected to have pink or a variegated type of flower. Mendel did not subscribe to this idea. Instead, he hypothesized that each parent contributes some heritable substance, which he called **elementen**.

Figure 1.1. Sir Francis Galton proposed the law on ancestral heredity which quantified the hereditary contribution of parents, grandparents, and the other ancestors to an offspring's trait. (Source: Eveleen Myers, Public Domain).

Figure 1.2. Gregor Mendel is touted as the father of genetics for his ground-breaking study on common pea plants that lead him to propose the three laws of inheritance. (Source: National Institutes of Health, Public Domain).

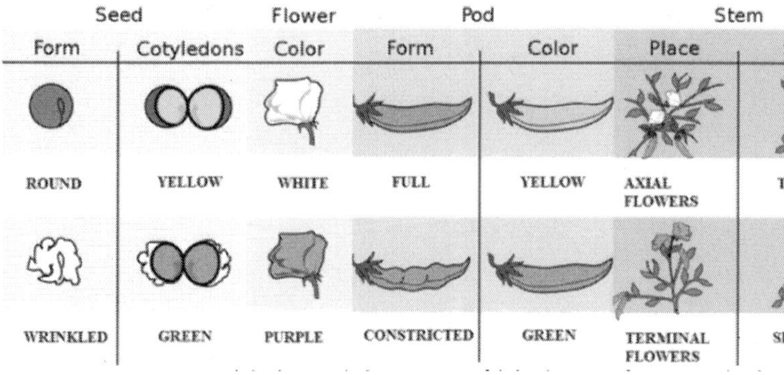

Figure 1.3. Gregor Mendel observed the pattern of inheritance of seven traits in the common pea plant. Each of these traits had two forms. (Source: Mariana Ruiz, Public Domain).

We now know that Mendel's elementen refers to the assembly of physical genes on the chromosome. Mendel came up with his laws after eight years of experimenting on the common pea plant (*Pisum sativum*). He performed controlled fertilization by using a small paintbrush to transfer pollen from one flower to the same flower (i.e., self-fertilization) or to another plant's flower (i.e., cross-fertilization). He observed seven traits of the pea plant, each of which had two types (see Figure 1.3). These included height (tall or short), pod shape (full or constricted), pod color (yellow or green), seed shape (round or wrinkled), cotyledon color (green or yellow), flower color (white or violet), and flower position (axial or terminal). For each trait that Mendel observed, he focused on how the elementen for that trait was distributed among the progeny (Miko, 2008a).

1.1.1. Law of Uniformity

Mendel self-fertilized pea plants for two years and observed that their characteristics remained constant in each generation. These parental lines could be considered pure-breeders, true-breeders or homozygous for the trait that they consistently show.

Mendel and his assistants eventually developed 22 varieties of pea plants with combinations of these consistent features (Miko, 2008a). Mendel did not only self-fertilize pure-breeding parents, but he also cross-fertilized plants that are pure-breeding for the two forms of a trait (e.g. tall crossed with short). The latter crosses produced hybrids or heterozygous types. The cross-fertilization between tall and short plants produced hybrid seeds that all grew to be tall, negating the blending theory of inheritance. Such cross-bred offspring constitute the first filial generation or F_1 generation. Mendel called the character (in this case, tallness) that appeared in the F_1 as dominant while the opposite (in this case, small height) that disappeared in the F_1, he called recessive (Bateson and Mendel, 2013).

This was also observed in other traits of the pea plants such as flower color (see Figure 1.3). From these results, Mendel proposed his first law, the law of uniformity. This states that all offspring of a cross between two pure-breds that differ by only one trait will have the identical appearance (Miko, 2008a).

1.1.2. Law of Independent Segregation

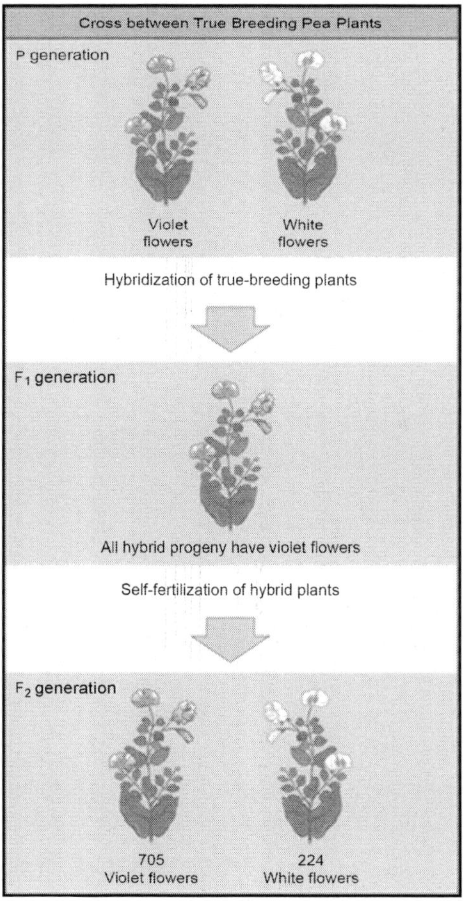

Figure 1.4. Mendel performed crosses between two true-breeding or homozy-

gous plants that differ by one trait producing a hybrid progeny (F_1 generation), which resemble one of the parents and are all identical in terms of that one trait. These offspring exhibit the so-called dominant trait. Mendel proceeded to self-fertilize the F_1s to produce the F_2 generation. The progeny in the F_2 showed a constant average of proportions, i.e. 75% exhibit the dominant trait while 25% show the recessive trait, which was not observed in the F_1. (Source: OpenStax Biology 2nd Edition, Creative Commons License).

Mendel self-fertilized two F_1 s=F1s to generate the **F_2 generation**, which consist of a mixture of tall and short plants. When accounting for the F_2 progeny, Mendel observed that the proportion of tall and short offspring was consistent, i.e. 75% were tall (dominants) while 25% were short (recessives) (Bateson and Mendel, 2013). Thus, the hidden recessive trait was carried down to the F_2 generation from the earlier parental generation. The technique employed by Mendel to observe the inheritance of a single trait is now called **monohybrid cross** (see Figure 1.5) (Miko, 2008a).

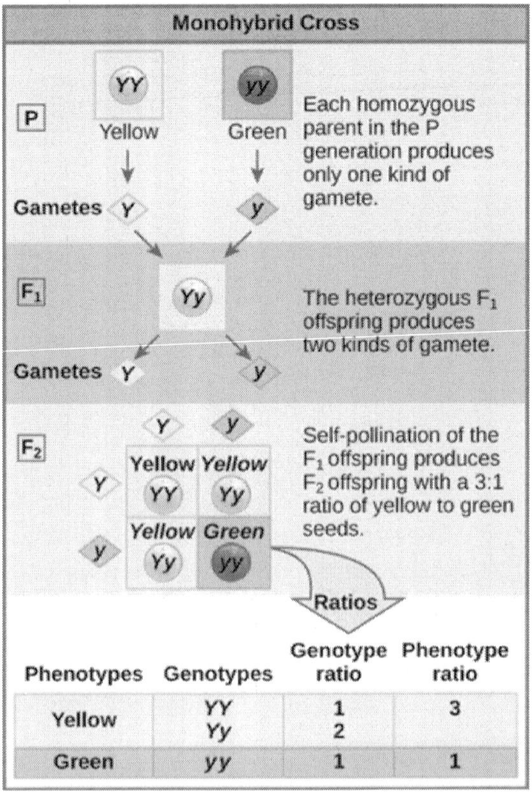

Figure 1.5. The crosses Mendel performed to produce the F_1 and F_2 progeny

constitute the monohybrid cross. The alleles of a gene are represented with the same letter with the capital letter representing the dominant allele while the lower-case letter represents the recessive one. Two alleles constitute a genotype for a single trait and these two get separated during meiosis and are distributed to the gametes. A Punnett square can be employed to determine all possible combinations of the gametes into genotypes. (Source: Lisa Bartee, Creative Commons License).

Mendel used capital and small letters to represent the heritable units. For example, he assigned *Aa* as the notation for the hybrid genotype, with each letter representing each parent's contribution. This representation of the genetic composition is now called the **genotype**. On the other hand, the observable trait that corresponds to the genotype is called the **phenotype**. Mendel's use of two letters for the genotype notation is impressive because despite not fully understanding the concept of a gene, his notation allowed for representing the various forms of a gene called the **allele**. For example, a **gene** is responsible for a trait such as seed color of the pea plant. One letter is used to represent the gene. There are two types of seed color, yellow and green, and so there are two alleles of the gene for seed color. The dominant allele is for yellow and so it is represented by the upper case letter *Y*, while the recessive allele is for green and represented by the lower case letter *y*.

Following this notation, the genotypes of the individuals involved in a monohybrid cross can be represented (see Figure 1.5). The true-breeding or homozygous parents have two copies of identical alleles. Thus, the parent with yellow seeds has the *YY* genotype while the other parent with the green seeds has the *yy* genotype. The F_1 hybrid offspring has the same phenotype as the parent with the yellow seeds but it has a different genotype. It is heterozygous with the genotype *Yy*. Therefore, Mendel inferred that the hybrid progeny carries the potential or allele for green seeds even though it does not express it. It also means that while a single copy of the dominant allele is sufficient for it to be expressed, the recessive allele needs to be present in two copies in order to be expressed. Mendel then proposed his second principle of heredity, the **law of independent segregation**, which states that the alleles in the genotype separate during meiosis and are distributed into gametes (Miko, 2008a). This means that the *YY* parent will produce gametes bearing the *Y* allele while the *yy* parent will produce gametes with *y* alleles. During the cross-fertilization, these two gametes combine to form the *Yy* offspring. In turn, the F_1 offspring undergoes meiosis and produces two types of gametes, 50% of the gametes bear *Y* while the other half bears *y*.

As previously mentioned, Mendel observed that self-fertilizing the F_1

produces F_2 progeny wherein 75% manifest the dominant trait (i.e., yellow seed in this case) and 25% display the recessive one (i.e., green seed), which can also be written as 3 dominant: 1 recessive. This constitutes the **phenotypic ratio**. Following the law of segregation, the **genotypic ratio** can also be determined, which provides all the possible genotypes and their respective proportions. One method for determining this in the F_2 is by making a Punnett square (see Figure 1.5). The various types of gametes (in terms of allelic composition) from one parent are written on the uppermost row while the gametes from the other parent are written in the leftmost column. All possible genotype combinations from the alleles can then be accounted for. The genotypic ratio is then determined by getting the proportion of each genotype. In this particular case, the expected F_2 genotypic ratio is 1 YY: 2 Yy: 1yy. This means that ¼th or 25% of the genotypes in the F_2 are expected to be YY, 50% are Yy, and 25% are gg.

1.1.3. The Law of Independent Assortment

Mendel also examined the inheritance of two traits by performing **dihybrid crosses**(see Figure 1.6). The homozygous parental plants utilized in this cross differ by two traits, i.e. seed color and seed shape. For instance, plants with yellow, round seeds were crossed with those which had green, wrinkled seeds. From his earlier monohybrid crosses, Mendel concluded that yellow seed color and round seed shape are dominant over green and wrinkled seeds. Thus, he expected that the F_1 hybrid would all be having yellow and round seeds, and that is exactly what he observed. He then self-fertilized the F_1 plants and observed a phenotypic ratio of 9 yellow, round seeds: 3, yellow, wrinkled seeds: 3, green, round seeds: 1 green, wrinkled seeds in the F_2 or it can be expressed as 9/16: 3/16: 3/16: 1/16(Miko, 2008a).

Mendel then examined the phenotypic ratio for each trait separately. He determined that the seed color showed 3 yellow: 1 green phenotypic ratio while the seed shape phenotypic ratio was 3 round: 1 wrinkled. From these data, Mendel proposed his third law, the law of independent assortment. According to this law, even though there were two traits in the cross the alleles for one trait segregate into the gametes during meiosis independently of the alleles in the other trait. Therefore, the inheritance of one trait did not affect the inheritance of the other trait (Miko, 2008a).

Using Mendel's notation for representing alleles in the genotype, the expected genotypic ratio in the F_2 of a dihybrid cross may be determined by the Punnett square method (see Figure 1.6). In the previously discussed cross involving the seed color and shape, the two true-breeding parents in the cross are represented as *YYRR* for the parent with yellow, round seeds and *yyrr* for the one with green, wrinkled seeds. Having in mind the law of assortment, the *YY* alleles separate from each other during meiosis in the same manner that *RR* alleles also get separated from each other. Hence, the double dominant parent produces gametes bearing the *YR* combination of alleles. The same principle applies to the double recessive parent so that its gametes contain the *yr* alleles. All the F_1 offspring are therefore expected to be double heterozygotes, i.e. *YyRr*, exhibiting yellow, round seeds.

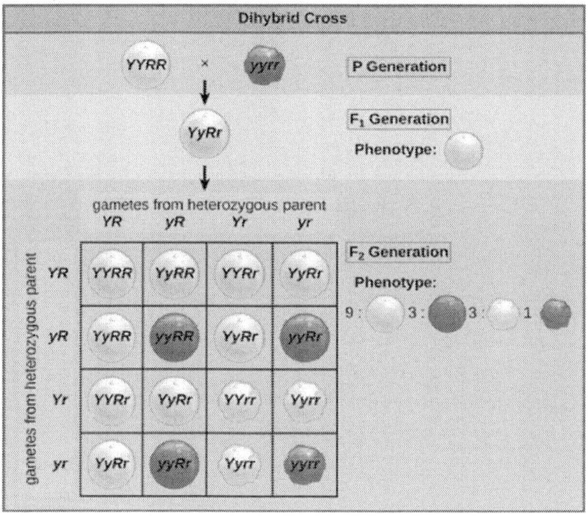

Figure 1.6. A dihybrid cross looks at the inheritance of two traits. Four types of gametes are expected from the dihybrid F_1 offspring. These four types are then combined in a Punnett square to predict the expected genotypes in the F_2. Accounting for the phenotypes of these genotypes results in a phenotypic ratio of 9 dominant in both traits: 3 dominant in the first trait, recessive in the second: 3 recessive in the first trait, dominant in the second one: 1 recessive in both traits. (Source: Lisa Bartee, Creative Commons License).

When the *YyRr* individual undergoes meiosis, the *Yy* and *Rr* gene pairs are expected to segregate independent of each other (i.e., *Y,y, R,r*). Then the separated alleles from the first trait combine with the separated alleles from the second trait. The types of gametes formed can be determined by utilizing the dichotomous branching method:

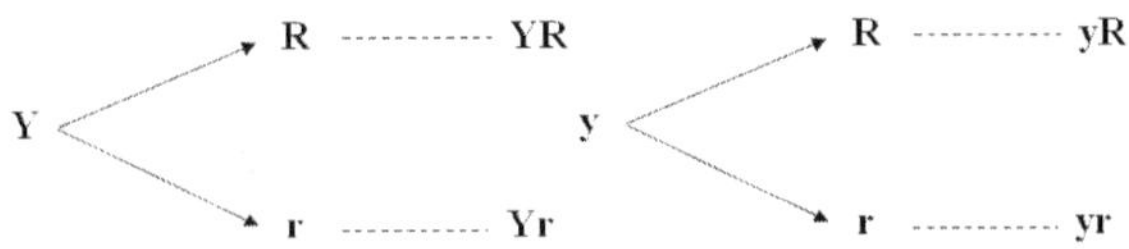

Then using the Punnett square , these gametes can be combined to determine the genotypes expected in the F2 generation after self-fertilizing the F_1 (see Figure 1.6). The genotypic ratio can be determined by accounting for the different genotypes and their corresponding proportion. The dichotomous branching method can also be employed to determine the F_2 ratios. In this method, the genotype of each trait is taken independently and then two monohybrid crosses are performed:

Yy x Yy →1 YY: 2 Yy: 1 yy (genotypic ratio)

3 yellow seeds: 1 green seed (phenotypic ratio)

Rr x Rr → 1 RR: 2 Rr: 1 rr (genotypic ratio)

3 round seeds: 1 wrinkled seed (phenotypic ratio)

The genotypes and phenotypes derived from each cross are then combined by dichotomous branching. In this method, the genotypes are combined and their corresponding proportions are multiplied to determine the genotypic ratio:

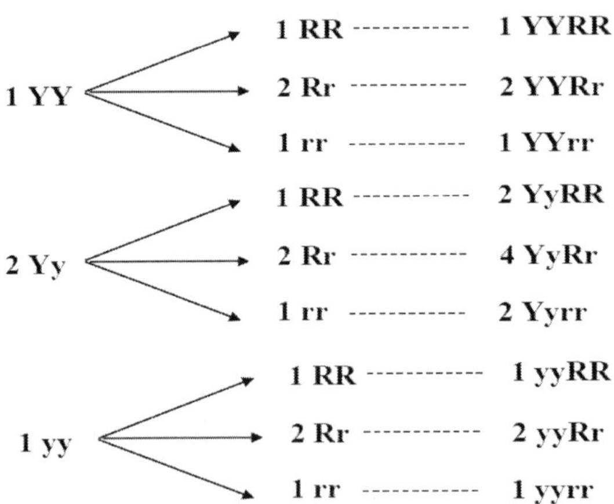

The same procedure is done to determine the phenotypic ratio, where the phenotypes are combined and the proportions are multiplied:

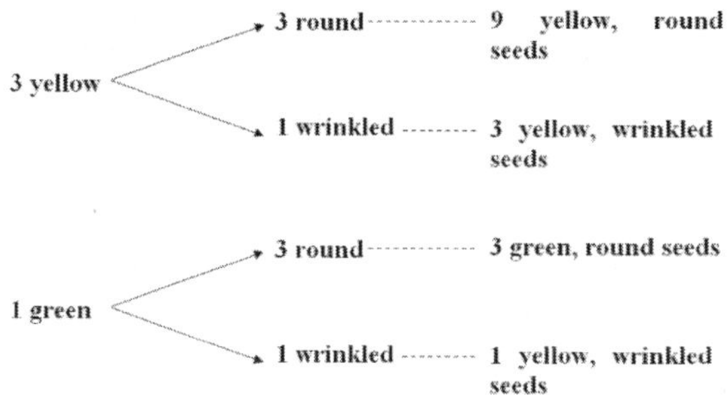

The dichotomous method is a faster and more efficient method of determining genotypic and phenotypic ratios unlike the Punnett square method, which entails counting similar genotypes in the square and listing the various genotypes with their corresponding proportions. There is also no need to determine the types of gametes when using this method.

More than 30 years later, when Mendel's heritable units were discovered to reside in cellular structures known as **chromosomes**, it became apparent that the alleles in a gene pair separate because chromosomes separate during meiosis and that the alleles of different traits show independent assortment because they reside in different chromosomes. A more detailed discussion of chromosomes will be provided in another section of this chapter.

1.1.4. Mendelian Inheritance in Humans

Figure 1.7. William Bateson rediscovered the work of Mendel and recognized its implication in human heredity. (Source: www.amphilsoc.org, Public Domain).

William Bateson, a British biologist, is credited for coining the term genetics(see Figure 1.7). Along with the botanists Hugo de Vries, Carl Correns, and Erich Tschermak von Seysenegg, they championed Mendel's work more than three decades after he published his experiment and so they are known as the rediscoverers of Mendel (Encyclopaedia Britannica, 2019b). Bateson supposed that Mendel's laws of heredity may be extended to the understanding of the inheritance of human traits.

In his book, Mendel's Principles of Heredity, he suggested that if parents were studied carefully, their contribution to their child's traits may be determined. However, Mendel's laws of inheritance are associated with statistical outcomes that require a large number of offspring. This presents a challenge in humans, since their mating does not usually produce a lot of offspring. Additionally, The more obvious human traits such as eye color, skin and hair color or height do not exhibit the expected transmission ratios as it was eventually determined that several genes are involved in these traits or that their inheritance is more complex than that proposed by Mendel (Ostrer, 1998). Still, Mendelian heredity is observed in humans. It was first demonstrated in unusual traits and rare diseases like brachydactyly (short hands), hereditary hematuria (blood in urine), and keratosis palmaris et plantaris (thickened palms and soles) (Ostrer, 1998). These conditions or diseases are called monogenic since these are caused by mutations or changes in a single gene. By constructing and analyzing a pedigree, which is a diagram showing a line of ancestors with their corresponding phenotype or genotype relative to a certain trait, the pattern of inheritance of such trait can be determined (see Figure 1.8). There is a standard key followed in order to represent male and female members of the pedigree, the relationship among the individuals, whether or not an individual is affected by the genetic condition, and also if certain individuals are deceased. Such pedigree analysis can help one determine if the gene for a trait is found in an autosome, which is a non-sex chromosome, or in a sex chromosome.

Furthermore, based on the pattern of inheritance observed in the pedigree, it can be inferred if a trait is due to a dominant or recessive allele (Chial, 2008).

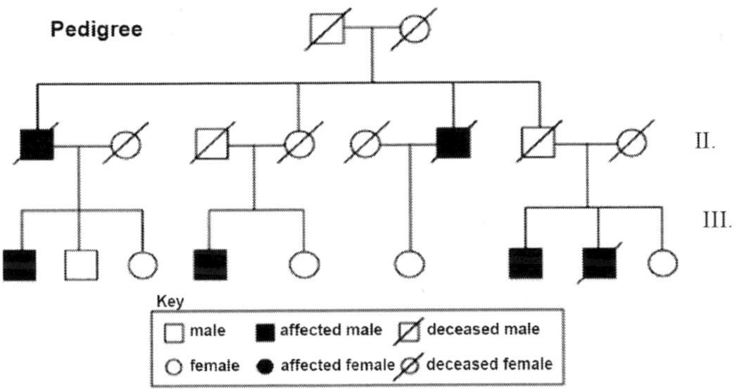

Figure 1.8. In a pedigree, the Roman numerals represent the generation number. Circles represent the female members of the family while squares are the males. Those who are unshaded do not show the trait while those who are shaded do. A horizontal midline connecting two individuals denotes marriage or mating between the two. A vertical line connects the parents to their children and these line branches out to connect to each child. (Source: Modified from National Institutes of Health, Public Domain).

1.1.4.1. Autosomal recessive inheritance

If the number of affected males and females in the pedigree is roughly similar, then the gene for the trait is likely to be located on an autosome. This is because both males and females of a species have the same number and types of autosomes. Therefore, all the genes that are in the autosomes are expected to be present in both sexes. A trait that is due to a recessive allele can only be expressed by an individual if there are two copies of the mutant allele. Therefore, individuals that are **heterozygotes** (i.e., with a heterozygous genotype) with one copy of the mutant allele are unaffected just like those that have a homozygous dominant genotype, which does not have any copy of the mutant allele. Nevertheless, heterozygotes are considered to be carriers of the allele for the trait. A clear pattern observed for autosomal recessive inheritance is that the trait may skip (i.e., affected individuals are not observed) one or more generations (see Figure 1.9). An affected offspring may be observed if the parents are both carriers or if both parents are affected. Unless a genetic test is available to determine the individuals' genotypes, it is difficult to differentiate the homozygous dominants from the heterozygotes since both genotypes manifest a normal phenotype. However, it can be concluded that the normal parents of the affected individuals have heterozygous genotypes.

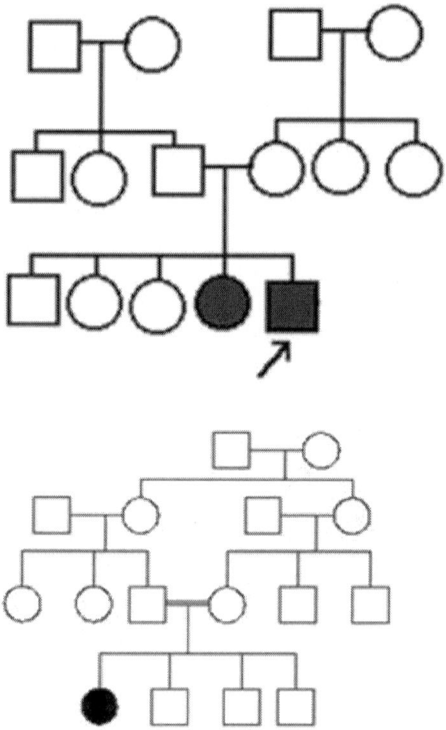

Figure 1.9. These two pedigrees show the classic skipping of generations which usually indicates that the trait is due to an autosomal recessive gene. Specifically, affected individuals were only observed in the third (for the upper pedigree) and fourth (for the lower pedigree) generations. Thus, affected individuals have a homozygous recessive genotype. The arrow indicates that the individual is the **proband**, or the starting point for the genetic study. It is interesting to note that the parents of the affected female in the lower pedigree are first cousins. Such consanguineous mating is denoted by a double horizontal midline. Mating between relatives increases the likelihood that two parents are carriers since they come from the same family, which has a history of the trait. (Source: AlěsPanczak, Creative Commons License).

Several human disorders are due to autosomal recessive alleles such as phenylketonuria (PKU), cystic fibrosis, sickle-cell anemia, and oculocutaneous albinism (Chial, 2008). A separate chapter will describe the molecular basis of these disorders.

1.1.4.2. Autosomal dominant inheritance

Disorders due to an autosomal dominant gene occur even if the individual only has a single copy of the mutant allele. Therefore, both homozygous dominants and heterozygotes manifest the condition. Only individuals with homozygous recessive genotype, i.e. with two copies of the non-mutant or **wild type** alleles, are unaffected. Affected mother or father having affected child/children is a pattern observed in the pedigree of this type of inheritance (see Figure 1.10). Thus, there are affected individuals in all generations.

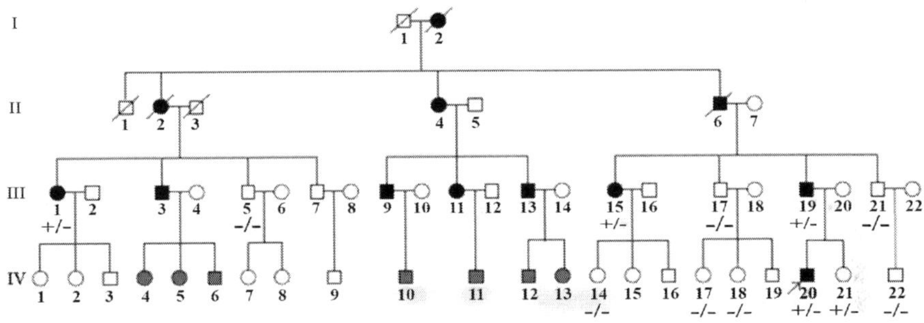

Figure 1.10. In this pedigree of a Chinese family with members suffering from autosomal dominant late- onset nonsyndromic hearing loss, there are six affected females and six affected males which are observed in all four generations. These patterns are characteristic of an autosomal dominant trait. Gray squares and circles represent members with unavailable status. All unaffected individuals are assumed to have homozygous recessive genotypes. Affected individuals may either have homozygous dominant or heterozygous genotypes. For the I2 female, she is likely to be a heterozygote since after marrying a normal male they had three affected children but one son (individual II1) was unaffected. Given these genotypes of the parents, all the affected children(generation II) are inferred to be heterozygotes. All the three affected siblings married normal mates and so it is expected to observe affected heterozygotes (+/-) as well as normal individuals (-/-) among their offspring(generation III). while - represents the recessive wild-type allele (Source: Zhang, C.; Wang, M.; Xiao, Y.; Zhang, F.; Zhou, Y.; Li, J.; Zheng, Q.; Bai, X.; Wang, H. Neural Plasticity1512831, 10 pp., 2016. Creative Commons License).

Apart from late-onset nonsyndromic hearing loss (see Figure 1.9), other diseases that are due to autosomal dominant alleles include Huntington's disease, myotonic dystrophy, familial hypercholesterolemia, neurofibromatosis, and polycystic kidney disease (Chial, 2008).

1.1.4.3. Sex-linked inheritance

The various traits of the pea plants studied by Mendel are all autosomal. However, in sexually reproducing organisms, the genes for certain traits are in the sex chromosome. These traits are called sex-linked and were first extensively studied in *Drosophila* (fruit fly) by Thomas Hunt Morgan, an American zoologist and geneticist (see Figure 1.11). In *Drosophila* and humans, there are two types of sex chromosomes, the X and Y chromosomes. Females have two copies of the X chromosomes, i.e. XX., a condition called homogametic since all female gametes are X-bearing. On the other hand, males are XY making them heterogametic with 50% of the sperm carrying X while the other 50% have Y. Males are also considered to be hemizygous, i.e. having a single copy of genes in the X chromosomes. The X and Y chromosomes are different from each other, so the genes present in the X chromosome are absent in the Y, and vice versa. Thus, the occurrence of a sex-linked trait depends on the sex of the individual. In the case of birds, males are homogametic with the ZZ genotype while females have ZW.

Figure 1.11. Thomas Hunt Morgan studied the inheritance of traits in *Drosophila* including those located in the sex chromosomes. The pattern of inheritance of these traits deviated from that of autosomal traits. (Source: National Institutes of Health, Public Domain).

In the case of a sex-linked trait, writing the genotype differ from the manner it is written for autosomal traits since the sex genotype is written and then the alleles are appended as superscripts. An example is eye color in *Drosophila*. The wild type allele, w+, is responsible for red eye color and it functions in a dominant manner. On the other hand, the mutant form, w, gives rise to white eye color. These alleles are in the X chromosome. Thus, homozygous dominant and heterozygous genotypes are written as $X^{w+}X^{w+}$ and $X^{w+}X^{w}$, respectively. For the female fruit fly, two copies of the mutant allele are necessary for the white eye to be expressed, i.e. $X^{w}X^{w}$. In the case of the male, the red-eyed genotype is $X^{w+}Y$ and since it is hemizygous, a single copy of the mutant allele results in a white eye phenotype, i.e. $X^{w}Y$.

When performing crosses involving sex-linked traits, the principles of segregation and assortment still apply with respect to the sex chromosomes (see Figure 1.12) so that the gametes formed and the expected genotypic and phenotypic ratios in the offspring can be predicted.

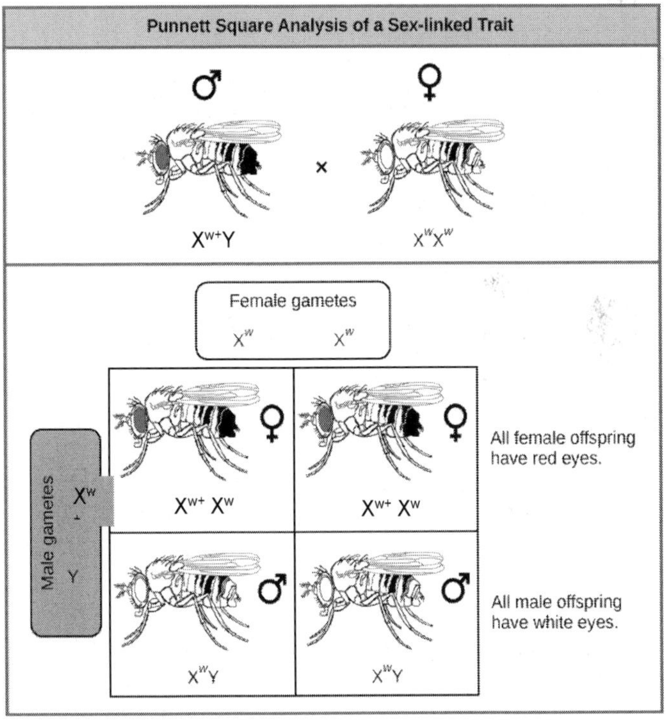

Figure 1.12. The allele for eye color in *Drosophila* is located in the X chromosome with the allele for red eye color (w+) being dominant and that for white color (w) being recessive. A red-eyed male is expected to produce X-bearing and

Y-bearing gametes. All the X-bearing sperms have the allele for red eye color. A white-eyed female is homozygous recessive and so all its gametes bear the X chromosome with the recessive allele for white eye color. A cross between these two is expected to produce red-eyed females and white-eyed males. (Source: Charles Molnar and Jane Gair, Creative Commons License).

The traditional method of differentiating whether the gene of a trait is autosomal or sex-linked is by performing reciprocal crosses. These involve performing crosses between true-breeding parents with opposite phenotypes, and hence genotypes, and then performing a reciprocal cross where the phenotypes of the parents are switched. An autosomal trait will yield the same type of offspring in reciprocal crosses, as demonstrated by the following diagram:

Female parent (♀)		**Male parent (♂)**
AA		aa
Yellow seed-bearing	x	Green seed-bearing

F_1 Aa (All yellow seed-bearing)

Reciprocal cross:

Female parent (♀)		Male parent (♂)
aa		AA
Green seed-bearing	x	Yellow seed-bearing

F_1 Aa (All yellow seed-bearing)

On the other hand, reciprocal crosses for sex-linked traits will produce different proportions of genotypes and phenotypes in the offspring (see Figure 1.13).

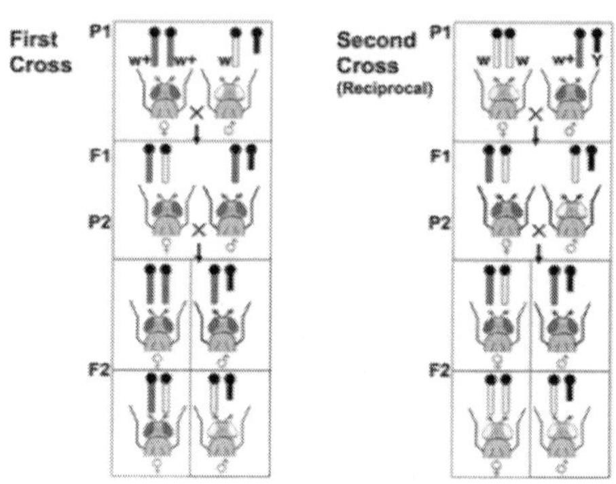

Figure 1.13. In the first cross where a true-breeding red-eyed female is crossed with a white-eyed male, the F_1 progeny is composed of heterozygote red-eyed

females and red-eyed males. When a reciprocal cross is performed involving a white-eyed female and a red-eyed male, the F_1 progeny consists of heterozygote red-eyed females and white-eyed males. (Source: Todd Nickle and Isabelle Barrette-Ng, Creative Commons License).

1.1.4.3.1. X-linked recessive inheritance

For traits that are due to an X-linked recessive allele, it can be observed from the pedigree that more males are affected than females because males need only a single copy of the allele to express the trait. This is because they only have one X chromosome, so they do not have a second allele to potentially cancel out the effect of the mutant one (Chial, 2008). On the other hand, females with two X chromosomes need to have two copies of the mutant allele in order to express the recessive phenotype. A pattern observed for this type of inheritance is that all the sons of an affected mother are affected since they will inherit an X chromosome with the recessive allele. Meanwhile, the daughters are normal because they will inherit another X chromosome from the normal father, making them heterozygous. Affected males may also be observed if their mother is a carrier (see Figure 1.14). On the other hand, affected daughters may be observed if the father is affected and if the mother is also affected or a carrier.

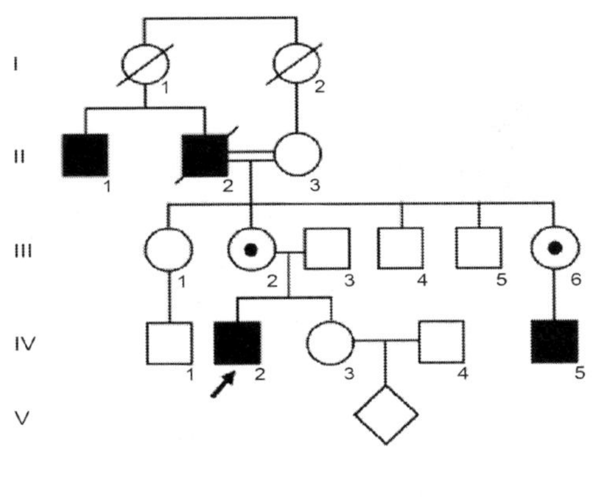

Figure 1.14. All the individuals affected with the retinal degenerative disease in this pedigree of a Chinese family are males which is one indication for an X-linked recessive inheritance. It is possible that individual I1, who is deceased, might have been affected or at least a carrier since both her sons are affected. The mothers of the affected individuals IV2 and IV5 are normal but they must be carriers of the mutant allele, which they inherited from their affected father.

This was confirmed by DNA sequencing in the case of individual III2. The dotted circles denote female carriers and the diagonal shape represents an unborn infant. (Source: Guo, H.; Li, J.; Gao, F.; Li, J.; Wu, X.; Liu, Q. BMC Ophthalmology 15, 85, 2015, Creative Commons License).

Genetic diseases that are due to an X-linked recessive allele include retinal degenerative disease (see Figure 1.14), red-green color blindness, hemophilia, and Duchenne muscular dystrophy (Guo *et al*., 2015; Chial, 2008). Due to its pattern of inheritance, more males are afflicted with these disorders than females.

1.1.4.3.2. X-linked dominant inheritance

For diseases that are due to an X-linked dominant allele, the pattern observed in the pedigree is that all the daughters of an affected father are affected, while none of the sons are. This is because the single X-linked dominant allele inherited by the daughters from the father is enough to express the condition while the sons inherit the Y chromosome from the affected father, which does not have the gene. For this reason, females are affected with X-linked dominant disorders at a higher frequency than males (see Figure 1.15). If the mother is affected and she has two copies of the dominant allele, all her children will be affected, whether sons or daughters. For heterozygote mothers, there is a 50% chance of having affected children, whether sons or daughters.

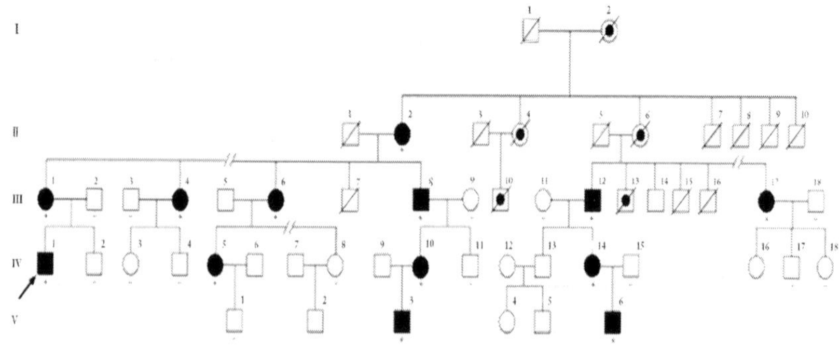

Figure 1.15. A family's pedigree for Charcot-Marie-Tooth disease type 1 (CMTX1) shows more affected females rather than males which is one indication for X-linked dominant inheritance. Dot symbols are used to represent affected individuals whose phenotype was deduced with a medical history. Most of the living family members were genetically tested and the symbol "+" means the individual has the dominant allele while "–" denotes absence of the domi-

nant allele. Individuals with no symbol were not genetically tested. It can be inferred that all affected females are heterozygous since they only had one affected parent while the other parent is normal. This was confirmed for most of the affected females because they were able to have normal children. (Source: Luo, S.; Jin, H.; Chen, J.; Zhang, L. Frontiers in Neurology10, 413, 2019, Creative Commons License).

Diseases that are inherited in an X-linked dominant manner are very rare. These include Charcot-Marie-Tooth disease type 1 (CMTX1) (see Figure 1.15), hypophosphatemic rickets, and Rett syndrome (Luo *et al.*, 2019; Chial, 2008).

1.1.4.3.3. Y-linked inheritance

Genetic disorders due to a Y-linked allele are extremely rare. Since only males have the Y chromosome, these alleles are inherited exclusively from fathers to sons (see Figure 1.16). A single copy of the mutated allele is sufficient to express the phenotype.

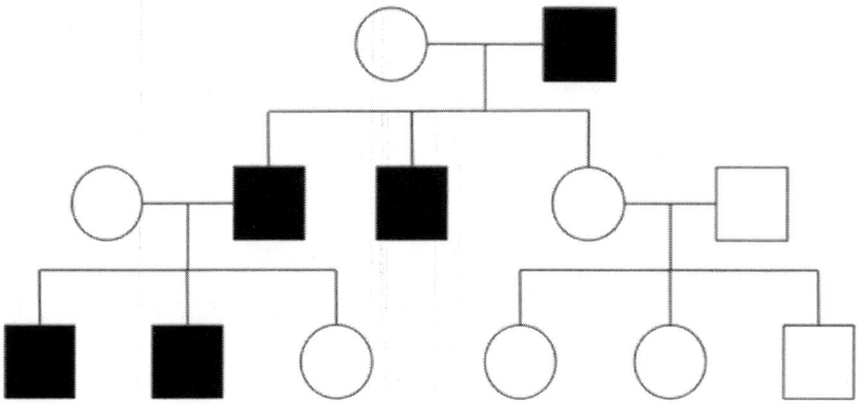

Figure 1.16. A Y-linked inheritance can be deduced from a pedigree where all affected members are males and all the sons of the affected father are affected while the daughters are normal. (Source: Madibc68, Creative Commons License).

An example of a condition due to a Y-linked gene is non-obstructive spermatogenic failure (Chial, 2008). The Y chromosome contains a small number of genes that are involved in the primary sexual development of males so mutations in these genes are expected to affect male sexual development.

1.2. CHROMOSOMES AND HEREDITY

Genes are located on chromosomes. This vital piece of information sheds greater light on explaining Mendel's laws of inheritance and advanced the field of genetics, particularly cytogenetics. Around the time that Mendel published his research, scientists already understood that cells are derived from other cells and that the hereditary information is located in the nucleus. Walther Flemming, a German anatomist, was able to improve his microscope's resolution to observe a fibrous network within the nucleus (see Figure 1.17A). He called this chromatin, meaning "stainable material". He noted that the chromatin formed into thread-like bodies during cell division. He called these mitosen (from the Greek word for thread). These were eventually called chromosomes, a term coined by Heinrich Waldeyer a few years later (O'Connor and Miko, 2008).

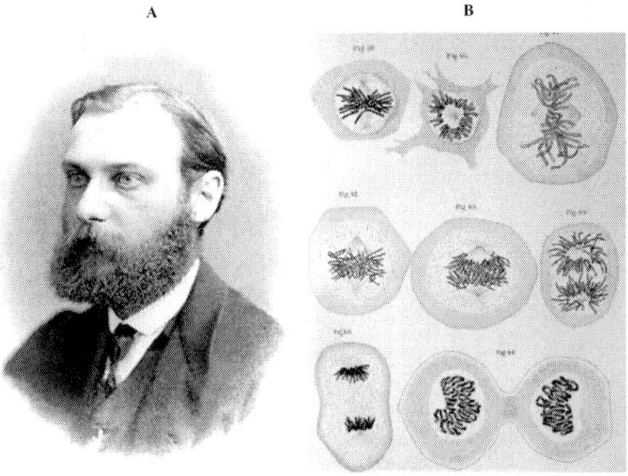

Figure 1.17. (A)Walther Flemming was a German anatomist who observed, illustrated, and correctly deduced the sequence of chromosome behavior during mitosis. (B) He illustrated how the chromosomes appeared under the microscope at various stages of mitosis. (Source: A: Public Domain; B: Walther Flemming, Public Domain).

Flemming made several microscopic observations of cells under various stages of cell division. Although he was observing fixed, and hence dead cells, he illustrated the correct sequence of chromosome movements during mitosis (see Figure 1.17B). From his observations, he noted that the chromosomes split along their length at one point of mitosis. He hypothesized that these separated chromosomes were distributed into different daughter cells during meiosis. When it was possible to observe live dividing cells, Flemming's findings were confirmed to be correct (O'Connor and Miko, 2008).

1.2.1. The Chromosome Structure and Number

The fibrous chromatin in the nucleus observed by Flemming condenses into thread-like, dark staining structures called chromosomes (from the Greek words *chroma*, meaning color, and *soma*, meaning body) once the cell starts dividing. Each chromosome has a constricted region called **centromere** (see Figure 1.18). This divides the chromosome into the upper p arm and the lower q arm. On the other hand, the ends of a chromosome are called the telomeres. Prior to cell division, the chromosome replicates so that it is made up of two sister chromatids. The location of the centromere along the chromosome gives it a characteristic shape, which can be used to classify chromosomes (see Figure 1.19). Metacentric chromosomes have centrally located centromeres so that the p and q arms are of equal length. Submetacentric chromosomes have p arms that are slightly shorter than the q arms because their centromeres are located off-center. The centromere in acrocentric chromosomes is located near the tip so that the p arm is significantly shorter than the q arm. The centromere is located at the tip of telocentric chromosomes so that they only have one arm.

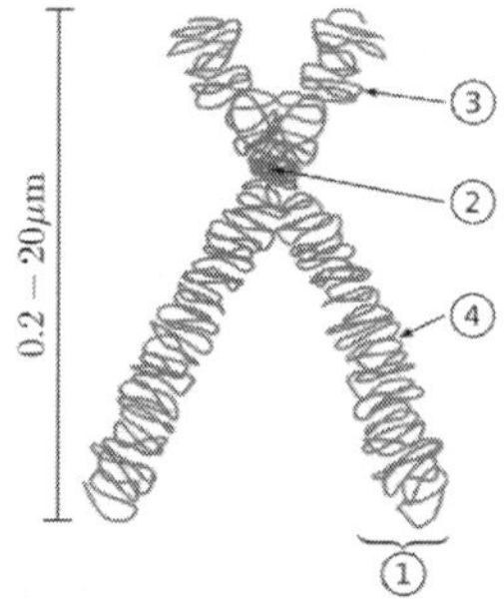

Figure 1.18. A replicated chromosome is made up of two sister chromatids joined by a centromere (2). The end of the chromosome arms is the telomere (1). The upper arm of the chromosome is called the p arm (3) while the lower one is the q arm (4). Source: Magnus Manske, Creative Commons License).

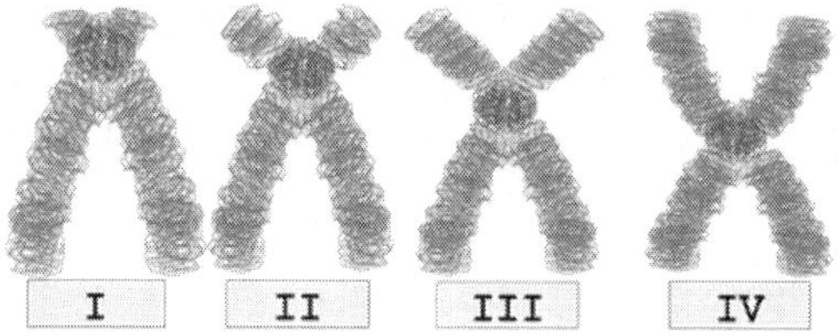

Figure 1.19. Chromosomes may be classified based on the placement of their

centromeres. A telocentric chromosome (I) has its centromere located at the tip of the chromosome. An acrocentric chromosome's(II) centromere is positioned near the end of the arms while that of a submetacentric chromosome (III) is positioned just off-center. Finally, a metacentric chromosome (IV) has its centromere in the middle. (Source:Fockey, 2003, Creative Commons License).

Advances in chromosome study revealed that the centromere is made up of numerous **centromere** proteins and serves as the site of kinetochore assembly. The kinetochore is a protein complex that is observed under the electron microscope to be made up of several layers of plate-like structures. The kinetochore complex is made up of 80 or more proteins (see Figure 1.20). These facilitate microtubule attachment, microtubule polymerization, and motor protein-directed movement during cell division (O'Connor, 2008a).

The number of chromosomes in a cell may be a haploid (n), which has one set of different chromosomes or diploid (2n), which has two sets of homologous chromosomes. A diploid chromosome number is usually achieved after fertilization where the chromosomes from the egg cell (i.e., the maternal chromosomes) combine with the chromosomes from the sperm (i.e., paternal chromosomes). The chromosome number may be more than diploid, like in cases of aneuploidy or polyploidy. Most organisms have a fixed chromosome number. For example, human cells are diploid with 46 chromosomes (2n=46) except the gametes which are haploid with 23 chromosomes (n=23).

Deviations from the chromosome structure and number may result in congenital malformations, intellectual disability, miscarriage, stillbirth, and risk of malignancy (Korf, 2017). An example is having an extra chromosome 21 in humans, which results in Down syndrome.

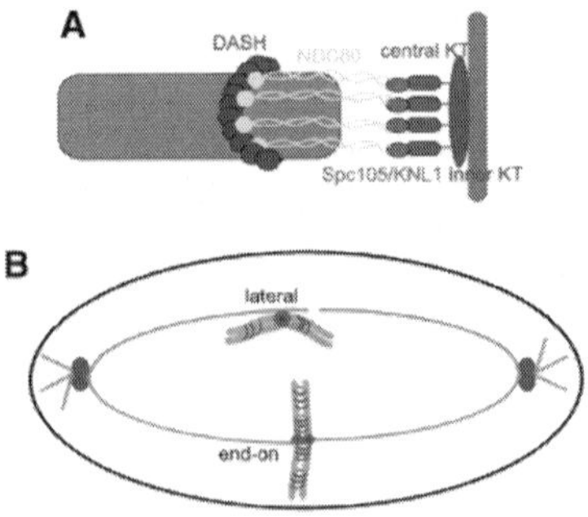

Figure 1.20. (A) The proteins of the inner kinetochore (purple) assemble on the

chromosome centromere (gray). The outer kinetochore proteins (SPC105 : light blue, NDC80: yellow, and DASH: blue) is where the microtubules bind while the proteins making up the central kinetochore (red) link the inner and outer kinetochore subcomplexes. (B) Kinetochores are involved in the attachment and polymerization of microtubules during cell division. (Source: Duro, E.; Marston, A.L. Genes and Development 29(2), 109-122, 2015, Creative Commons License).

1.2.2. Chromosome Theory of Inheritance

At the end of the 19th century, there were marked improvements in specimen preparation and staining, as well as microscopy, which enabled closer observation of chromosomes and their behavior during cell division, fertilization, and embryonic development. This was what **Theodor Boveri**, a German embryologist, did with the roundworm *Parascaris equorum* (formerly known as *Ascaris megalocephala*) and sea urchin (see Figure 1.21). Boveri is mainly remembered for his identification of chromosomes as the vectors of Mendelian factors or genes. His use of *Parascaris* embryo is ideal for studying early development because its embryonic cells have only two pairs of chromosomes and they develop into distinct somatic and germ cells. **Somatic cells** are those that eventually make up the various non-reproductive parts of the organisms while **germ cells** give rise to the gametes. Meanwhile, sea urchin egg cells are also ideal for study because they are transparent and available in large numbers (O'Connor and Miko, 2008; Satzinger, 2008).

A B

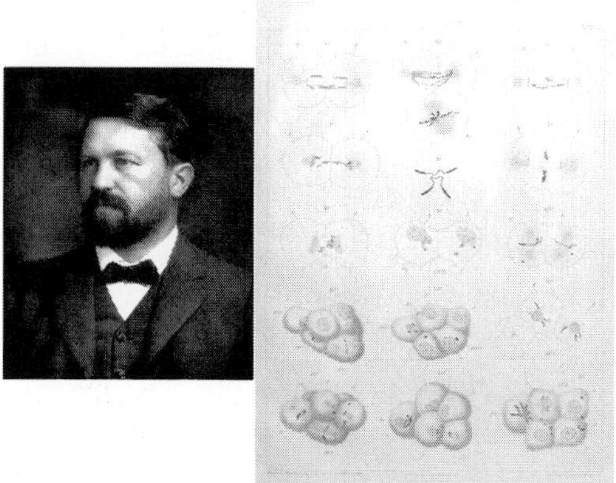

Figure 1.21. (A) Theodor Boveri inferred that Mendel's hereditary units are located in the chromosomes. (B) Boveri made drawings illustrating the various

stages of cell division that occur during the embryonic development of *Parascaris equorum*. (Source: A: Hugo Freund and Alexander Berg (ed.), Public Domain; B: Theodor Boveri, Public Domain).

Boveri observed that the number of chromosomes was reduced during the development of germ cells specifically the egg cell, into gametes, (a process that is eventually called meiosis). However, since both the egg and sperm contribute equal sets of chromosomes to the next generation of cells, the cells acquired a complete set of chromosomes after fertilization. He was able to monitor the chromosomes as the nuclei of the *Ascaris* embryo cells underwent cleavage (Satzinger, 2008; Maayan, 2011). Boveri further demonstrated the contribution of each chromosome during development by observing a sea urchin egg cell that was fertilized by two sperm cells. This meant that the fertilized cell contained three sets of chromosomes, i.e. one set of maternal and two sets of paternal chromosomes. This resulted in an unequal distribution of chromosomes during subsequent cell divisions so that the resulting embryos contain different sets of chromosomes. These embryos were allowed to develop further. Some died while others generated sea urchin larvae exhibiting various abnormal or undeveloped parts. Boveri performed another experiment where he split the embryo from polyspermic eggs into its individual cells or blastomeres. He allowed each cell to develop further into an embryo and observed that most of the resulting larvae are malformed. These results led Boveri to conclude that only the normal full set of chromosomes can support the development and that each chromosome contained a specific hereditary property for the organism (Satzinger, 2008; Maderspacher, 2008).

Around the same time that Boveri worked on chromosomes, **Walter Sutton**, an American geneticist who was a graduate student at that time, confirmed and expanded on Mendel's and Boveri's conclusions. He studied meiotic cells in the testes of the lubber grasshopper (*Brachystola magna*). It is possible to distinguish individual chromosomes in these cells. He was able to identify 11 pairs of chromosomes that could be distinguished by their sizes and one sex chromosome. He observed the behavior of these chromosomes through the various stages of meiosis, including synapsis, where homologous paternal and maternal chromosomes pair up, and determined that the chromosomes maintained their stable structure throughout meiosis. In his paper about this study that was published in 1902, he concluded: "I may finally call attention to the probability that the association of paternal and maternal chromosomes in pairs and their subsequent separation during the reducing division as indicated above may constitute the physical basis of the Mendelian law of heredity" (O'Connor and Miko, 2008).

He also noted that the alignment of synapsed chromosomes along the cell's equator during metaphase is random. This means that the paternal or maternal chromosome does not favor one side or pole during cell division. Thus, each chromosome is independent of the others so that the gametes formed after meiosis do not necessarily contain the same mixture of chromosomes. Moreover, he correctly assumed that each chromosome may contain more than one trait. This provided the basis for the observed variation in heritable traits (O'Connor and Miko, 2008).

Less than a decade after Boveri and Sutton published their findings that supported the chromosome theory of inheritance, Thomas Hunt Morgan and his colleagues at Columbia University identified hundreds of fruit fly (*Drosophila*) genes, like the gene for eye color that is located on the X chromosome (see Figure 1.11) (Section 1.2.2.2 discusses further the concept of sex linkage). They determined that a recessive mutation of the gene, which is for red eye color, is responsible for the unusual white eye color that they observed in their cultures (see Figure 1.12). Since female flies usually have two X chromosomes while the males have one X chromosome and a smaller Y chromosome, they hypothesized that male flies will always express the gene in their single X chromosome. Meanwhile, female flies would show white eyes only if they inherit two copies of the recessive gene. They performed thousands of mating that confirmed this hypothesis. However, they observed that on rare occasions, a heterozygous female fly crossed with a normal male produces white-eyed females. This contradicted with what was expected, which was all females produced should be red-eyed. It was Morgan's wife, **Lilian Vaughan Morgan**, who correctly surmised that these females might have an exceptional chromosome composition. In fact, these females had two X chromosomes and a Y chromosome, which was a result of defective meiosis. Specifically, the X chromosome containing the mutant allele replicated, then failed to separate so that the resulting offspring had two copies of the recessive allele. These results provided experimental evidence that genes are actual entities that are located on chromosomes (O'Connor and Miko, 2008). This association of chromosome with heredity gave rise to a special field called **cytogenetics**.

1.2.2.1. Chromosome segregation via cell division

When it has become clear that genes are indeed located on chromosomes, the bases of Mendel's laws of inheritance also became clear. The segregation and assortment of Mendelian factors (or genes) result from the separation and assortment of chromosomes during cell division.

1.2.2.1.1. The cell cycle

During their lifetime, cells enter the cell cycle, which is made up of the **interphase** and the **M or division phase** (see Figure 1.22). Interphase occurs prior to division and is divided into three stages. While this is a non-dividing phase of the cell, it is characterized by certain metabolic processes. The first stage of interphase is $\mathbf{G_1}$, the primary phase where the cells grow through the synthesis of proteins. This is followed by the **S phase** in which DNA is replicated so that each chromosome is made up of two identical fibers known as **sister chromatids**, which are joined to each other at the centromere. Then $\mathbf{G_2}$ occurs ,which is the second growth phase where the organelles replicate, chromosomes condense, and the spindle apparatus starts to form.

Following the interphase is cell division or the M phase where the replicated chromosomes are distributed to the daughter cells. There are two types of cell division: mitosis and meiosis. Each type has its own role in an organism's development and life cycle.

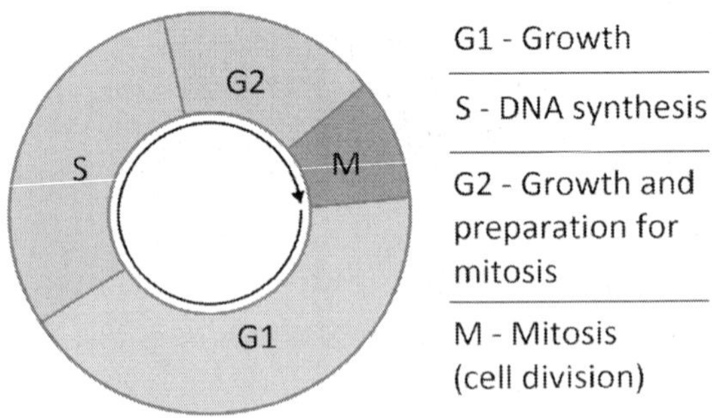

Figure 1.22. The cell cycle is made up of the interphase, which prepares the cell prior to cell division, and the M or division phase. Interphase is subdivided into three stages, i.e. G_1, S, and G_2. (Source: Simon Caulton, Creative Commons License).

1.2.2.1.2. Mitosis preserves the genetic information

Mitosis is a type of cell division that produces two daughter cells that are genetically identical to each other and to their parental cell. It is essential for the development of a single-celled zygote into an embryo and finally into a multicellular adult,as well as for replacing aged and damaged cells. Unicellular organisms reproduce by a process similar to mitosis called binary fission. Mitosis has four stages (see Figure 1.23).

The first stage of mitosis is **prophase** where the nucleoli and nuclear membrane disintegrate and the chromosomes continue the condensation process that began during G_2 so that they become visible under the light microscope. The assembly of the spindle fiber apparatus also continues at this stage. In animal and human cells, this involves the replication of centrioles and one of them moves to the opposite side. These two centrioles are soon connected by spindle microtubules.

Metaphase soon follows where the sister chromatids align at the cell's equator with each chromatid facing each pole of the cell. At **anaphase**, the centromeres divide and the sister chromatids separate from each other and move to their respective poles. Because each chromatid has a centromere, it is now considered a chromosome. Thus, there are now two sets of chromosomes, one set in each of the two poles of the cell. The last stage is telophase where the spindle apparatus is disassembled and a nuclear membrane is formed around each of the chromosome sets.

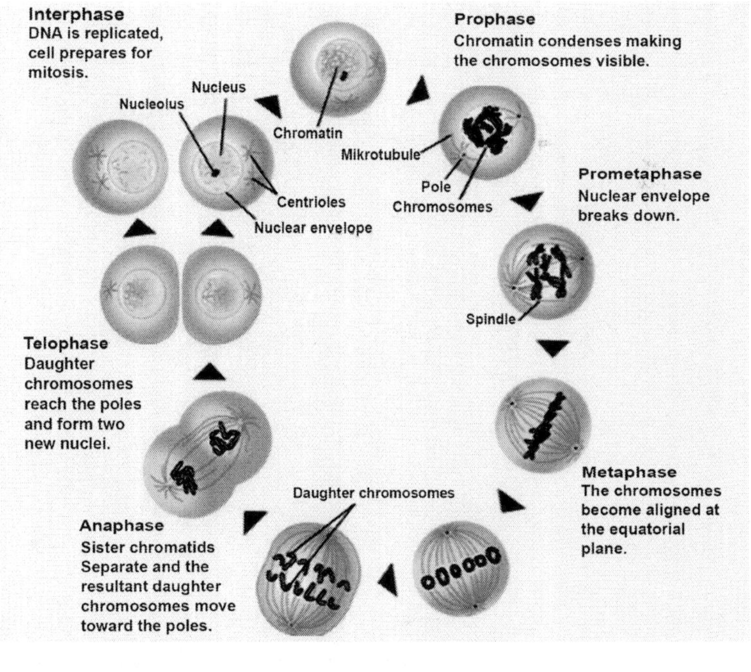

Figure 1.23. Mitosis distributes the four replicated chromosomes into the two

daughter cells so that the daughter cells also have four chromosomes. Apart from having the same number of chromosomes, the resulting cells are genetically identical to each other and to the parental cell. (Source: JPablo Cad, Juliana Osorio, Matthias M, and M3.dahl, Creative Commons License).

After **telophase**, the cell has two nuclei containing identical sets of chromosomes. It then undergoes cytokinesis where a new membrane (and a new cell wall in plants) is formed to divide the cell into two distinct daughter cells. Mitosis is also known as equational division since the chromosome number of the daughter cells is equal to that of the parental cell. Thus, the somatic or body cells in humans are expected to contain the same set of 46 chromosomes.

1.2.2.1.3. Meiosis generates genetic variation

A special type of diploid cells known as germ cells or meiocytes undergo meiosis in order to produce the haploid gametes. The major differences of this type of cell division from mitosis are the occurrence of genetic recombination and the fact that it consists of two division phases: meiosis I and meiosis II. However, interphase precedes only the first phase. Both phases have the same four stages as mitosis.

Meiosis I is also known as reductional division because the starting diploid cell gives rise to haploid daughter cells after this phase. Figure 1.24 shows the stages of meiosis I.

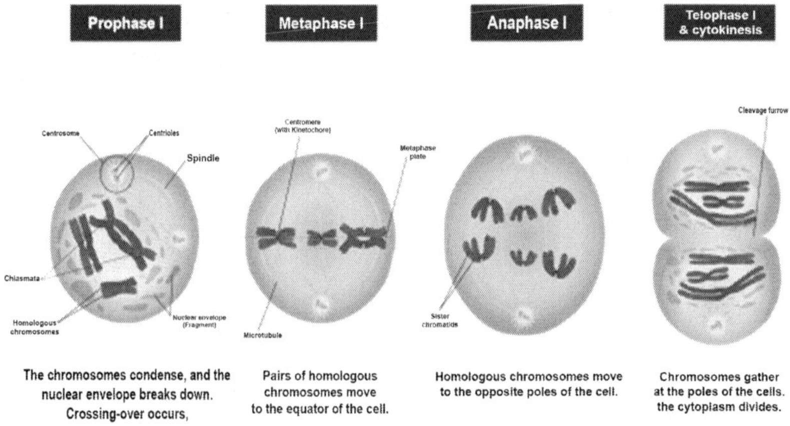

Figure 1.24. From a diploid parent cell with four chromosomes, the stages of meiosis I produce haploid daughter cells (n=2) that are genetically different from each other. (Source: Ali Zifan, Creative Commons License).

Prophase I deviates significantly from the prophase of mitosis as there are events that are exclusive to this stage. It is made up of five sub-stages. The first one is called **leptotene** where the chromosomes condense and become visible. This is followed by **zygotene** characterized by the pairing or synapsis of homologous chromosomes, which is facilitated by proteins of the synaptonemal complex (see Figure 1.25). Homologous chromosomes refer to the maternal and paternal chromosomes that are similar in structure and genetic composition. Synapsis leads to the formation of a **bivalent** which is made up of four chromatids that are joined in a zipper-like fashion. In human germ cells, the homologous autosomes form 22 bivalents. Among females, their two X chromosomes can fully synapse. On the other hand, the X and Y chromosomes are not fully homologous, but they can synapse in the short homologous pseudoautosomal regions they share (Blamire, 2002; van der Heijiden, 2011). After synapsis, crossing-over or recombination ensues where certain segments from the same region of two chromatids in the bivalent break off and are inserted after exchanging places add: (see Figure 1.26). This occurs during **pachytene**. If the segments exchanged during crossing over are from non-sister chromatids, this may produce genetic variations since the chromosomes in the bivalent are now made up of a mixture of paternal and maternal genes. During **diplotene**, the synaptonemal complex breaks down so that the chromatids in the bivalent eventually disengage from each other except at the **chiasma** (plural: chiasmata) which is the point where crossing-over took place. Finally, at **diakinesis**, the chromosomes condense further so that they are at their densest. The chiasmata also move along the length of the chromosomes until they reach the end in a process called **terminalization**. It is also at this stage that the nucleoli and nuclear membrane disintegrate (Blamire, 2002).

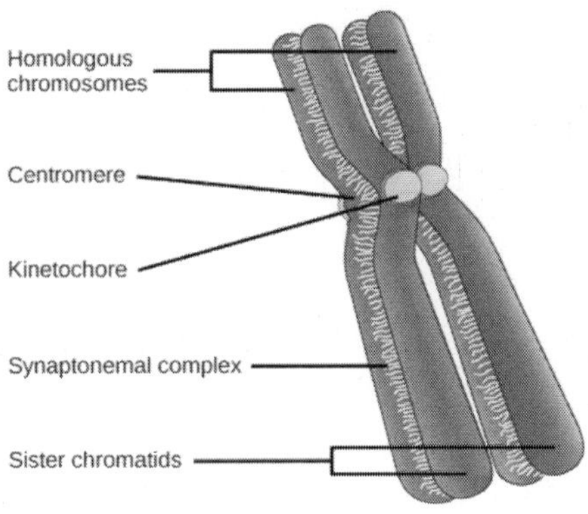

Figure 1.25. At the zygotene stage of prophase I, homologous chromosomes

pair together along their length with the aid of proteins in the synaptonemal complex. (Source: OpenStax Biology 2nd Edition, Creative Commons License).

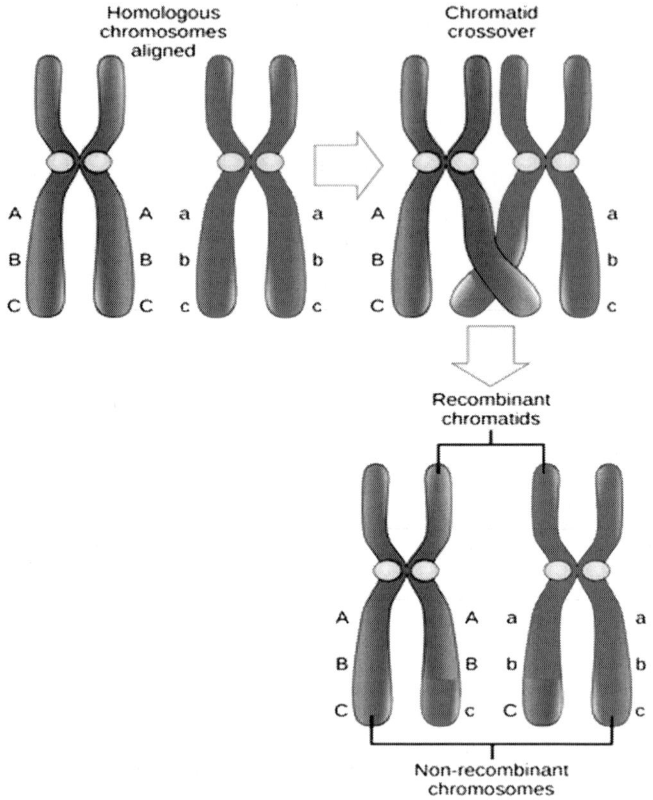

Figure 1.26. Crossing over is the exchanging of segments between the chromatids in the bivalent. If this occurs between non-sister chromatids, then there is genetic recombination of maternal and paternal genes. (OpenStax Biology 2nd Edition, Creative Commons License).

The paired homologous chromosomes joined at the chiasma line up at the center of the cell at metaphase I. The manner of alignment is such that the paternal chromosome in the bivalent is facing one pole while the maternal chromosome is oriented toward the other pole. Mendel's law of independent assortment is observed since the alignment of bivalents at metaphase I is independent of each other (see Figure 1.27). This means that the orientation of the maternal or paternal chromosomes in a bivalent is random relative to the orientation of paternal and maternal chromosomes in other bivalents. Thus, there are several possible combinations of paternal and maternal chromosomes that will move to one pole in the next stage. The next stage is anaphase I where the paired chromosomes in the bivalent separate into two **univalents** that move to opposite poles. This is the step when the diploid chromosome number gets reduced to haploid per pole. This also forms the basis for Mendel's law of independent segregation. Specifically, the maternal and paternal forms of a gene separate cleanly during meiosis and eventually get distributed to different gametes. At telophase I, a nuclear envelope forms around each set of chromosomes in the two poles. This may be followed by cytokinesis or not. In any case, the final product is either two cells or a binucleated cell with a haploid chromosome number per nucleus.

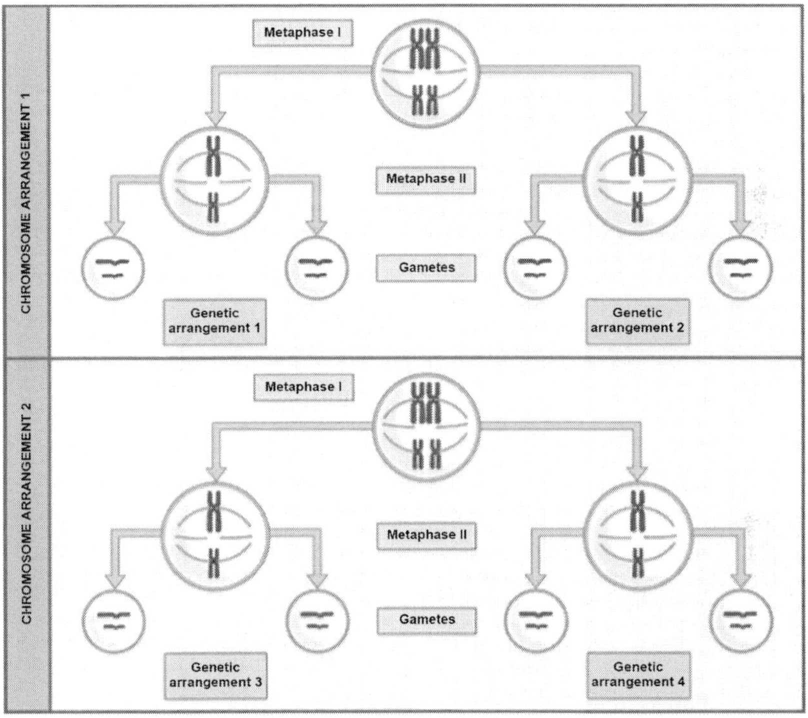

Figure 1.27. There are various ways that bivalents may align themselves along the equator at metaphase I. These result in gametes with assorted combinations of paternal and maternal chromosomes. These provide the explanation underlying Mendel's law of independent assortment. (Source: OpenStax Biology 2nd Edition, Creative Commons License).

Meiosis II, just like mitosis, is an equational division. The haploid cells from meiosis I enter meiosis II to produce cells that are also haploid in chromosome number. However, unlike mitosis, meiosis II proceeds without the cells undergoing interphase. This means that the DNA content of the starting cells is not replicated. The events in the four stages of this phase are very similar to those of mitosis (see Figure 1.28).

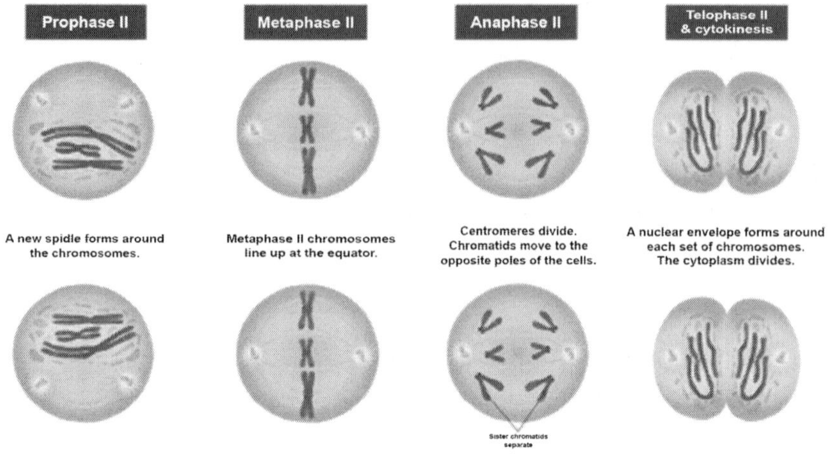

Figure 1.28. The events of meiosis II are very similar to mitosis which distributes the genetic material of the haploid parent cell (n=2) equally to the haploid daughter cells (n=2). (Source Ali Zifan, Creative Commons License).

At prophase II, the nuclear envelopes disintegrate. Then the univalents made up of chromatids with crossed-over segments align at the center of the cell at metaphase II. Each chromatid is oriented to face a certain pole. Then at anaphase II, the chromatids separate from each other and move to opposite poles. At this point, the chromatids are already considered chromosomes. This is followed by telophase II where nuclear envelopes form around the four sets of chromosomes. Finally, cytokinesis follows producing haploid gametes that are genetically different from each other. In the case of sperm formation, the cytoplasm is distributed equally after meiosis I and II so that the four sperm cells formed are of the same size and are all functional. On the other hand, cytokinesis during egg formation is unequal so that the products after meiosis I are one large **oocyte** and a smaller, non-functional polar body. After meiosis II, the unequal separation is repeated in the oocyte while the polar body is divided equally. As a result, the final products are one large ovum or egg cell and three polar bodies (see Figure 1.29).

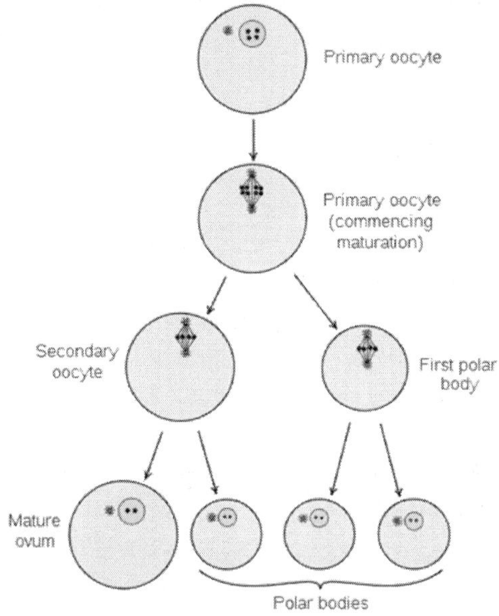

Figure 1.29. Cytokinesis during egg cell formation is characterized by the unequal distribution of cytoplasmic content so that after meiosis I, one large oocyte and a much smaller polar body are produced. Both cells undergo meiosis II but the large oocyte exhibits unequal cytokinesis while the first polar body's cytoplasmic content is divided equally. In the end, only the single large cell can function as an egg cell while the three polar bodies are non-functional. (Source Henry Vandyke Carter and Mysid, Public Domain).

If the sperm fuses with the egg during fertilization, then the characteristic diploid chromosome number of the organism is restored. The single-celled diploid zygote formed after fertilization will then undergo a series of mitoses in order to grow and develop into a multicellular form.

1.2.2.2. Inheritance of linked genes

As discussed in the previous sections, several researchers were able to confirm and explain the bases of Mendel's laws of inheritance. However, further studies on inheritance produced results that deviated from the expected Mendelian inheritance. An example was a study by **William Bateson**, **Edith Rebecca Saunders**, and **Reginald Punnett**. They performed dihybrid crosses in sweet pea plants similar to those done by Mendel. Specifically, they wanted to observe the inheritance of flower color and the pollen grain shape. Purple flowers (P) are dominant over red (p) while long pollen grains (L) are dominant over round ones (l). They crossed plants that were homozygous for purple flower and long pollen grains (PPLL) with plants homozygous for red flowers and round pollen grains (ppll). Therefore, the F_1 plants had purple flowers with long pollen grains and they were expected to be heterozygous for both traits (PpLl) (Lobo and Shaw, 2008).

Next, they crossed the F_1 plants with each other to produce the F_2 generation. The researchers were expecting that the phenotypic ratio would follow the Mendelian F_2 ratio in a dihybrid cross where the two genes show independent assortment. This means that there should be 9 plants with purple flowers and long pollen grains; 3 with purple flowers, round pollens; 3 with red flowers, long pollens; and 1 with red flowers, round pollens. However, they observed that there were more plants showing the parental phenotypes (i.e. purple flower, long pollen and red flower, round pollen) and fewer plants showing the recombinant phenotypes (i.e., purple flower, round pollen and red flower, long pollen). These deviations from the expected proportion were statistically significant which means they cannot be due to chance (Lobo and Shaw, 2008; Bateson *et al*., 1905).

The researchers hypothesized that the parental alleles for flower color and pollen shape are coupled or connected so that they do not exhibit independent assortment. However, much like Mendel, they were not able to fully explain the "coupled" inheritance they observed (Lobo and Shaw, 2008; Bateson *et al*., 1905). It was a few years later when Morgan's studies on *Drosophila* demonstrated that coupled inheritance is observed when genes are located near each other on the same chromosome. Thus, a chromosome contains not just one but several genes, which constitute a **linkage group**. For his part, Mendel only observed independent assortment or non-linked inheritance because the pea plant genes he chose to study were either located in different chromosomes or were distantly located in the same chromosome so that he would have been unable to detect linkage (Lobo and Shaw, 2008).

It was previously discussed (in sections 1.1.4.3 and 1.2.2) that Thomas Hunt Morgan and his team discovered a form of linkage in their extensive study of inheritance in *Drosophila*. This is **sex linkage**, which refers to the inheritance of genes that are located on the sex chromosomes. In a sense, the inheritance of sex and another trait is coupled. In the case of *Drosophila,* the gene for its eye color is located on the X chromosome. This conclusion was drawn by Morgan and his colleagues after they performed a cross between a white-eyed male fly and a red-eyed female. Since the allele for red eyes is dominant over the mutant allele for white eyes, all the F_1 flies were red-eyed as expected. The F_1 flies were crossed to produce the F_2 offspring. Assuming that sex and eye color are independently assorting, It is expected that 75% of the progeny are red-eyed males and females while 25% are white-eyed males and females. However, they observed 50% red-eyed females, 25% red-eyed males, and 25% white-eyed males but no white-eyed females. Since white eye color seemed to occur only in males, Morgan initially referred to it as a **sex-limited** trait (Morgan, 1910; Miko, 2008b).

However, when Morgan crossed the original white-eyed male with some of the red-eyed F_1 females, he obtained approximately equal numbers of red-eyed females, red-eyed males, white-eyed females, and white-eyed males (i.e. 1:1:1:1 ratio). Therefore, Morgan rejected his initial assumption that the gene for white eye color is only carried by males. Still the results made Morgan infer that the gene for eye color must be sex- related. He then illustrated the crosses that may explain the deviation of the results that he observed in the F_2 from what was expected. He represented the allele for red eye color with R and the allele for white eye color with W. Moreover, he represented the female genotype as XX while the males were simply X. Table 1.1 presents the Punnett square that represents the crosses proposed by Morgan (Morgan, 1910).

Table 1.1. The crosses proposed by Thomas Hunt Morgan to explain the proportion of phenotypes he observed in the F2 when he mated a red-eyed female *Drosophila* with a white-eyed male:

Parents of 1st cross	RR XX (red-eyed female) x WWX (white-eyed male)		
Gametes		*½ WX*	*½ W*
F_1 Genotypes	*RX*	RW XX	RW X
F_1 Phenotypic ratio		½ or 50% red-eyed female	½ or 50% red-eyed male
Parents of 2nd cross	RW XX	x	RW X
Gametes		*½ RX*	*½ W*
F_2 Genotypes	*½ RX*	¼ RRXX (red-eyed female)	¼ RW X (red-eyed male)
	½ WX	¼ RW XX (red-eyed female)	¼ WW X (white-eyed male)
F_2 Phenotypic ratio		½ or 50% red-eyed females	¼ or 25% red-eyed male; ¼ or 25% white-eyed male

Source: Morgan, T.H., (1910). Science, 32, 120-122.

Although Morgan used separate symbols for the X chromosome and for the alleles for eye color, he concluded that during the formation of sperms in the F1 red-eyed male (RW X), "the R and X go together". He further noted that "R and X are combined and have never existed apart" (Morgan, 1910). In this sense, Morgan concluded that the factors or genes for femaleness and eye color are coupled (Morgan, 1911).

A year after publishing his results on sex-linked traits of *Drosophila*, Morgan noted that while coupling was observed in some traits, other traits showed little or no coupling. He correlated this with the distance of the genes with respect to each other in the chromosome and the occurrence of crossing-over during gamete formation (although Morgan did not use the term crossing-over, he described it in his paper). Specifically, Morgan described that genes that are near each other are less likely to be separated during crossing-over while genes that are distant from each other could be separated by crossing-over. For these distant genes, although they are in the same chromosome, their linkage is broken when crossing-over occurs between them (see Figure 1.30) (Morgan, 1911).

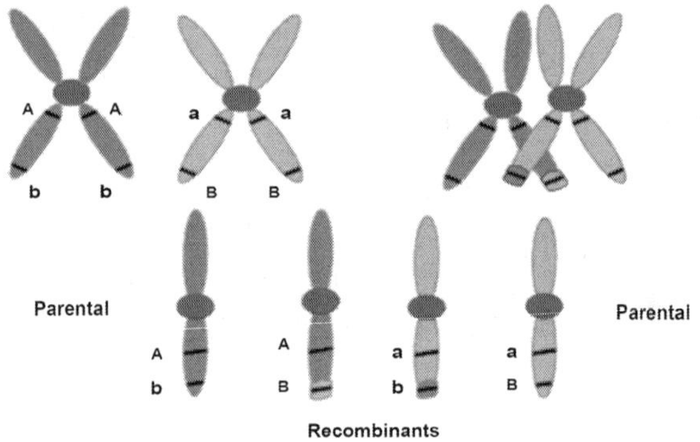

Figure 1.30. Crossing-over during meiosis can separate linked genes producing recombinant or new combination of genes in the gamete. On the other hand, if chromosome regions did not undergo crossing-over, the original or parental combination of genes is retained. (Source: Abby Provenzano, Creative Commons License).

A student of Morgan, **Alfred Sturtevant**, was able to infer that the differences in the strength of linkage as determined by the distance between genes in the chromosome could be used to determine the sequence and distance of genes in the chromosome. A greater proportion of recombinant phenotypes observed in F_2 meant that a greater frequency of crossing-over took place between the two genes involved. Thus, these two genes must be farther from each other than the genes for phenotypes that exhibited a lesser proportion of recombinants in the progeny. Over the course of one night, Sturtevant produced the first **genetic** or **linkage map** showing the order and relative spacing of six sex-linked genes in *Drosophila* (see Figure 1.31).

He made use of the percent or frequency of cross-overs or recombinants observed between two genes to map out their distances. Specifically, one unit of genetic distance in his map is equivalent to one cross-over out of every 100 gametes formed (i.e., 1% cross-over or recombination frequency). For instance, gene P (for red eyes) and gene M (for miniature wing size) exhibited 109 crossovers in 405 gametes. This gives a 26.9% recombination frequency, which is equivalent to 26.9 map units. The P and M genes exhibit **incomplete linkage**. This means that they are mostly inherited together (i.e., with >50% frequency of parental types) but occasionally get unlinked by recombination (i.e., with 26.9%, which is <50%, of recombinant types). On the other hand, C and O, which are two genes necessary for red eye color show **complete linkage**, i.e. they are always inherited together and are never separated by crossing over. Hence, no recombinants for these two genes were observed in the progeny. If the recombination frequency between two genes is equal to or greater than 50%, this indicates an absence of linkage and that the two genes are independently assorting. It is likely, therefore, that these genes are located in different chromosomes. Considering that the methodology determines genetic statistical linkage, Sturtevant recognized that his map units do not represent actual or physical distances between genes (Sturtevant, 1913). Still, Sturtevant's methodology was widely used in mapping genes along the chromosomes of various organisms.

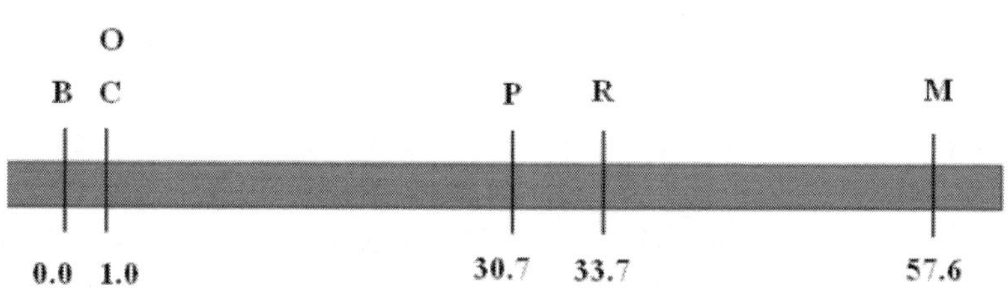

Figure 1.31. Alfred Sturtevant made the first genetic or linkage map showing the arrangement and relative distances of six genes in the X chromosome of *Drosophila* (fruit fly). The distances indicated are equivalent to the percent recombination or crossing-over between two adjacent genes. (Source: Adapted from Sturtevant, A.H., 1913. Journal of Experimental Zoology, 14, 43-59).

1.3. THE AGE OF THE DEOXYRIBONUCLEIC ACID (DNA)

The year 1953 was a major turning point in the field of genetics when the biochemical structure of the hereditary factor spoken about by Mendel and the other classical geneticists was revealed. This ushered in a new era where the genetic material is now studied as a biomolecule, the deoxyribonucleic acid (DNA), that lends itself to modification and manipulation. This groundbreaking discovery, however, was built upon a number of previous discoveries by various researchers.

1.3.1. Unlocking the Biochemical Identity of the Genetic Material

Nearly a century earlier in 1869, **Friedrich Miescher**, a Swiss physician, isolated a substance from leucocytes that had properties different from the protein that he was studying (see Figure 1.32). He reported that unlike proteins, this substance precipitated from the solution when acid was added and dissolved again when alkali was added. He named this substance **nuclein** as it was located in the nuclei of the cells. **Albrecht Kossel**, a German biochemist, subsequently discovered that nuclein was made up of four bases and sugar molecules. Also, **Eduard Zacharias**, a German botanist, showed that nuclein was an integral part of chromosomes(Dahm, 2004).

Figure 1.32. Friedrich Miescher was a young Swiss physician when he first isolated DNA from the nuclei leucocytes in pus, He called what he isolated nuclein. (Source: PBS, Public Domain).

Nevertheless, most scientists, long after the death of Miescher,believed that the protein components of chromosomes constituted the genetic material because these have 20 different amino acids while DNA is made up only of four different nucleotides. Since the cell contains a great quantity of genetic information, these scientists thought that it can only be stored in a complex molecule like protein (Dahm, 2004).

Then in 1928, **Frederick Griffith**, a British microbiologist, performed a study which initiated a chain of studies that unlocked the chemical identity of the genetic material. He experimented on the pneumonia-causing bacterium, *Streptococcus pneumoniae*, which has two strains, the non-infective type (also called the R strain since it forms rough- looking colonies) and the infective type (also called the S strain because it forms colonies with a smooth surface due to the gelatinous capsule present in this strain). He observed that when a mixture of non-infective R strains and heat-killed infective S ones were injected in mice, the mice died due to pneumonia (see Figure 1.33). Analysis of the mice's blood samples revealed that they contain the S strain. Since the injected S strain was already dead, the only explanation for this observation is that the living R bacteria were converted to S type by a substance or a "transforming principle" from the dead S bacteria. The process involved is called **transformation**, which in genetics refers to the transfer of genetic material from one cell to another to alter the genetic make-up of the recipient cell (Raven and Johnson, 2002). Still, some bacteriologists suspect that the transforming substance was a protein.

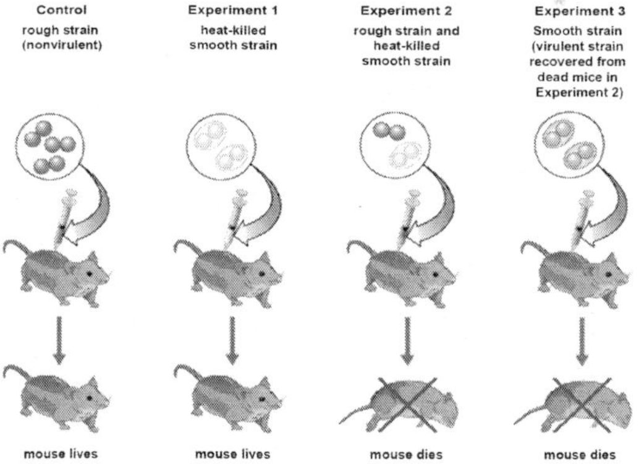

Figure 1.33. The series of experiments performed by Frederick Griffith involved injecting the nonvirulent rough (R) strain, the virulent smooth (S) strain, or a combination of either dead or living R and S strains of *Streptococcus pneumoniae* in mice. His results showed that the mouse died when a mixture of

heat-killed S strain and viable R strain were injected, and viable S strain was recovered from the dead mouse. This indicates that the R strain was transformed to become a virulent S strain by a certain substance from the dead S strain. (Source: OpenStax: Microbiology, Creative Commons License).

The identity of the transforming substance or transforming principle was determined through an experiment performed by a trio of Canadian and American researchers, **Oswald Avery**, **Colin Macleod**, and **Maclyn McCarty** in 1944. They were able to isolate almost pure and high molecular weight transforming principle from the S strain of *S. pneumoniae* and added it to the cultures of R cells (see Figure 1.34). They observed transformation where a few S cells were produced in the cultures. They also observed that enzymes which digest lipids, proteins, and ribonucleic acid (RNA) did not affect the transforming activity while treatments (i.e., enzyme and temperature) that destroy DNA stopped the activity of the transforming principle. All these observations led to the conclusion that the transforming substance is DNA (Raven and Johnson, 2002; Cobb, 2015; National Human Genome Research Institute, 2013).

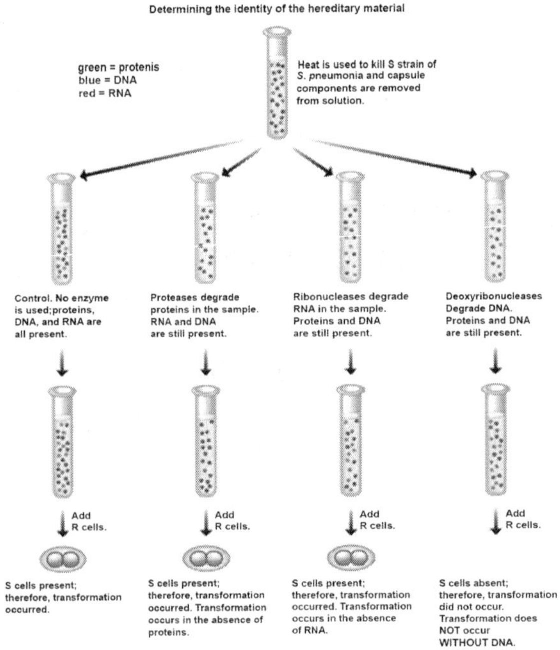

Figure 1.34. Oswald Avery, Colin Macleod, and Maclyn McCarty provided the experimental proof that the transforming substance in Frederick Griffith's

experiment is DNA. They isolated the transforming principle from the S strain of *Streptococcus pneumoniae* and observed that it converted R cells to become the S strain. The transforming activity was not affected by the destruction of proteins and RNA, but it stopped when DNA was digested. (Source: OpenStax: Microbiology, Creative Commons License).

This significant study was followed by the so-called "blender experiment" by the American geneticists **Alfred Hershey** and **Martha Chase**, in 1952 (see Figure 1.35). Their results confirmed that the transforming substance is DNA. They infected the bacterium *Escherichia coli* with the bacteriophage T2. A **bacteriophage** is a virus that infects bacteria and is generally made up of DNA that is enclosed in a protein coat (Klug *et al.*, 2006). Apart from the protein-enclosed DNA that constitutes the head, bacteriophage T2 has a collar, tail, and tail fibers. T2 infects bacteria by attaching its tail on the bacterial cell surface allowing the entry of phage material into the host cell. Then this material multiplies to form hundreds of progeny bacteriophages that eventually cause the host cell to burst open, so the progenies are released.

Hershey and Chase's experiment aimed to determine which between the DNA and protein components of T2 is inherited by the progeny bacteriophage. They were able to monitor the DNA and protein components by labeling these radioactively. Labeling of DNA was done by growing *E. coli* in a medium with ^{32}P, a radioactive form of phosphorus, since phosphorus is a component of DNA but not of protein. On the other hand, the phage protein was labeled by growing *E. coli* in a medium with ^{35}S, a radioactive form of sulfur. Sulfur is a component of proteins and not of DNA.

E. coli were infected with bacteriophage T2 labeled with either ^{32}P or ^{35}S. After infection, the bacteriophages were detached from the host cell by centrifugation and sheared by being swirled by a kitchen blender. Removing and destroying the T2 did not affect the subsequent infection process indicating that the genetic material for making bacteriophage T2 was injected into the host cell shortly after bacteriophage attachment.

Analysis of the infected bacteria revealed that ^{32}P radioactivity was observed in *E. coli* with only a little of ^{35}S radioactivity. This indicates that it is DNA that was injected by bacteriophage T2 during the infection process, leaving out most of the protein component. Another pivotal finding is that 30-50% of the observed ^{32}P radioactivity but less than 1 percent of ^{35}S radioactivity was inherited by the progeny bacteriophage. These observations led Hershey and Chase to conclude that the genetic material in T2 bacteriophage is DNA. This conclusion is eventually extended to living organisms whether bacteria, protists, fungi, plants, animals, and humans except in some viruses with RNA as their genetic material (Lewin, 2004).

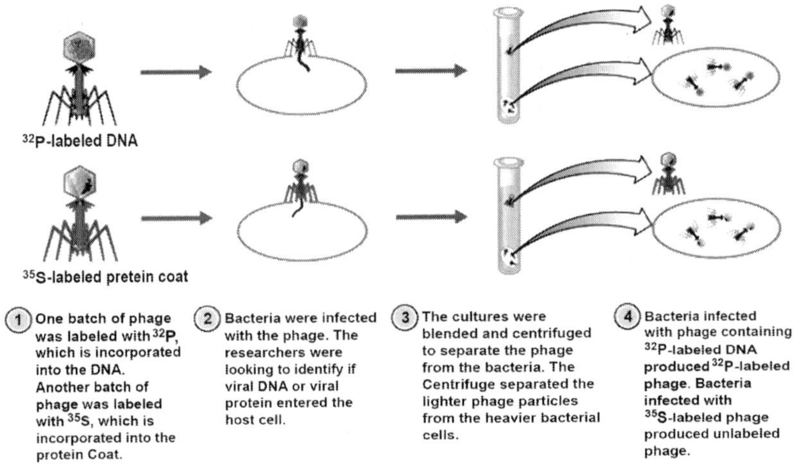

Figure 1.35. Alfred Hershey and Martha Chase performed an experiment where they separately labeled the DNA and protein components of T2 bacteriophage with radioactive isotopes of phosphorus (^{32}P) and sulfur (^{35}S), respectively. Their results proved that it is mostly the viral DNA that entered the host *Escherichia coli* during infection and that this is responsible for producing new bacteriophages. (Source: OpenStax: Microbiology, Creative Commons License).

1.3.2. The DNA Chemical Structure

DNA is one of the two types of nucleic acids. The other type is RNA, which is the genetic material of some viruses. Both molecules are polymers made up of building blocks known as **nucleotides** (see Figure 1.36). A nucleotide is actually made up of three components, a phosphate group, a 5-carbon sugar (also known as a pentose), and a nitrogenous base. It was the Russian biochemist **Phoebus Levene** who discovered the order of the nucleotide components and how nucleotides are put together to form nucleic acids (Pray, 2008). The nitrogenous base is linked to the first carbon of pentose (1' carbon) while phosphate is bound to the fifth carbon (5'carbon). The number of carbon atoms in the pentose is primed in order to differentiate them from the carbon atoms of the base. DNA and RNA have the same phosphate groups. However, Levene discovered that these differ in their pentose component, DNA has **2-deoxyribose** while RNA has **ribose** (Pray, 2008). The nitrogenous part of the nucleotide can be any one of the five types, which could either be a single-ringed pyrimidine or a double-ringed purine (see Figure 1.36). The purine bases are **guanine** and **adenine**. The pyrimidine bases include **cytosine** and **thymine**, which is unique to DNA, or **uracil**, which is found only in RNA. Uracil has a chemical structure similar to thymine except that it lacks a methyl group. The phosphate group is linked to the pentose, which is then joined to a nitrogenous base to form a nucleotide.

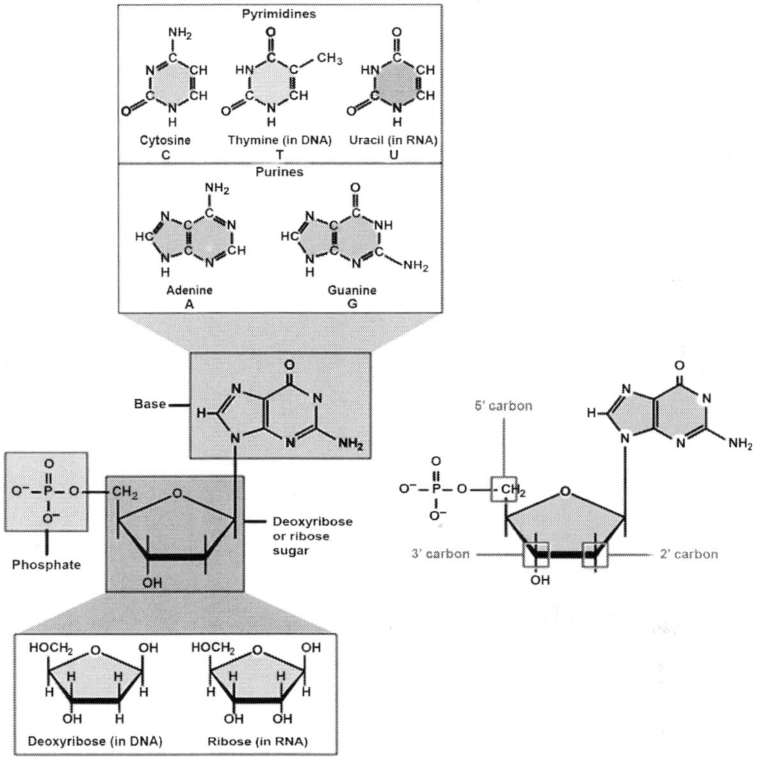

Figure 1.36. A nucleotide is formed when a nitrogenous base is joined to the pentose via the 1' carbon and a phosphate group is linked to the 5' carbon. The pentose in the nucleotide of DNA is deoxyribose while in RNA, it is ribose. The nitrogenous bases are of two types, pyrimidines and purines. Pyrimidine bases include cytosine, thymine (found only in DNA), and uracil (found only in RNA). Purine bases are adenine and guanine. (Source: OpenStax Biology 2e, Creative Commons License).

Several nucleotides are joined together by phosphodiester linkages formed between the phosphate group of one nucleotide and the oxygen (from the hydroxyl group) attached to carbon 5' of the pentose of the succeeding nucleotide. The resulting polynucleotide strand exhibits polarity, i.e. one end has the phosphate group, which is called the 5'-end or the phosphate end while the opposite end which has the sugar is the 3'-end or the hydroxyl end. RNA is generally single-stranded, but DNA is made up of two polynucleotide strands that are joined together by means of a series of hydrogen bonds (see Figure 1.37). The two strands of DNA are **antiparallel** to each other, meaning one strand runs from 5' to 3' and the opposite strand runs from 3' to 5'.

Since the nucleotides differ only by their nitrogenous base, the sequence of the nucleotides making up the nucleic acid is represented by the sequence of bases. The first letter of the bases of the nucleotides is written to denote the sequence of nucleotides (e.g. GTTCAAACTG). Also, the pair of nucleotides that are linked together in the DNA is called a base pair (bp) instead of nucleotide pair.

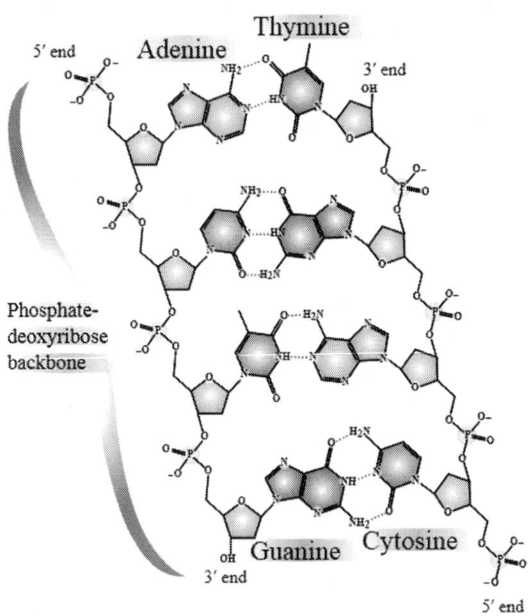

Figure 1.37. DNA is made up of two polynucleotide strands with the phosphate-deoxyribose outer backbone forming two antiparallel strands. The nitrogenous bases exhibit specific pairing (i.e., adenine is always paired with thymine while guanine pairs with cytosine) in the inner portion and are linked by hydrogen bonds. (Source: Madeleine Price Ball, Public Domain).

1.3.2.1. The DNA is a double helix

Native DNA is made up of two polynucleotide strands that form a right-handed double helix. This was first observed by **Rosalind Franklin**, a British chemist, in 1953. She was able to obtain an X-ray diffraction photograph of DNA fibers showing helical patterns that makes a complete turn every 3.4 nm and has a diameter of 2 nm (see Figure 1.38). Franklin's photograph provided crucial experimental data that allowed American biologist **James Watson** and English physicist **Francis Crick** to propose the double helix DNA model that is still accepted today. This proposed structure was published in *Nature* in 1953 (Passarge, 2001).

A B

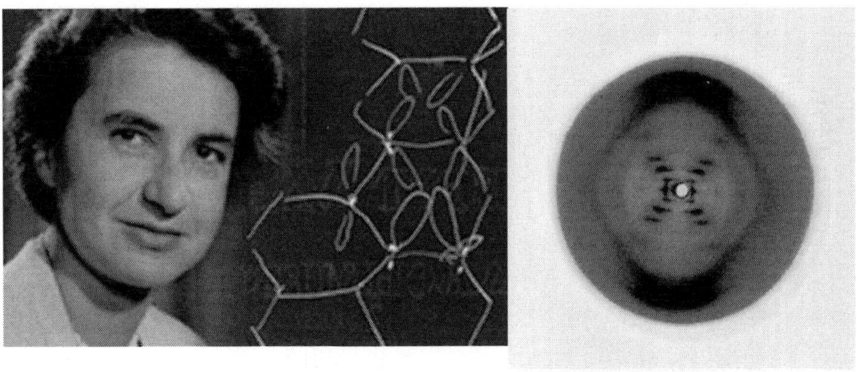

Figure 1.38. (A) Rosalind Franklin was a British chemist and X-ray crystallographer who took X-ray diffraction photographs of DNA in an effort to determine its structure. (B) Franklins' famous photo 51 clearly showed that the DNA is a helix. (Source: Wellcome Library, Creative Commons License).

In the Watson-Crick model (see Figure 1.39), the two antiparallel strands making up the DNA make a series of 360° turns following a clockwise or right-handed path to form a helix. There are approximately 10 pairs of nucleotides in each complete turn. The coiling of the strands also leads to the formation of a major groove and a minor groove and based on their names, these two differ in depth and width (Schleif, 1993).

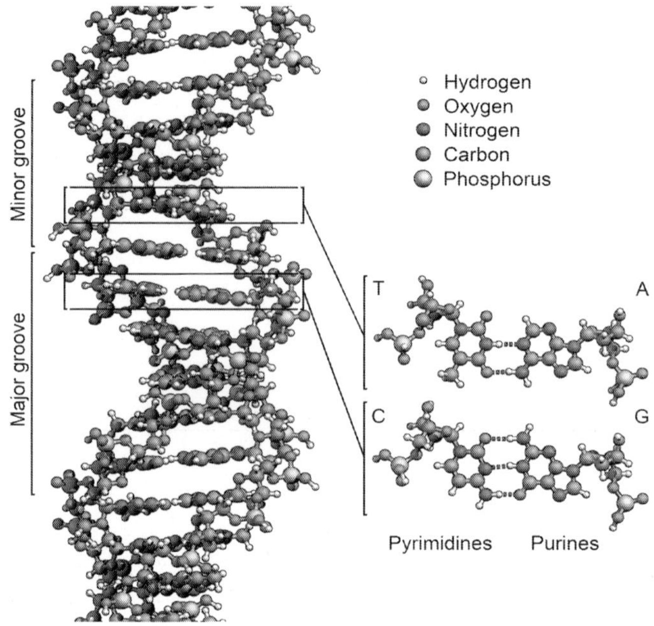

Figure 1.39. The DNA model proposed by James Watson and Francis Crick shows a double-stranded helix that winds clockwise and forms a minor and major groove per complete turn. The sugar and phosphate components of the nucleotide form the outer backbones while the nitrogenous bases are stacked inside the molecule, exhibiting specific base pairing. (Source: Zephyris, Creative Commons License).

The external backbones of the helix are made up of the sugar and phosphate groups of the nucleotides composing the two strands. On the other hand, the nitrogenous bases of the two strands are facing each other in the interior part of the helix and are connected by hydrogen bonds (see Figure 1.39). These base pairs are stacked inside the helix. The type of nitrogenous bases that face each other in the DNA molecule is specific and complementary. It was **Erwin Chargaff** who discovered specific base pairing by determining the nucleotide composition of DNA extracted from various organisms, ranging from a bacterium (e.g. *E. coli*) to a human. He found out that the four types of nucleotides are not present in equal proportions in the DNA. However, the amount of adenine is always equal to the amount of thymine and the amount of guanine is always equal to the amount of cytosine. Therefore, adenine pairs only with thymine (or vice versa) through the formation of two hydrogen bonds while cytosine pairs only with guanine (or vice versa) by forming three hydrogen bonds (see Figure 1.39).

These findings came to be known as **Chargaff's rules** and were used by Watson and Crick in deducing the DNA structure (Raven and Johnson, 2002). The complementary base pairing is always between a two-ringed purine and one-ringed pyrimidine. This results in a DNA molecule with a consistent diameter of 2 nm. The multiple hydrogen bonds that hold together the interior base pairs form a collectively strong linkage so that the two DNA strands in the helix are not easily separated from each other. The base pairs can stack over one another without collapsing into themselves through hydrophobic interactions (Schleif, 1993). The genetic information is encoded by the sequence of bases in the DNA molecule. There are limitless ways of sequencing the bases which enable the DNA to store the vast biological information that is needed to build an organism.

1.3.3. Packaging of the Eukaryotic DNA into Chromosomes

Chromosomes of eukaryotes are linear and are made up of chromatin, which is composed of DNA and protein. Chromatin can be of two types: the **euchromatin** and the **heterochromatin** (see Figure 1.40).

Euchromatin is a less densely packed region and appears dispersed in the nucleus. On the other hand, heterochromatin is a very densely packed region usually found in the area of the centromere.

An average chromosome is made up of a DNA strand that is 5 centimeters long when stretched out. Considering that the average size of the nucleus in mammalian cells is about 6 micrometers (0.0006 centimeters) and there is more than one chromosome in a eukaryotic cell, chromosomes have to be tightly folded in order to fit inside the nucleus.

At the same time, the highly condensed chromosome should allow for uncoiling during the processes of DNA replication and transcription when proteins and enzymes need to have access to the DNA molecule (Raven and Johnson, 2002).

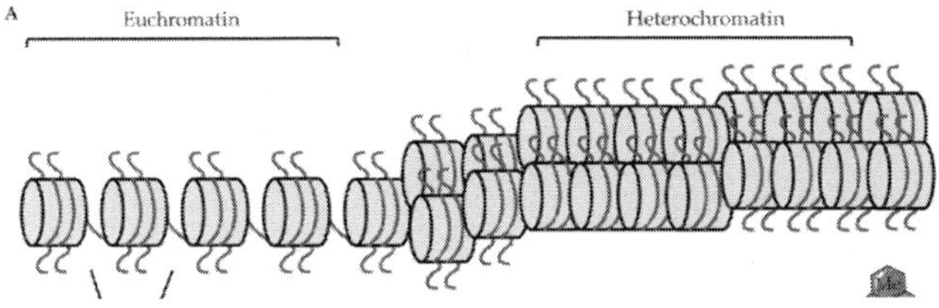

Figure 1.40. The linear eukaryotic chromosome is made up of the loosely packed euchromatin, which is easily accessed by the DNA binding proteins for gene

expression, and the condensed heterochromatin, which is less transcriptionally active. (Source: Modified from Wickramasekara, R.N.; Stessman, H.A.F. Biology, 8(1), 11, 2019; Creative Commons License).

The first level of coiling is when the DNA fiber is coiled every 200 base pairs (on average) around eight histone proteins, forming a bead-like structure that is made up of various types of histone proteins, i.e. two histone 2A (H2A), two H2B, two H3, and two H4. This forms the **nucleosome**, which is the fundamental subunit of chromatin (see Figure 1.41). The **histone** proteins are made up of basic, positively charged amino acids and these promote and guide the coiling of the DNA. This first level of packaging condenses the DNA six-fold. Each octameric bead is connected to another bead by a linker DNA that is associated with H1. A continuous string of nucleosomes may form a 10-nm fiber which is observed *in vitro* under low ionic strength conditions. The nucleosome coils further to form a 30-nm fiber or **solenoid** which has six nucleosomes per turn and condenses the DNA approximately 40-fold. The 30-nm fiber is the basic constituent of both the euchromatin and heterochromatin. With the aid of non-histone proteins, the 30-nm fiber itself undergoes coiling which condenses DNA about 1, 000-fold to form the euchromatin or a total of 10, 000-fold to form the heterochromatin or the mitotic chromosomes (Lewin, 2004).

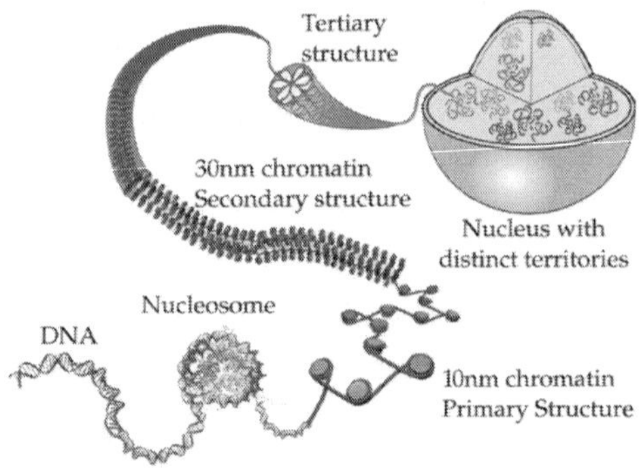

Figure 1.41. The DNA is condensed several thousand times in order to fit inside the microscopic nucleus. The DNA double helix is coiled into nucleosomes to form the 10-nm chromatin primary structure. This is further coiled into the secondary structure, the 30-nm chromatin, and then finally the tertiary structure of the chromosome. (Source: Iyer, B.V.S.; Kenward, M.; Arya, G. BMC Biophysics 4:8, 2011, Creative Commons License).

1.3.4. Replication of the DNA Molecule

In their published paper, Watson and Crick stated that "the specific base pairing we have postulated immediately suggests a copying mechanism" (Watson and Crick, 1953). Indeed, specific base pairing during DNA replication ensures that the base sequence is accurately copied so that two identical DNA molecules are produced after replication. These daughter DNA molecules are also identical to the original DNA molecule that served as the template. DNA is replicated during the S phase of interphase in the cell cycle.

1.3.4.1. Replication is semi-conservative

There were three possible modes of DNA replication, conservative, semi-conservative or dispersive (see Figure 1.42). In the **conservative model**, the parent DNA remains intact and it is copied to produce entirely new strands of daughter DNA molecules. The two strands of the helix separate from each other in the **semiconservative model**. Each strand serves as a **template** or pattern in forming a new polynucleotide strand. The third mode of replication, the **dispersive type**, is called such because the parental DNA is dispersed throughout the daughter DNA molecules so that these are made up of a mixture of old and new DNA. Among these three, DNA was proven to replicate in a semiconservative manner (Raven and Johnson, 2002).

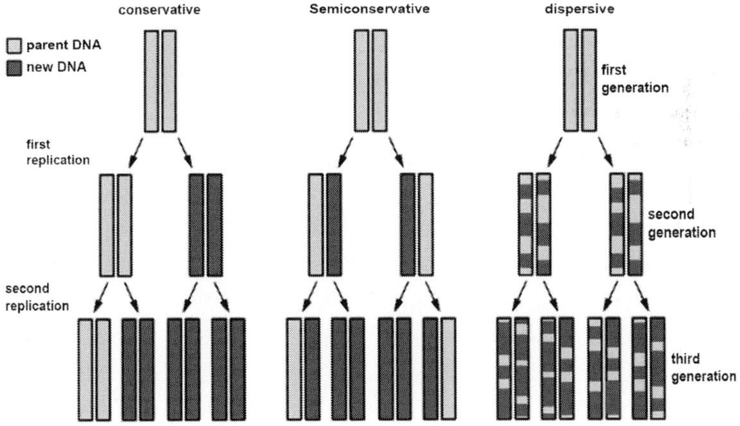

Figure 1.42. There are three proposed modes of DNA replication. In the conservative model, the parental strands remain together after replication and two new strands make up the daughter DNA molecule. In the semiconservative model, the parental strands separate from each other and serve as templates for the

synthesis of new strands so that after replication, there are two DNA molecules made up of an old and new strand joined together. In the dispersive model, the products of replication have regions of double-stranded parental DNA and regions of double-stranded daughter DNA. (Source: OpenStax: Microbiology, Creative Commons License).

1.3.4.2. DNA replication in eukaryotes

Both prokaryotic and eukaryotic cells generally follow similar steps when they replicate their DNA but they also exhibit certain differences. This discussion focuses on the replication process in eukaryotes, which occurs in the nucleus. The steps involved in DNA replication are divided into initiation, elongation, and termination.

1.3.4.2.1. Initiation of replication

Replication starts at a specific point in the DNA molecule called the origin of replication or the ori site. When replicating chromosomes are observed, a Y-shaped region of active replication eventually moves away from the ori site and progressively along the length of the parental DNA. This region is called a replication fork (Alberts et al., 2014). From each ori site, replication proceeds in both directions forming a replication bubble, which is essentially two adjacent replication forks that are oriented in opposite directions. Each human chromosome is made up of a considerable length of DNA such that there are thousands of ori sites along each chromosome (Valenzuela, 2012) (see Figure 1.43). Therefore, several replication bubbles simultaneously move on each human chromosome. The base sequences at the ori sites vary among organisms but these are usually rich in adenine (A)-thymine (T) base pairs. The origin recognition complex, made up of six subunits of protein, binds to each ori site and facilitates the binding of other proteins. Among these is a pair of enzymes called CMG helicases, which are positioned opposite each other, that break the hydrogen bonds between the base pairs at the ori site (see Figure 1.44). As a result, a region of parental DNA is unwound exposing the bases in the interior. Since the ends of the chromosome are fixed, the local unwinding of the double-strands creates torsional stress, which is relieved by the formation of supercoils.

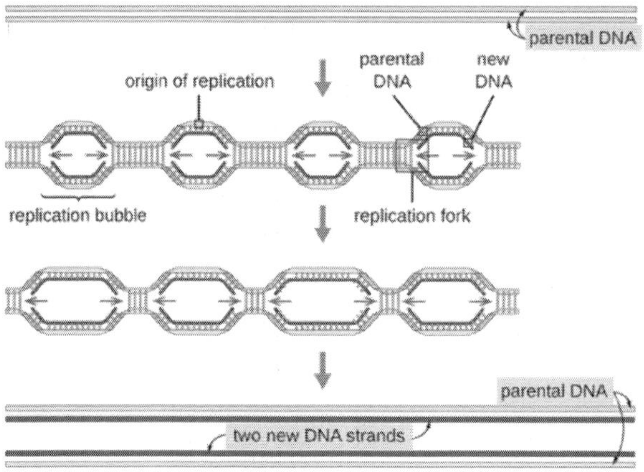

Figure 1.43. Human chromosomes, or eukaryotic chromosomes in general, are made up of linear and very long DNA so that there is a the need for multiple origins of replication and replication bubbles in order for DNA synthesis to be completed quickly. (Source: OpenStax: Microbiology, Creative Commons License).

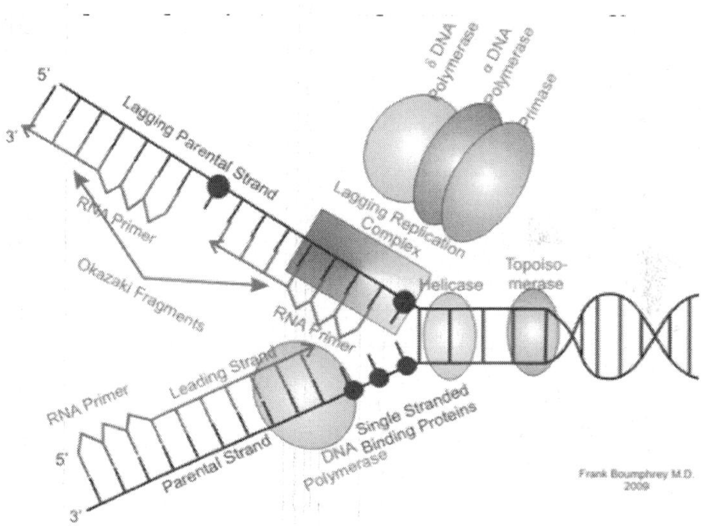

Figure 1.44. At the replication fork, the helicase enzyme unwinds the double-stranded parental DNA to generate single- stranded templates for replication. The supercoils produced by this localized unwinding of the DNA are untangled by topoisomerase. Single- stranded binding proteins associate with the template strands to stabilize them. A primase-DNA polymerase α complex synthesizes a short RNA primer followed by a short DNA segment. This primer is elongated

by the DNA polymerase δ. The leading daughter strand is extended continuously. On the other hand, the lagging strand has several primers that are extended by DNA polδ producing Okazaki fragments. (Source: Frank Boumphrey, Creative Commons License).

Enzymes called topoisomerases bind the DNA ahead of the helicase and cut either one or both strands so that the supercoils swivel and are removed. The cleaved strands are then resealed. Multiple copies of single -stranded binding proteins called replication protein A (RPA) bind the single-stranded templates produced. These maintain the uniform conformation of the DNA template making the replication process more efficient. (Raven and Johnson, 2002; Klug et al., 2006; Lodish et al., 2003).

1.3.4.2.2. Elongation of the daughter DNA

The main enzyme for synthesizing a new DNA strand is **DNA polymerase**. It catalyzes the addition of a new nucleotide by joining the 5' end of that nucleotide to the 3' end of a previous nucleotide. This means that it synthesizes new strands in the 5'→3' direction (see Figure 1.44). There are several types of DNA polymerases in eukaryotes with distinct functions (see Table 1.2).

Table 1.2. The Different Types of DNA Polymerase in Eukaryotes

Type of DNA Polymerase	Function
DNA polymerase α	Forms a complex with primase, adds DNA nucleotides to the short RNA primer synthesized by the primase
DNA polymerase β	Repairs DNA
DNA polymerase δ	Synthesizes daughter DNA strands
DNA polymerase ε	Synthesizes mainly the lagging strand
DNA polymerase γ	Synthesizes mitochondrial DNA
DNA polymerase ζ	Repairs DNA

DNA polymerase cannot synthesize new strands *de novo*, requiring an existing strand, called a **primer**, with a 3' end that DNA polymerase then eloWngates. A primer is a short RNA strand made up of about 10 nucleotides and is synthesized by an enzyme, **DNA primase**, by pairing the first few nucleotides in the template with their complementary RNA nucleotides. For example, if the template strand has the sequence ATCGTCCTAA, then the primer produced will have the sequence UAGCAGGAUU. This creates a hybrid DNA-RNA duplex in the replication fork. In eukaryotes, primase and

DNA polymerase α(alpha)form a complex (primase-DNA pol α), with the primase synthesizing a short RNA primer and the polymerase extending it with deoxynucleotides so that the final primer is a mixture of both RNA and DNA (Alberts *et al.*, 2014; Lodish *et al.*, 2003).

Using the template strand that runs from 3' to 5', a single primer that is oriented at the 5' to 3' direction is synthesized (see Figure 1.44). **DNA polymerase δ** (delta)then elongates the primer into a daughter DNA strand observing specific base pairing with respect to the template. As it adds new nucleotides, the enzyme is presented with a free 3' end so elongation occurs continuously in the same direction as the helicase unwinds the double-stranded template. This daughter strand is called the **leading strand**. The opposite parental strand oriented from 5' to 3', serves as the template for the synthesis of the **lagging strand**. The elongation of this strand is opposite to the direction of template unwinding. Unlike the leading strand, the lagging strand requires several primers that are generated every time the template is unwound further. This means that DNA pol δ in the lagging strand extends the primers from their 3' end and away from the replication fork until it reaches the previous primer. It then proceeds to extend a new primer that is synthesized further along the template strand. The discontinuous elongation of the lagging strand produces DNA segments that are separated by primers. These are called **Okazaki fragments** (named after their discoverer **Reiji Okazaki**) and are about 100-200 nucleotides long (Alberts *et al.*, 2014; Lodish*et al.*, 2003).

Ribonuclease H and **FEN I** eventually remove the RNA primers in the daughter strands, and these are replaced with DNA nucleotides by DNA pol δ. Finally, **DNA ligase** catalyzes the formation of phosphodiester bonds in order to link together the newly formed DNA segments to the Okazaki fragments (Lodish*et al.*, 2003)

There are other types of DNA polymerases in eukaryotes including **DNA polymerase ε** (epsilon) that is also involved in elongating the daughter DNA similar to the DNA polymerase δ but is thought to function under different cellular conditions or to elongate only the lagging strand (O'Connor, 2008b). However, in yeast, DNA pol **ε is said to elongate mainly the leading strand while DNA** pol δ is for the lagging strand (Doublié and Zahn, 2014). Then there are **DNA polymerase β** (beta) and ζ (zeta), which function for DNA repair, and **DNA polymerase γ** (gamma), which is involved in the synthesis of mitochondrial DNA (mtDNA) (O'Connor, 2008b).

1.3.4.2.3. Termination of replication

In eukaryotes, the DNA that makes up the **telomere**, the end portion that stabilizes the linear chromosomes, is not fully replicated in the lagging strand. This is because when the primer at the end is removed, there is no longer an Okazaki fragment after the gap which can provide the 3' end to enable a DNA polymerase to insert DNA nucleotides. This means that the telomere may shorten every time the chromosome undergoes replication. To solve this dilemma, an enzyme known as **telomerase**, which has a catalytic part and a built-in RNA sequence, replenishes the telomere sequence after every replication process (see Figure 1.45).

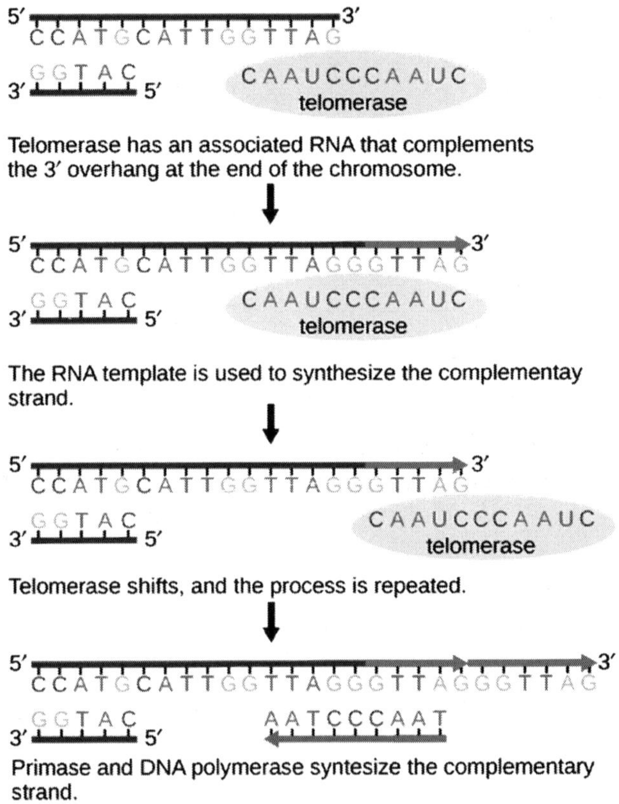

Figure 1.45. The length of telomeres of eukaryotic chromosomes is maintained by the activity of the telomerase enzyme, which uses its built-in RNA as a template in order to extend the DNA sequence of the telomere in one strand. It then repeats until the template of the lagging strand has been lengthened considerably. The newly synthesized segment can then be used as a template

by primase to make a new primer in the lagging strand. The primer is then extended by DNA polymerase so that the full telomere sequence is replicated. (Source: OpenStax: Microbiology, Creative Commons License).

A part of the RNA sequence is complementary to the sequence of the 3' overhang in the template of the lagging strand, which is known as the **G-tail.** In humans, the sequence is GGGTTA, which is repeated thousands of times in the telomere. The rest of the RNA sequence is used as a template by telomerase to extend the G-tail of the template DNA in the 5'→3' direction. Thus, telomerase is distinct from DNA polymerase because it synthesizes DNA from an RNA template. This process is called **reverse transcription**. The telomerase shifts to the newly-elongated region to continue providing a template for lengthening the G-tail. This cycle may occur multiple times. Then during the next round of DNA replication, primase synthesizes a primer for the telomere of the daughter strand by complementing the G-tail. The primer is extended by DNA polymerase to complete the telomere (O'Connor, 2008b; Alberts *et al*., 2014).

1.3.4.3. Expression of genetic information

The association of genotype with a human disorder started with the observation of a British physician, **Archibald Garrod**, that his patients with inherited diseases lacked certain enzymes needed to break down certain metabolites (Raven and Johnson, 2002). In his book entitled *The Inborn Errors of Metabolism*, published in 1902, Garrod described the hereditary alkaptonuria to be due to an anomaly in the catabolism of aromatic amino acids resulting in increased levels of homogentisic acid. These, in turn, lead to the "black urine" observed in affected children (Piro *et al*., 2010).

Garrod's conclusion was confirmed by **George Beadle** and **Edward Tatum** who performed an experiment on the biosynthetic processes in the fungus *Neurospora*. They induced mutations in the fungi by X-ray bombardment. They discovered three mutant strains that could not synthesize certain growth factors required by *Neurospora*, such as vitamin B_6(or pyridoxine), vitamin B_1, and para-aminobenzoic acid. They called the strain that could not synthesize vitamin B_6 as a pyridoxine-less mutant. They concluded that a gene responsible for the formation of an enzyme involved in the biosynthesis of vitamin B_6 was mutated by the X-rays (Griffiths, 2000; Chhetri, 2014). This gave rise to the **one gene: one enzyme hypothesis**. This means that a gene contains the code or instructions necessary to produce a corresponding enzyme. Enzymes were eventually discovered to be proteins and so the hypothesis was expanded to indicate that a gene contains the code for the synthesis of proteins in general. A further modification to the hypothesis became necessary when certain proteins were discovered to be made up of more than one subunit. If the protein is a homomultimer which has the same type of subunits, then it is coded for by one gene. In contrast, if the protein is made up of different types of subunits, or it is a heteromultimer, then more than one gene is responsible for it. Thus, the hypothesis was modified to become **one gene: one polypeptide** (Lewin, 2004). In the phenotypic level, this means that certain traits are due to more than one gene. It must also be pointed out that some genes code for RNAs that are no longer translated into a polypeptide.

The synthesis of polypeptides from DNA involves the processes of transcription and translation. This unidirectional flow of genetic information is known as the central dogma of molecular genetics, which is summarized in the following diagram:

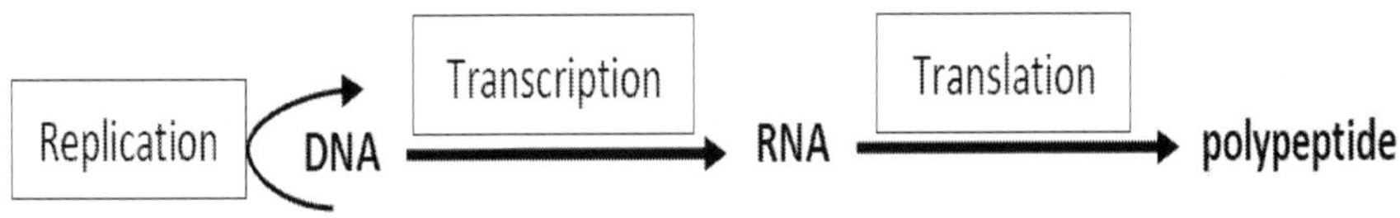

1.3.4.3.1. RNA synthesis via transcription

The entire DNA molecule does not contain codes for RNA or polypeptide synthesis. Along the length of the DNA strand, certain segments contain these codes and these segments are the genes. **Transcription** is the process of copying these genetic codes stored in the very large DNA molecule into a much smaller, short-lived RNA molecule which is also known as the **transcript** (see Figure 1.46). It occurs in the nucleus of eukaryotes.

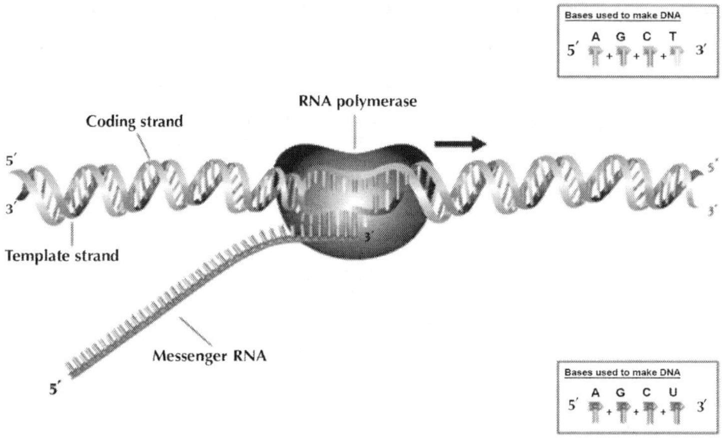

Figure 1.46. The main enzyme of the transcription process is RNA polymerase which opens the double- stranded DNA and copies the gene sequence from the DNA template to produce RNA in a 5'→3' direction. (Source: Genomics Education Programme, Creative Commons License).

1.3.4.3.1.1. Initiation of transcription

The principal enzyme for transcription is RNA polymerase. It can open up the helix as well as synthesize RNA *de novo*, without requiring a primer. In eukaryotes, there are three types of RNA polymerase, which are structurally similar to each other but produce the three types of RNA (see Table 1.3). **RNA polymerase I**, which occurs in the nucleolus, is responsible for producing most of the **ribosomal RNA (rRNA)**. This type of RNA is called such because it associates with certain proteins to form the ribosomes. **RNA polymerase II**, which is found in the nucleoplasm, is responsible for the synthesis of most **messenger RNA (mRNA)**. This is the RNA that is used as a template for polypeptide synthesis. This enzyme also produces non-coding RNAs that regulate gene expression such as microRNA (miRNA), small nuclear RNA (snRNA), small nucleolar RNA (snoRNA), small interfering RNA (siRNA), and long noncoding RNA (lncRNA). Finally, **RNA polymerase III** in the nucleoplasm produces 5S rRNA, some snRNAs, and **transfer RNA** (tRNA). Transfer RNA brings specific amino acids to the ribosomes during polypeptide synthesis (Klug *et al*., 2006; Schleif, 1993; Alberts *et al*., 2014).

Table 1.3. The Three Types of RNA Polymerase in Eukaryotes

Type of RNA Polymerase	Location in the cell	Type of RNA synthesized	Promoter
RNA polymerase I	Nucleolus	18S, 5.8S, and 28S ribosomal RNA	Located upstream of ribosomal RNA gene, with an AT-rich region near the start site
RNA polymerase II	Nucleoplasm	messenger RNA, microRNA, small nuclear RNA, small nucleolar RNA, small interfering RNA, long non-coding RNA	TATA box, usually located upstream of the start site
RNA polymerase III	Nucleoplasm	5S ribosomal RNA, small nucleolar RNA, transfer RNA	Maybe located upstream or downstream of the start site

Unlike in DNA replication where both strands of the DNA helix serve as the template and the entire DNA molecule is replicated, the template for transcription is one of the two DNA strands. The strand serving as the template for the transcription of a gene is called the **anti-coding** or **anti-sense strand** while the complementary **coding** or **sense** strand contains the genetic sequence (see Figure 1.46). Considering that several genes are contained along the DNA of a chromosome, several anti-coding template segments may be located in either strand.

Transcription of the gene starts at the **initiation** or **start site** (or the **+1 base**). This site corresponds to the first nucleotide in the DNA template that is paired with an RNA nucleotide. However, RNA polymerase recognizes and binds a certain region of the DNA called the **promoter** before starting to produce RNA. A promoter sequence is considered a **cis-acting element**. This means that it is located in the same DNA molecule as the gene it regulates. The three types of RNA polymerases have their own promoters (see Table 1.3). The core promoter of RNA polymerase I is located at approximately -50 bp upstream (i.e., 50 bp on the 5' side or opposite the direction of transcription) of the start site. It usually overlaps with the start site in the rRNA gene. This promoter contains a conserved AT-rich region, which surrounds the start site. RNA polymerase II promoters are usually located upstream of the start site. Similar to prokaryotes, a consensus or preserved sequence in promoters of RNA polymerase II is present in eukaryotes. This is known as the **TATA** or the **Goldberg-Hogness** box because it is rich in thymine and adenine. This is located in the middle of -30 and -100 (i.e., 30 and 100 nucleotides upstream from the start site). Additional promoter elements such as the

CAAT and **GC box**, which are located between -40 and -150, are needed to successfully start transcription. Meanwhile, some RNA polymerase III promoters are also located in the upstream position while the majority are downstream of the start site (i.e., on the 3' side making it a part of the gene sequence) (Klug *et al.*, 2006; Berg *et al.*, 2002; Paule and White, 2000). The TATA promoter is bound, not only by RNA polymerase II, but also by proteins called **general transcription factors (GTFs)**, which vary in their number of subunits as well as their function (see Table 1.4). These assist in correctly orienting RNA polymerase at the promoter, facilitate unwinding of the double-stranded DNA, and disassociate RNA polymerase from the promoter allowing it to proceed to the start site.

The binding of RNA polymerase II and the GTFs around the TATA promoter form the **preinitiation complex** (see Figure 1.47). **TFIID** is the first one to bind the TATA box via its subunit called the TATA box-binding protein (TBP). Apart from its TBP component, TFIID has 13-14 TBP-associated factors (TAFs) some of which directly interact with transcriptional activators (Dadoune *et al.*, 2004). TBP interacts with the minor groove of the DNA molecule causing a distortion in the DNA molecule. The distortion is proposed to serve as a landmark for the promoter location. It also brings closer DNA sequences on both sides of the promoter allowing the binding of other GTFs. The binding of TBP to DNA is proposed to be facilitated by the dimeric **TFIIA** (Liu *et al.*, 1999). It is followed by the monomeric **TFIIB**, which associates with TBP and DNA on either side of the TATA box via its C-terminal core domain. The N-terminal zinc-ribbon domain recruits RNA polymerase/TFIIF complex to the promoter site and bridges it with TBP. The two domains of **TFIIB** are joined by a flexible linker, which stimulates the initiation of transcription possibly by increasing the affinity of RNA polymerase to DNA (Lodish *et al.*, 2003; Alberts *et al.*, 2014; Weinzieri and Wiesler, 2011).

Table 1.4. Transcription Factors Involved in Eukaryotic Transcription

Transcription Factor	Number of Subunits	Functions
TFIIA	2	Facilitates and stabilizes the binding of TFIID to the TATA box, enhances TFIID function by counteracting the effects of TFIID repressors (Liu *et al.*, 1999)
TFIIB	1	Recruits RNA polymerase to the promoter site, acts as a bridging factor between TBP and TFIIF, stimulates transcription initiation (Weinzieri and Wiesler, 2011)

TFIID	1 TATA-binding protein (TBP) 13-14 TBP-associated factors	TBP recognizes and specifically binds the TATA box, some TAFs interact with activators of transcription (Dadoune *et al.*, 2004)
TFIIE	4	Recruits and regulates the enzymatic functions of TFIIH, possibly stabilizes the single-stranded DNA template and work with TFIIH to remove the general transcription factors from the initiation complex, possibly facilitates the recruitment of elongation factors (Okamoto *et* al., 1998)
TFIIF	2	Positions RNA polymerase II at the start site,stabilizes RNA polymerase interaction with TBP and TFIIB (Lodish *et al.*, 2003; Alberts *et* al., 2014)
TFIIH	10	Its ATPase, helicase/translocase activities enables the ATP-dependent opening of the promoter region and the transcription start site. Its kinase domain phosphorylates RNA polymerase II, which possibly leads to dissociation of the preinitiation complex (Rimel and Taatjes, 2018).

Then a complex of the dimeric TFIIF and RNA polymerase II binds, positioning RNA polymerase II at the start site. **TFIIF** stabilizes the interaction of RNA polymerase with the TBP and TFIIB. This is followed by the binding of the tetrameric TFIIE. **TFIIE** provides a docking site for and regulates the enzymatic activity of the next TF, which is TFIIH. **TFIIH** has ten subunits and it has an ATPase and helicase activity so that it hydrolyzes ATP in order to unwind the double-stranded DNA at the promoter region and start site forming an **open complex**. The helicase domain can also function as 5'→3' DNA translocase, where it translocates in a 5'→3' direction along the non-template strand. Since upstream DNA is bound to TBP, the translocation activity downstream generates a torsional stress, which can be relieved by the opening of the double-stranded DNA around the start site.

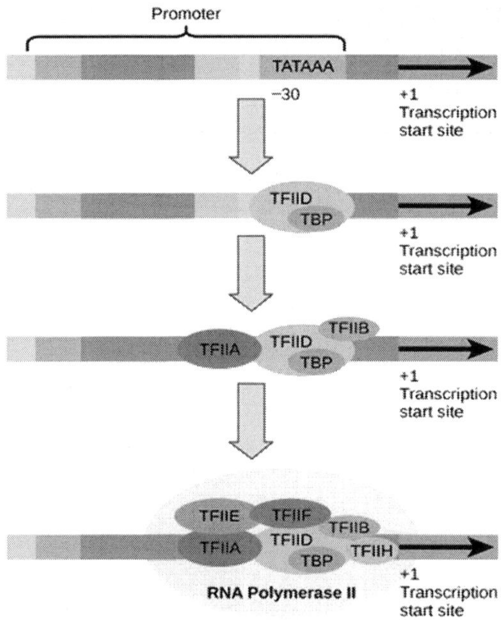

Figure 1.47. The preinitiation complex for transcription is formed by the sequential assembly of transcription factors (TFs) and RNA polymerase II at the TATA box. The first to bind the TATA promoter is TFIID via its TATA-binding protein (TBP) subunit. TFIID is assisted by TFIIA. The next to join the assembly is TFIIB. This is followed by the binding of the preformed complex of TFIIF and RNA polymerase II. Then TFIIE and TFIIH join, respectively, completing the complex. (Source: OpenStax Biology 2nd Edition, Creative Commons License).

The single-stranded template is then bound to the RNA polymerase active site. It is prevented from re-annealing with its complementary strand by the bound TFIIB, TFIIE, and TFIIF. TFIIH also has a kinase domain, which phosphorylates the C-terminal domain (CTD) of the largest subunit of RNA polymerase II. This kinase domain is thought to be involved in scanning for the start site and in dissociating the preinitiation complex (Lodish *et al.*, 2003; Alberts *et al.*, 2014; Okamoto *et al.*, 1998; Rimel and Taatjes, 2018). The phosphorylation of RNA polymerase II is also involved in facilitating **RNA processing**, which refers to the various modifications undergone by the precursor mRNA (the immediate product of transcription) to become a mature mRNA. For instance, the phosphorylated CTD recruits and activates 5'-capping enzymes and regulates splicing of the mRNA (Rimel and Taatjes, 2018).

The activity of eukaryotic promoters (also in prokaryotic promoters but less frequently) is influenced by one or more sequences known as **enhancers** (see Figure 1.48). These are 8-30 bp long cis-regulatory sequences, which are located 100 to 10,000 bp upstream or downstream from the starting point of transcription. These enhancers serve as binding sites for activator proteins that increase the efficiency of initiating transcription by five to a thousand-fold. It is proposed that the DNA segment between the promoter and enhancer forms a loop to enable the physical interaction of the RNA polymerase in the promoter region with the proteins bound to the enhancer (Schleif, 1993) (see Figure 1.46). A large complex of proteins called **Mediator** transmit the signals from the transcription activators in the enhancer regions to the preinitiation complex (see Figure 1.49) (Soutourina, 2018).

In contrast, **silencers** are cis-regulatory sequences that reduce or abolish transcription (see Figure 1.49). Other classes of regulatory sequences are **insulators**, which have two types (see Figure 1.49). Enhancer-blocking (EB) insulators maintain the specificity of the action of cis-regulatory sequences by preventing these from impinging on neighboring genes. Barrier-form insulators, on the other hand, set up regulatory boundaries by preventing chromatin condensation in active regulatory regions.

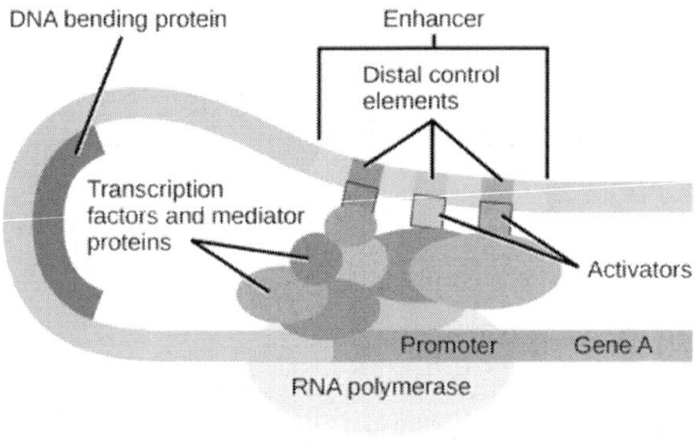

Figure 1.48. Promoters and enhancers are both cis-acting elements that are distant from each other in the eukaryotic DNA. RNA polymerase and transcription factors bind the promoter in order to initiate transcription. On the other hand, activators bind the enhancer and interact with the transcription factors in order to increase the efficiency of transcription initiation. DNA bending protein facilitates the looping of the DNA to enable the physical interaction of proteins in the promoter and enhancer regions. (From OpenStax Biology 2nd Edition, Creative Commons License).

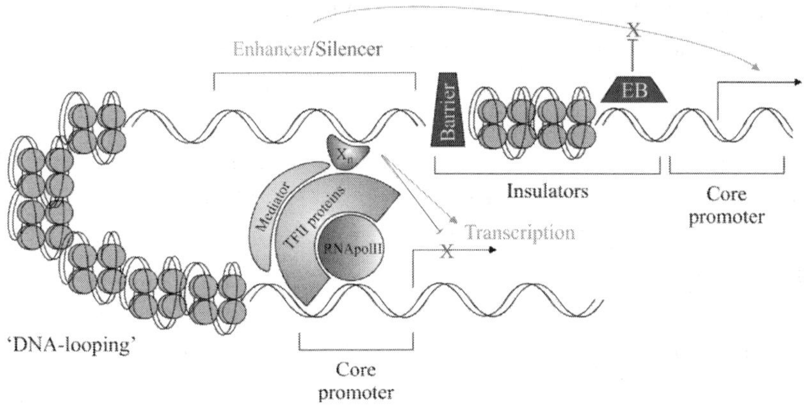

Figure 1.49. Eukaryotic transcription starts with the formation of the preinitiation complex, which includes various types of transcription factor II (TFII proteins) and RNA polymerase II. This complex is bound to the core promoter region, which consists of binding sites of the TFII proteins and RNA polymerase II as well as the transcription start site. In addition, distant cis-regulatory sequences modulate the transcription process. Enhancers increase transcription rate with the aid of a Mediator complex that transmits the signal from the activator bound to the enhancer to the preinitiation complex(green arrow) while silencers decrease or stop transcription (red T bar). Barrier-form insulators demarcate the boundary of active regulatory regions and enhancer-blocking (EB) insulators block regulatory sequences from influencing adjacent genes. (Source: Cowie, P., Ross, R., Mackenzie, A. Biology 2(1), 64-84, 2013, Creative Commons License).

1.3.4.3.1.2. Elongation of the RNA transcript

When the open complex is formed, RNA polymerase II starts to synthesize a short segment of RNA from the start site by pairing the nucleotides in the DNA template with the complementary RNA nucleotide. This means that adenine in the template is paired with uracil, thymine with adenine, guanine with cytosine, and cytosine with guanine. This results in the formation of RNA-DNA hybrid duplex. After synthesizing a 12-13-base RNA molecule, RNA polymerase II needs to dissociate from the promoter/start site region and proceed downstream to transcribe the rest of the gene. A key step enabling the RNA polymerase II to escape from the promoter is the presence of sufficient RNA nucleotides as well as the addition of phosphate groups, by TFIIH, to specific serine residues in its C-terminal domain. As a result,

RNA polymerase II dissociates from the complex of GTFs (Alberts *et al*., 2014; Rimel and Taatjes, 2018).

After escaping from the promoter, RNA polymerase II is observed to pause after producing 20-80 bases long transcript. This is possibly regulated by the kinase activity of TFIIH. This pausing is also observed in the other types of eukaryotic RNA polymerases as well as in prokaryotic ones (Sims III *et al*., 2004). RNA polymerase II also undergoes a series of structural changes to enhance its processivity. This refers to the capacity of the enzyme to transcribe for long distances without detaching from the template. This is necessary since newly initiated RNA polymerase II tends to slip off from the template. After the transition from initiation to elongation, most of the GTFs are released from the open complex. Thus, these are available to initiate a new round of transcription with another RNA polymerase (Alberts *et al*.; Rimel and Taatjes, 2018). Succeeding rounds of initiation may proceed at a faster rate than the first round (Sims III *et al*., 2004).

RNA polymerase II is assisted by proteins known as **elongation factors** to increase its processivity as well as to overcome pause in transcription. Among these elongation factors, **TFIIS** is the most well-known. *In vitro*, it has been observed to stimulate RNA polymerase II to cleave the synthesized RNA that has dissociated from the template (see Figure 1.50). This generates a new 3' end and allows realignment of the RNA polymerase II active site so that the enzyme can continue elongating the transcript (Alberts *et al*., 2014; Kim *et al*., 2007; Sims III *et al*., 2004). **TFIIF** can also function as an elongation factor. After being released after the initiation process, it re-associates with RNA polymerase II, especially when the enzyme pauses, and induces a conformational change in the enzyme to increase the elongation rate. Once this is accomplished, TFIIF is released from RNA polymerase II. Other elongation factors include those belonging to the Elongin complex, the ELL (eleven-nineteen lysine-rich leukemia) family, and the NELF complex (Sims III *et al*., 2004).

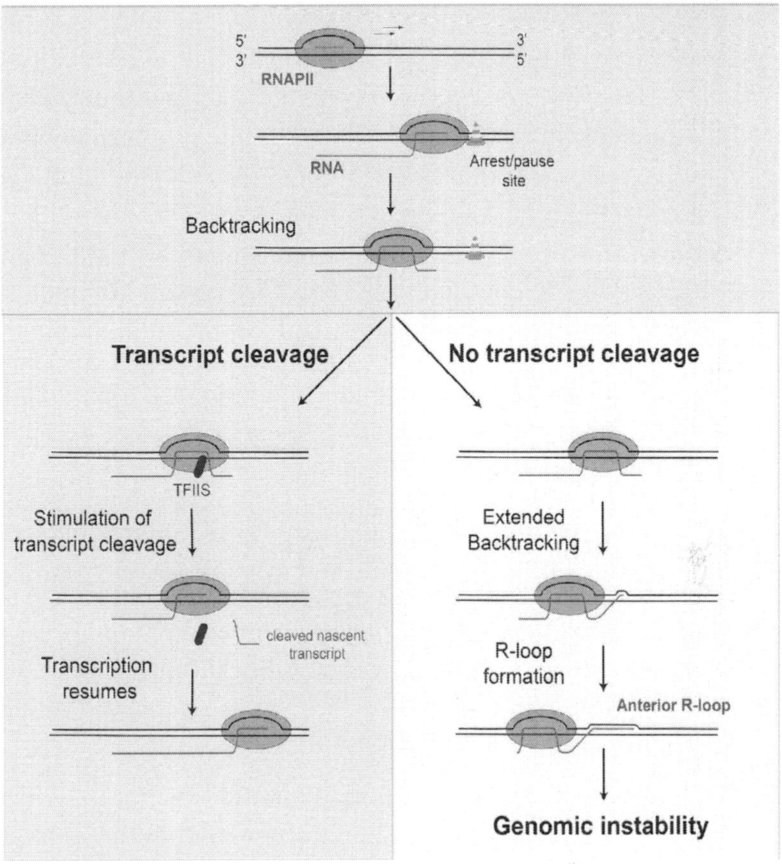

Figure 1.50. TFIIS is an elongation factor for eukaryotic transcription. It recognizes backtracked or paused RNA polymerase II then stimulates cleavage of the transcript. This allows RNA polymerase II to resume elongation of the transcript. Without transcript cleavage, transcript elongation is disturbed. (From Zatreanu, D., Han, Z., Mitter, R., Tumini, E., Williams, H., Gregersen, L., Dirac-Svejstrup, A.B., Roma, S., Stewart, A., Aguilera, A., Svejstrup, J. Molecular Cell 76(1), 57-69.e9, 2019, Creative Commons License).

Since eukaryotic DNA is associated with proteins, the continued synthesis of RNA requires **ATP-dependent chromatin remodeling complexes**. These enable RNA polymerase to overcome protein blocks as it moves along the chromatin. Moreover, **histone chaperones** facilitate nucleosome disassembly in front of a moving RNA polymerase and assemble them behind (Alberts *et al.*, 2014).

1.3.4.3.1.3. Termination of transcription

Termination of transcription varies depending on the type of RNA polymerase. Transcription of the rRNA gene by RNA polymerase I is terminated with the help of a **termination factor**. This is a DNA-binding protein that recognizes and associates to a specific DNA sequence downstream of the transcribed sequence (Lodish*et al.*, 2003).

RNA polymerase II recognizes a **termination signal** in the pre-mRNA transcript, which directs cleavage and polyadenylation (or the addition of several adenylate residues) of the pre-mRNA at its 3' end. The site of polyadenylation is preceded by a consensus sequence, i.e. AAUAA. Polyadenylation initiates termination but the actual termination occurs at multiple RNA polymerase pause sites, which are approximately 0.5-2 kilobases (kb) beyond the termination signal. The pause site may be a hairpin structure and/or a certain sequence that is bound by a specific protein factor (Lodish*et al.*, 2003; Eun, 1996).

RNA polymerase III terminates at a region of the template rich in adenine residues. This results in a stretch of DNA-RNA hybrid with A-U base pairs, which can easily be separated due to its instability (Lodish *et al.*, 2003).

1.3.4.3.2. Post-transcriptional mRNA modification

In eukaryotes, the **precursor mRNA** initially produced after transcription is a much longer molecule compared to the **mature mRNA** that is eventually used for protein synthesis. This is because the precursor mRNA undergoes RNA processing where it is modified to produce the mature mRNA used as a template for translation (see Figure 1.51).

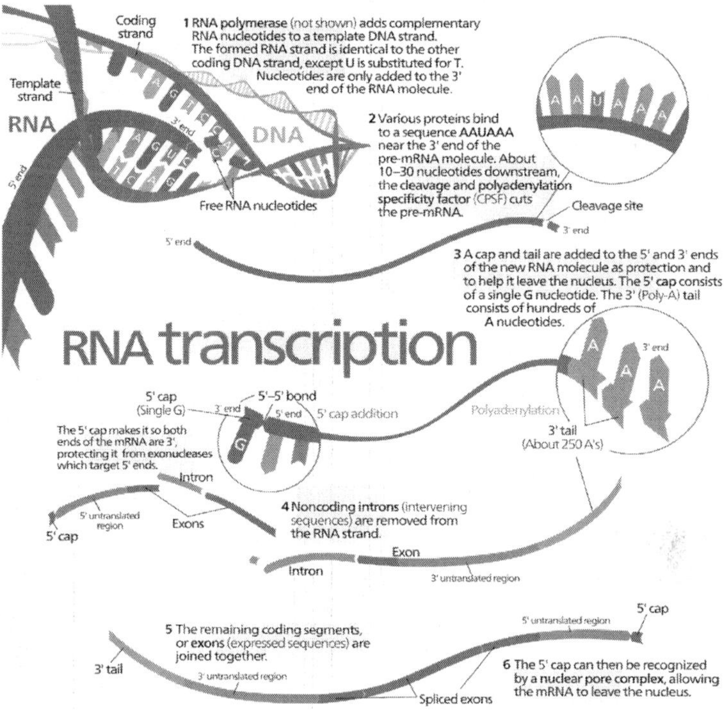

Figure 1.51. The immediate product of transcription in eukaryotes is a precursor mRNA that undergoes modifications such as the addition of a cap at the 5'end and a poly-adenylate tail at the 3' end (step 3), the removal of non-coding introns (step 4), and joining together the exons (step 5). (Source: Kelvin Song, Creative Commons License).

1.3.4.3.2.1. Capping of mRNA

Even before the mRNA is completed, a 7-methylguanosine (7-mG) cap is added to its 5' end (see Figure 1.51). The cap possibly protects the mRNA from being digested by nucleases. It may also be involved in transporting the mature mRNA from the nucleus into the cytoplasm in order to initiate translation (Klug *et al*., 2006). Ribosomes translate cap-containing mRNAs more efficiently than those without a cap (Schleif, 1993).

1.3.4.3.2.2. Poly-A tailing of mRNA

After the addition of the 7-mG cap to the 5'end of the mRNA, the precursor mRNA is cleaved at a certain point, which is 10-35 nucleotides from a highly conserved AAUAAA sequence. Then a series of adenine nucleotides, as many as 250 residues, are added to the 3' end (see Figure 1.51). Poly-A tails are also observed in some prokaryotic mRNA. On the other hand, eukaryotic mRNAs that code for histone proteins seem to lack this modification. Similar to the 5' cap, the 3' poly-A tail is possibly important in protecting the mRNA from degradation and is involved in exporting the mRNA to the cytoplasm (Klug *et al.*, 2006).

1.3.4.3.2.3. Splicing of the RNA

One significant difference between prokaryotic and eukaryotic genes is the presence of intervening sequences. These are called **introns**, which do not code for amino acids. These sequences are removed from the precursor mRNA and are no longer part of the mature mRNA that serves as a template for polypeptide synthesis (see Figure 1.51). The expressed segments in between the introns are known as **exons**.

In mRNA, the process of **splicing** involves the removal of the introns and the rejoining of the exons (see Figure 1.51). There are several types of splicing mechanisms that are observed in eukaryotes.

Introns that are part of the rRNA primary transcript are classified as **group I** introns. These introns themselves have the needed enzymatic activity for their own removal. These RNAs that have autocatalytic activity and can splice themselves are called **ribozymes**.

Group II introns are also autocatalytic and can perform self-excision. This type of introns is observed in the primary mRNA and tRNA transcripts produced in the mitochondria and chloroplasts.

The introns of nuclear mRNA are much longer and greater in number. Their splicing involves a more complex mechanism. These introns have GU at their 5' end, called the splice **donor sequence**, and AG at the 3' end, called the splice **acceptor sequence**. These consensus sequences attract certain molecules that form the **spliceosome**, a molecular complex essential for splicing mRNA introns. A very important component of the spliceosome is the **small nuclear RNAs (snRNAs)**, which are 100-200 nucleotides long and associate with proteins to form **small nuclear ribonucleoproteins (snRNPs)**. There are different types of snRNPs that are designated as U1, U2, and so forth due to the presence of several uracil nucleotides in the snRNAs.

The nucleotide sequence of U1 snRNA is complementary to the 5' splice donor sequence of the intron. After the pairing of these sequences, other types of snRNPs are added in sequence (U2, U3, U4, U5, and U6) to complete the spliceosome which will then excise the introns starting from the 5' end of the intron (see Figure 1.52). The freed guanine nucleotide residue forms a phosphodiester bond at the 2' site of an adenine nucleotide residue upstream from the 3' end of the intron. The intron is removed as a lariat (i.e., a looped structure), then the exons are joined together and the snRNPs are released (Merchant, 2018).

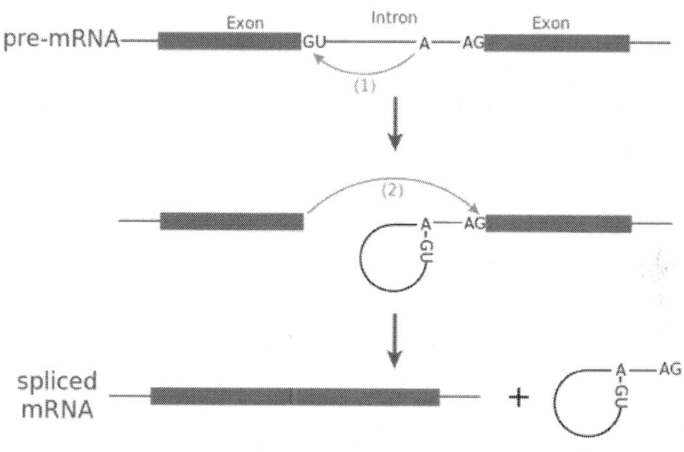

Figure 1.52. The eukaryotic precursor mRNA needs to have its introns removed. This is accomplished by the spliceosome complex which cuts at the 5' GU end of the intron. The freed guanylate is linked to an adenylate residue within the intron to form a lariat. Another cut is made at the 3' AG end of the intron, which removes the intron and allows the exons to be joined together. (Source: BCSteve, Creative Commons License).

Precursor tRNAs also contain introns that are spliced. However, tRNAs need to be folded correctly into their cloverleaf form before they are processed. Unlike mRNA splicing, no lariat intermediate is generated when introns are excised from the tRNA (Alberts *et al.*, 2014).

1.3.4.3.2.4. RNA editing

RNA editing involves a change in the sequence of the precursor mRNA before the translation process. The two main types of RNA editing are **substitution editing**, wherein the nucleotides in the mRNA are changed, and **insertion/ deletion editing**, in which nucleotides are added to or subtracted from the mRNA. RNA editing is also observed in mitochondrial RNAs of certain eukaryotes. RNA editing seems to play an important role in regulating gene expression. This is illustrated by the effect of the mutation on the *ADAR* enzymes which are necessary for RNA editing in mammals. In one study involving mice embryos, those with one copy of the *ADAR1* gene that is defective because of mutation die due to defective hematopoietic system. In another study, mice with two defective copies of the *ADAR2* gene showed normal embryonic development but after birth, they are prone to epileptic seizures and eventually die during the weaning stage (Klug *et al*., 2006).

Approximately 10% of the nucleotides in tRNAs are also altered via more than 50 different types of chemical modifications. For instance, the amino group may be removed from adenosine to produce inosine. This affects the conformation of the tRNA enhancing its recognition of the appropriate mRNA during the translation process. Other modifications influence the capacity of the tRNAs to attach to the correct amino acid (Alberts *et al*., 2014).

1.3.4.3.3. Protein synthesis via translation

The RNA products of the transcription process are necessary for manufacturing proteins (see Figure 1.53). The rRNAs are assembled with ribonucleoproteins in the nucleus to form ribosome subunits and these are exported into the cytosol. Likewise, mRNAs and tRNAs are translocated into the cytosol to form a translation complex with the ribosome. Every nucleotide triplet in the mRNA is called a **codon** and it corresponds to an amino acid during translation. As previously discussed, the mature tRNA has a cloverleaf structure with a series of base- paired stems and unpaired loops. The 3'end of the tRNA is the amino acid binding site. Opposite this site is a loop containing three nucleotides that make up the **anticodon**, which is complementary with the codon in the mRNA. This means that each type of amino acid is bound to a tRNA whose anticodon is complementary to the codon for that amino acid.

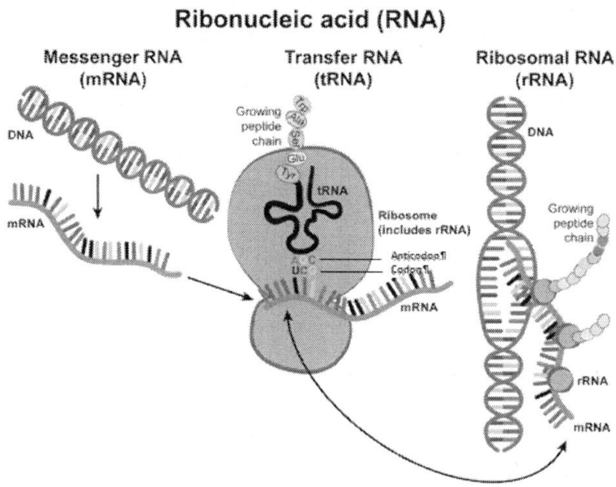

Figure 1.53. The three types of RNA produced during transcription are necessary for protein synthesis. The ribosomal RNA (rRNA) associates with ribosomal proteins to form the ribosome. The messenger RNA (mRNA)contains the codes (or the codons) for the amino acid sequence of the protein to be synthesized while the transfer RNA (tRNA) brings the amino acid to the site of protein synthesis. The anticodon in the tRNA is complementary to the codon in the mRNA, which specifies the amino acid carried by that tRNA. (Source: Modified from WebStockReview, Creative Commons License).

1.3.4.3.3.1. The genetic code

As previously mentioned, the mRNA contains a series of codons that are equivalent to amino acids. A total of 64 possible triplet combinations may be derived from the four types of nucleotides in the mRNA. Each of these codons is unambiguous, i.e. it specifies only one type of amino acid. The series of codons in the mRNA is "read" in a non-overlapping manner. There are only 20 types of amino acids which means that more than one type of codon may specify an amino acid. This is called the **degeneracy** of the code. The code is also nearly **universal** which means that it is the same for all organisms, with the mtDNA being the exception. The correspondence of codons to amino acids is summarized in the genetic code chart. As can be seen in the chart (see Figure 1.54), three of the codons (i.e. UAA, UGA, UAG) do not correspond to any amino acid. These are called **stop** or **termination** or **nonsense codons**. These serve as signals to terminate the translation process (Klug *et al*., 2006).

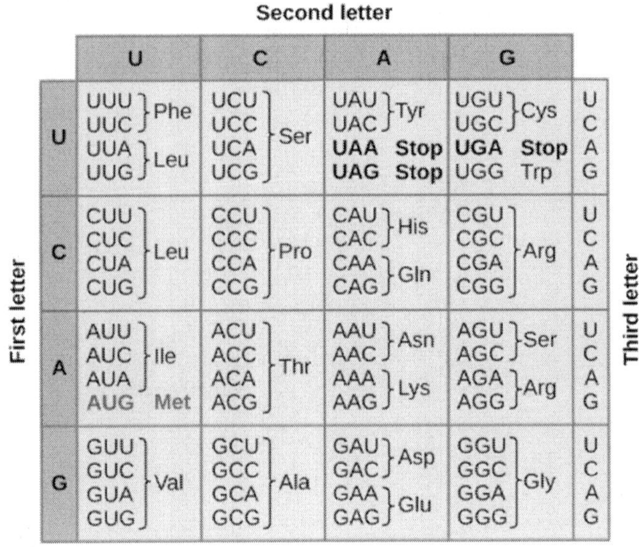

Figure 1.54. The genetic code chart lists the 64 possible codons in the messenger RNA. One of these is the start codon (AUG in red), three are stop codons (in bold fonts), and the rest correspond to any of the 20 amino acids. (Source: National Institutes of Health, Public Domain).

1.3.4.3.3.2. Activation of tRNAs

Prior to translation, tRNAs must be linked to their corresponding amino acid (see Figure 1.55). Since there are 20 types of amino acids, there must be at least 20 types of tRNAs. If the basis is the number of amino acid-coding codons to be complemented by the anticodons, then there should be 61 tRNAs. However, due to the **wobble hypothesis**, which proposes that the 3rd nucleotide in the codon can pair with a variety of nucleotides in the anticodon, a certain tRNA anticodon may correspond to more than one type of codon. Thus, it is thought that there are at least 32 types of tRNAs.

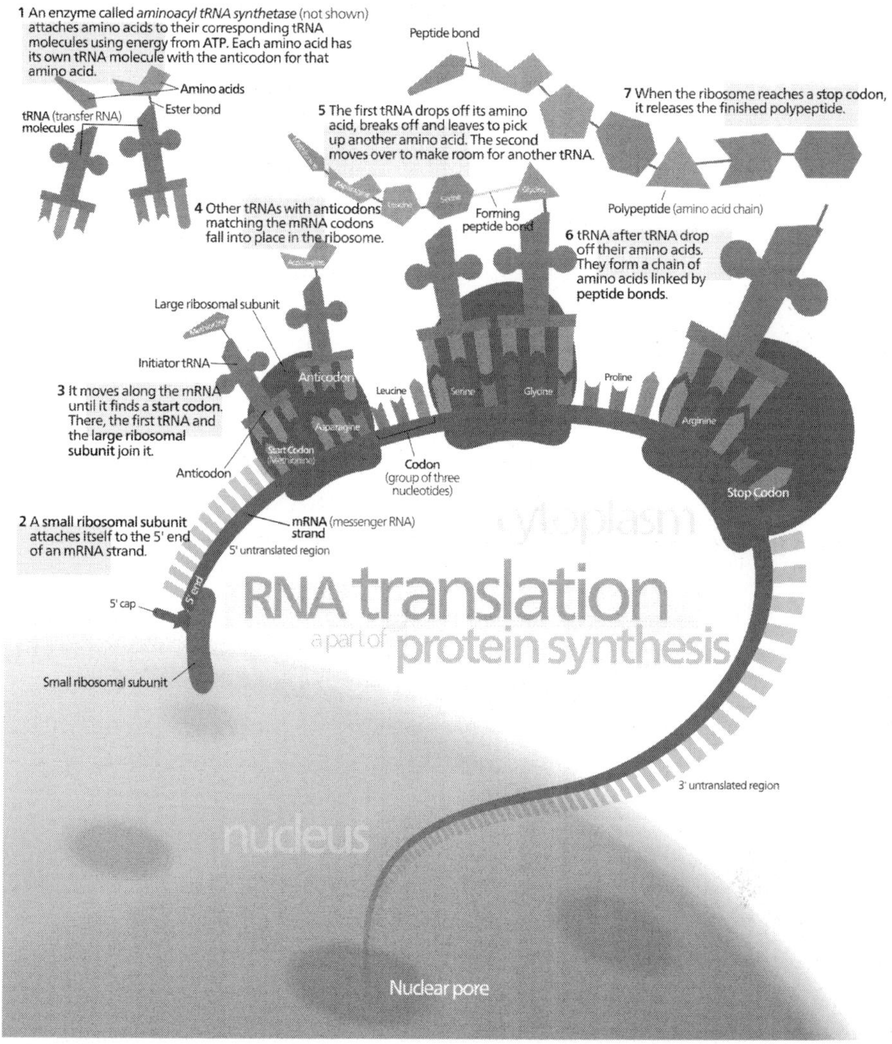

Figure 1.55. The translation process requires the activation of tRNAs (step 1) catalyzed by aminoacyl tRNA synthetases. At the initiation phase (steps 2 and 3), the tRNA bearing methionine is positioned opposite the start codon. The tRNA carrying the amino acid that corresponds to the next codon is then put in place. The previous tRNA leaves the translation complex and the ribosome moves every three nucleotides to expose the next codon. These steps occur repeatedly resulting in elongation of the amino acid chain (steps 4, 5, and 6). Finally, when the stop codon is encountered by the ribosome, termination of translation ensues (step 7). (From Kelvin Song, Creative Commons License).

The attachment of an amino acid to its tRNA is catalyzed by the enzyme, aminoacyl tRNA synthetase. There are 20 of these enzymes, one that is specific for each amino acid. The amino acid reacts with and is linked to a molecule of adenosine triphosphate (ATP) via a high energy bond. Hence, the amino acid is said to be activated. The resulting intermediate binds to an aminoacyl tRNA synthetase that recognizes the amino acid and the tRNA. This is essential in maintaining the accuracy of the translation process. The enzyme then catalyzes the transfer of the amino acid to the 3' end of the tRNA (Klug *et al.*, 2006). The aminoacyl tRNA synthetase is able to identify its correct tRNA molecule not only through the anticodon but also by recognizing certain structural details of the tRNA (Schleif, 1993; Lodish *et al.*, 2003).

1.3.4.3.3.3. Initiation of protein synthesis

The ribosome is made up of a small and a large subunit which are usually separate from each other when the organelles are not involved in polypeptide synthesis. Ribosomes in eukaryotes may either be free-floating in the cytoplasm or bound to the membrane of the endoplasmic reticulum (ER). The translation process is similar in these ribosomes; they differ in the types of proteins they synthesize. Ribosomes on the ER produce proteins that are secreted by the cell while cytosolic ribosomes produce the proteins used by the cell itself.

In eukaryotes specifically, the small subunit is designated as the **40S** (S stands for svedberg, which measures how fast molecules move when centrifuged) and the large subunit is **60S** and they associate with each other to form the **80S** ribosome. When not engaged in protein production, these two subunits are kept separated by two eukaryotic initiation factors, eIF3 and eIF6. Apart from having a binding site for the mRNA, there are also three binding sites of tRNA in the ribosome. The **peptidyl (P) site** is where the tRNA attached to the growing polypeptide chain is bound while the **aminoacyl (A) site** is where an incoming amino acid-tRNA is bound. The third one, the **exit (E) site** is where a tRNA devoid of any amino acid is bound prior to being released from the ribosome (Lodish *et al.*, 2003; Alberts *et al.*, 2014).

The 40S subunit-eIF3complex is bound by eIF1A and by a complex of Met-$tRNA_i$ (i.e., methionine bound to an initiator tRNA, which initiates protein synthesis), eIF2, and GTP. These form the **preinitiation complex**. The **initiator tRNA ($tRNA_i$)** has a nucleotide sequence distinct from other tRNAs that carry methionine so that it is recognized by initiation factors. Additionally, the Met-tRNA$_i$ is the only charged tRNA that can bind directly to the P site and to the small ribosome subunit, even without the presence of the large subunit to complete the ribosome. The eIF4 has an eIF4E subunit that binds the 5' cap of the mRNA (see Figure 1.55). Then eIF4G and eIF3 mediate the association of the mRNA-eIF4 complex with the preinitiation complex to form the **initiation complex**.

Another subunit of eIF4, eIF4A, has a helicase function so that it hydrolyzes ATP to unwind secondary structures in the mRNA and, along with the 40S subunit, scans the mRNA from 5'→3'until the Met-tRNA$_i$ anticodon recognizes the start or initiation codon at the P site. This codon is usually the first AUG downstream from the 5' end of the mRNA.

In addition, this is surrounded by the so-called **Kozak sequence** (named after **Marilyn Kozak**), i.e. 5' ACC<u>AUG</u>G 3'. The A before the AUG and the G after it are the most important nucleotides that influence the selection of the correct AUG as the start codon as well as the efficiency of translation initiation. Therefore, if the first AUG does not have this surrounding sequence, it is sometimes skipped by the scanning complex, which then proceeds to scan along the mRNA until it finds the AUG surrounded by the Kozak sequence. This allows for the production of different protein molecules from the same mRNA (Lodish *et al.*, 2003; Alberts *et al.*, 2014).

Once the start codon has been recognized, the GTP associated with eIF2 is hydrolyzed, a reaction that stops the scanning of the mRNA. Then, the initiation factors dissociate from the initiation complex. The 60S large subunit then associates with the small subunit, with the assistance of eIF5, to form the 80S ribosome. At this point, the Met-tRNA$_i$ remains at the P site, leaving the A site empty (Lodish *et al.*, 2003; Alberts *et al.*, 2014).

1.3.4.3.3.4. Elongation of the polypeptide chain

After the initiation of protein synthesis, the next phase involves the addition of amino acids to the initial methionine in order to produce a polypeptide (see Figure 1.55). Elongation factors (EFs) make this process more accurate and efficient. In eukaryotes, there are two of these, EF1 and EF2 (Alberts *et al.*, 2014).

The incoming tRNA to bind to the A site should have an anticodon that is complementary to the codon exposed at the A site. Furthermore, it should be bearing the amino acid that corresponds to the codon. This amino acid-tRNA associate with EF1α-GTP before binding to the A site. At this point, both P and A sites are occupied by tRNAs with their respective amino acids. The base pairing of the anticodon and the codon at the site results in the hydrolysis of the GTP associated with EF1α.

This, in turn, induces the ribosome to shift in conformation so that the EF1α -GDP is released but the amino acid-tRNA remains tightly bound to the A site. Meanwhile, at the P site, the high energy bond between methionine and the initiator tRNA breaks to drive the linking of methionine to the free amino group of the adjacent amino acid. The enzyme **peptidyl transferase**, which is located at the large ribosome subunit, catalyzes the formation of a peptide linkage between these two amino acids. As a result, the A site now contains a tRNA carrying a dipeptide. The ribosome then translocates to the 3' direction so that the next codon in the mRNA is exposed at the A site and the tRNA-dipeptide is shifted to the P site. On the other hand, the empty tRNA is now bound to the E site. This translocation step is promoted by the hydrolysis of another GTP associated with EF2. The tRNA at the E site dissociates from the complex and is now free to be activated anew with another amino acid. The ribosome returns to its previous conformation where its A site is open. The previous steps of entry of a new amino acid-tRNA/EF1α-GTP complex to the A site, dissociation of the carboxyl end of the peptide from the tRNA at the P site to form a peptide bond with the new amino acid, ribosome translocation, and ejection of the empty tRNA are repeated in a cycle of reactions. These result in the elongation of the polypeptide by adding one amino acid at a time as the translation complex moves along the mRNA (Alberts *et al*., 2014; Lodish *et al*., 2003).

1.3.4.3.3.5. Termination of the translation process

The synthesis of the polypeptide comes to a halt when any one of the stop codons (i.e., UAA, UAG or UGA) is reached (see Figure 1.55). The stop codons are no longer recognized by tRNAs but serve as a molecular signal for **eukaryotic release factors (eRFs)** to facilitate the termination process. One of these is eRF1 which is structurally similar to tRNA. When a stop codon is at the A site, it binds to the A site. Another one is eRF3, which binds to GTP. The eRF3-GTP complex works with eRF1 to induce the cleavage of the polypeptide-tRNA so that the polypeptide chain is released from the translation complex. Other release factors mediate the dissociation of the ribosome to its subunits, as well as the release of mRNA and terminal tRNA (Lodish *et al*., 2003; Alberts *et al*., 2014; Klug *et al*., 2006).

1.3.4.3.4. Posttranslational modification of polypeptide

The freed polypeptide chain then folds to its three-dimensional conformation before it can perform its activity. Other polypeptides need to bind **co-factors**, to be modified enzymatically or to assemble with other protein subunits in order to be functional (Lodish *et al.*, 2003).

1.3.4.3.4.1. Protein folding

There are eight bond angles that are stereochemically allowed in the polypeptide backbone. This means that if a certain polypeptide chain has 20 amino acids, it can theoretically fold into 8^{20} conformations. However, a protein generally adopts a single conformation called the **native state**, which is the most stable form for most proteins.

It was observed that after denaturation in vitro, in which a protein loses its native conformation due to extreme pH, high temperature, or exposure to organic solvents, proteins return to their native state spontaneously.

This indicates that the information needed for protein folding is encoded in the amino acid sequence (Lodish *et al.*, 2003; Koolman and Roehm, 2005).

The cell shows remarkable efficiency in protein folding so that more than 95% of the cellular proteins are folded in their native state. This may be due to **chaperones**, a class of proteins found in all organisms that plays an important role in the general protein-folding mechanism of cells. There are two families of chaperones, **molecular chaperones** and chaperonins. Molecular chaperones bind and stabilize unfolded or partly folded proteins. This prevents these proteins from aggregating and being degraded. Some molecular chaperones, such as the heat shock protein 70 (Hsp70), initiate the degradation of misfolded or damaged proteins (see Figure 1.56). On the other hand, **chaperonins** directly facilitate protein folding (Lodish *et al.*, 2003).

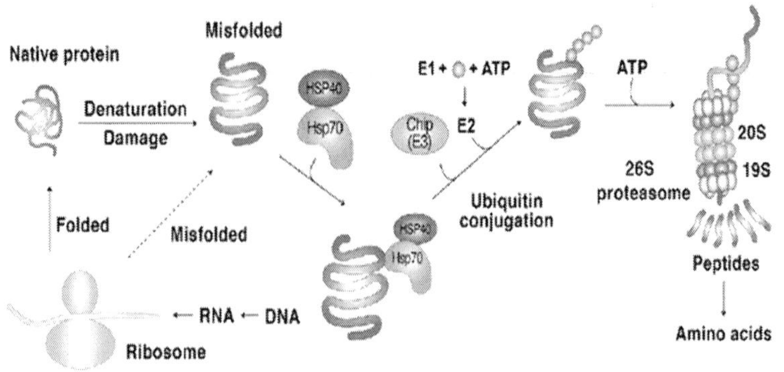

Figure 1.56. Heat shock protein 70 (Hsp70), a class of molecular chaperones, recognizes the hydrophobic sequences of misfolded or damaged protein. This initiates a cascade of reactions in the ubiquitination pathway that eventually results in the degradation of the protein. This is one of the ways of protein quality control since the accumulation of misfolded and damaged proteins leads to a host of diseases. (Source: Ciechanover, A.; Kwon, Y.T. Frontiers in Neuroscience11, 185, 2017, Creative Commons License).

1.3.4.3.4.2. Removal of certain amino acids

The enzymatic cleavage of the polypeptide backbone to remove one or more amino acid residues is known as **proteolysis**. In eukaryotes, the first methionine is often removed. Moreover, the N-terminus of the polypeptide contains up to 30 amino acids that compose the **signal sequence**. This is important in directing where the resulting protein should be transported in the cell. For example, there are specific signal sequences that target a polypeptide to the nucleus, the mitochondria, the ER, and so forth. Once the polypeptide reached its final location, the signal sequence is removed (Klug *et al*., 2006).

Other polypeptide chains are trimmed to obtain their final, active form. This is the case for the enzymes involved in blood coagulation, digestion, and apoptosis. For example, certain digestive enzymes known as endoproteases and carboxypeptidases are manufactured and secreted as inactive zymogens. This is to protect the secretory cell from degrading itself. Once the zymogens are released in the digestive tract, these undergo proteolytic cleavage to activate the enzymes. Proteolysis of larger polypeptides also generates active peptide hormones such as insulin (Lodish *et al*., 2003) (see Figure 1.57).

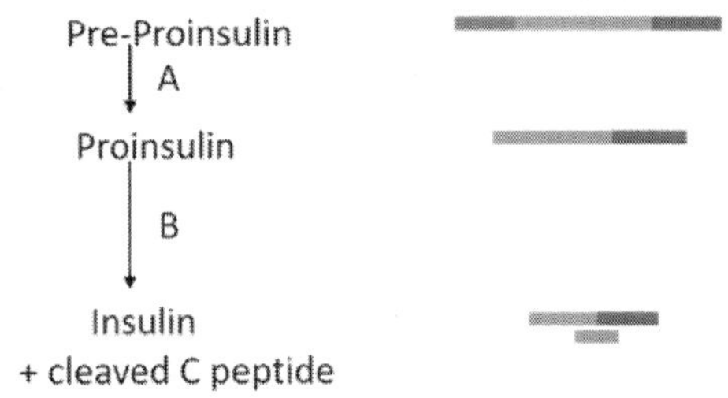

Figure 1.57. The immediate product of translation is called pre-proinsulin. Its signal peptide (blue segment) is eventually cleaved forming proinsulin. Proinsulin is folded in the ER then transported to the Golgi apparatus where the C-peptide (light green segment) is cleaved to form a free C- peptide and mature insulin. (Source: Anoel8, Creative Commons License).

In bacteria and some eukaryotes, **protein self-splicing** occurs where an internal segment of the polypeptide is cleaved and then the ends of the remaining polypeptide are rejoined. This process is autocatalytic, i.e. it occurs on its own without the need for enzymes (Lodish *et al*., 2003).

1.3.4.3.4.3. Modification of the amino acids

Most proteins are chemically modified after being synthesized in the ribosomes. These modifications include the addition of a chemical group to the amino group or carboxyl group on either end of the polypeptide or to a side chain within the polypeptide. Some of the chemical modifications undergone by amino acids are discussed in this section.

Acetylation

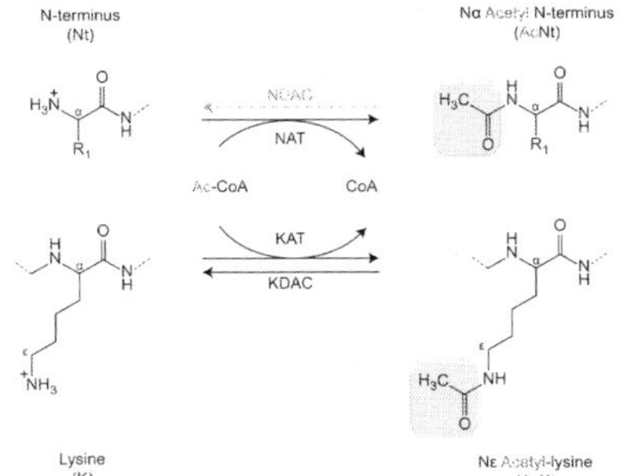

Figure 1.58. Acetylation at the N-terminus of the protein involves N-terminal acetyltransferase (NAT), which catalyzes the transfer of an acetyl group from

acetyl-CoA (Ac-CoA) to the free α-amino group. An N-terminal deacetyltransferase (NDAC) is yet to be discovered.

Another type of enzyme for acetylation is lysine acetyltransferase (KAT) that **catalyzes** the transfer of acetyl group to the ε-amino group of the lysine (K) side chains.

The attached acetyl group may be removed by the activity of lysine deacetyltransferase (KDAC). (Source: Ree, R.; Varland, S.; Arnesen, T. Experimental & Molecular Medicine 50:90, 2018, Creative Commons License).

Acetylation occurs in approximately 80% of all proteins, making it the most common chemical modification (see Figure 1.58).

This is thought to regulate the life span of proteins as non-acetylated proteins are quickly digested by proteases inside the cell. An acetyl group may also be added to an internal amino acid within the protein (Lodish *et al.*, 2003).

Phosphorylation

Phosphate groups may be added to the polypeptide. Specifically, amino acids with hydroxyl (-OH) groups such as serine, threonine, tyrosine, and histidine are phosphorylated via the activity of kinases. This type of modification is reversible, which means that the phosphorylated amino acid may be dephosphorylated by phosphatases. Alternating between the phosphorylated and dephosphorylated state is critical for regulating the activity of kinase enzymes (Klug *et al.*, 2006; Lodish *et al.*, 2003).

Glycosylation

Carbohydrate chains, either linear or branched, may also be added to the polypeptide to form a **glycoprotein**. The side chains of asparagine, serine, and threonine are the common sites for glycosylation (see Figure 1.59). Several secreted proteins and membrane proteins have glycosylated amino acids (Klug *et al.*, 2006; Lodish *et al.*, 2003). The addition of carbohydrate to these proteins occur in the ER and the Golgi apparatus. Glycosylation is required for protein folding (Buxbaum, 2015). Glycoproteins have an important role in cell-to-cell as well as cell-to-environment interactions.

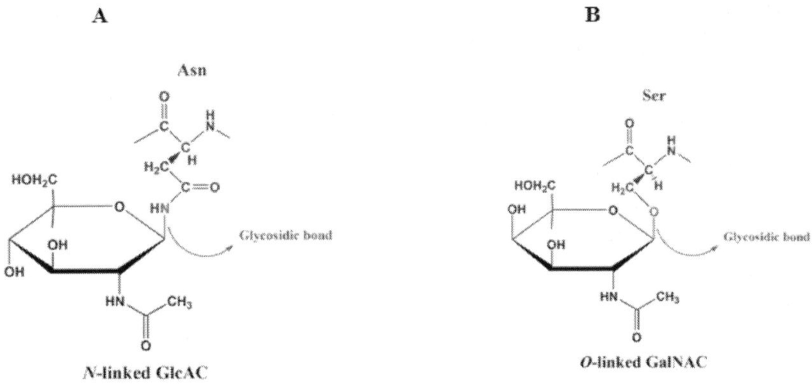

Figure 1.59. (A) If the carbohydrate group is attached to the nitrogen atom in the side chain of asparagine, the protein modification is called N-linked glycosylation. (B) If it is attached to the oxygen atom of serine or threonine, then it is called O-linked glycosylation. (Source: Tpirojsi, Public Domain).

Hydroxylation

Hydroxylation is the addition of a hydroxyl group to an amino acid in the polypeptide as catalyzed by iron-dependent hydroxylases. This occurs most commonly on proline, lysine, and asparagine. Hydroxylation is essential in the maturation of collagen fibers (Green and Garneau-Tsodikova, 2010) (see Figure 1.60).

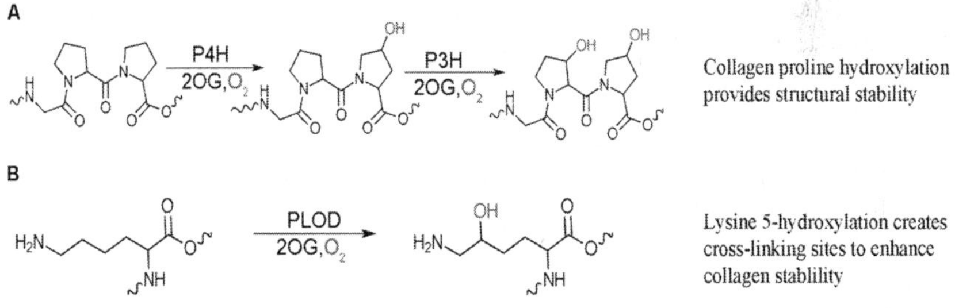

Figure 1.60. (A) Hydroxylation of proline residues of collagen is catalyzed by procollagen prolyl 4-hydroxylase (P4H) and prolyl 3-hydroxylase (P3H). This improves the structural stability of collagen. (B) Procollagen lysine 5-hydroxylase (PLOD) hydroxylates the lysine residues of collagen creating cross-linking sites, which provide further stability. (Source: van Staalduinen, L.M.; Jia, Z. Frontiers in Microbiology, https://doi.org/10.3389/fmicb.2014.00798, 2015, Creative Commons License).

Methylation

Figure 1.61. Methylation occurs in specific arginine (R) and lysine (K) residues of histone 3 (H3) and histone 4 (H4) tails. Histone methylation, depending on its location, is associated with various functions such as heterochromatin formation, gene silencing, alternative splicing, DNA damage response, transcription regulation, cell-cycle checkpoint, and cellular development. (Source: Kim, J-H; Lee, J.H.; Lee, I-S.; Lee, S.B.; Cho, K.S. International Journal of Molecular Sciences, 18(7), 1404, 2017, Creative Commons License).

Protein methylation entails the addition of methyl at either the N- or C-terminus, on the side chain nitrogen atoms of arginine, histidine, and lysine residues or on the carboxyl groups of glutamate, leucine, and isoprenylated cysteine. Examples of proteins that are methylated are histones and certain membrane receptors. Histone methylation regulates DNA replication, DNA repair, and transcription (Cheng *et al.*, 2012; Sawan and Herceg, 2010) (see Figure 1.61).

Carboxylation

The enzyme γ-glutamyl carboxylase catalyzes the addition of a carboxyl group to the glutamate residues of vitamin K-dependent proteins modifying these to γ-carboxylglutamate. Carboxylation of glutamate in prothrombin is essential for the blood clotting process in vertebrates (Bandyopadhyay *et al.*, 2002; Lodish *et al.*, 2003).

REFERENCES

1. Alberts, B., Johnson, A., Lewis, J., Raff, M., Roberts, K. & Walter, P., (2014). Molecular Biology of the Cell (6thedn.). Garland Science: New York.
2. Bandyopadhyay, P.K., Garrrett, J.E., Shetty, R.P., Keate, T., Walker, C.S. & Olivera, B.M., (2002). γ-Glutamyl carboxylation: An extracellular posttranslational modification that antedates the divergence of molluscs, arthropods, and chordates. Proceedings of the National Academy of Sciences of the United States of America, 99(3), 1264-1269.
3. Bateson, W. & Mendel, G., (2013). Mendel's Principles of Heredity (Dover edn.). Dover Publications, Inc.: New York.
4. Bateson, W., Saunders, E.R. & Punnett, R.C., (1905). Experimental Studies in the Physiology of Heredity. Reports to the Evolution Committee of the Royal Society, 2, 1-99.
5. Berg, J.M., Tymoczko, J.L. & Stryer, L., (2002). Biochemistry (5thedn.). W.H. Freeman: New York.
6. Blamire, J., (2002). Prophase I. http://www.brooklyn.cuny.edu/bc/ahp/LAD/C9/C9_m_prophase.html (Accessed on 28 January 2020).
7. Bulmer, M., (1998). Galton's law of ancestral heredity. Heredity, 81(Pt 5),579-585. DOI:10.1046/j.1365-2540.1998.00418.x (Accessed on 11 December 2019).
8. Buxbaum, E., (2015). Fundamentals of Protein Structure and Function, 2nd Ed.; Springer International Publishing: Switzerland.
9. Cheng, D., Vemulapalli, V. & Bedford, M.T., (2012). Methods Applied to the Study of Protein Arginine Methylation. Methods in Enzymology, 512,71-92.
10. Chhetri, D., (2014). "Genetic Control of Biochemical Reactions in Neurospora" (1941), by George W. Beadle and Edward L. Tatum. https://embryo.asu.edu/pages/genetic-control-biochemical-reactions-neurospora-1941-george-w-beadle-and-edward-l-tatum (Accessed on 14 February 2020).
11. Chial, H., (2008). Mendelian Genetics: Patterns of Inheritance and Single-Gene Disorders. Nature Education, 1(1),63.
12. Cobb, M., (2015). Oswald Avery, DNA, and the transformation of biology. Current Biology, 24(2)R55-R60.
13. Dadoune, J.-P., Siffroi, J.-P. & Alfonsi, M.-F., (2004). Transcription

in Haploid Male Germ Cells. International Review of Cytology, 237, 1-56. https://doi.org/10.1016/S0074-7696(04)37001-4 (Accessed on 24 February 2020).

14. Dahm, R., (2004). Friedrich Miescher and the discovery of DNA. Developmental Biology, 278(2), 274-288. https://doi.org/10.1016/j.ydbio.2004.11.028 (Accessed on 5 February 2020).
15. Doublié, S. & Zahn, K.E., (2014). Structural insights into eukaryotic DNA replication. Frontiers in Microbiology, 5, 444. doi: 10.3389/fmicb.2014.00444 (Accessed on 13 February 2020).
16. Encyclopaedia Britannica, Inc., (2019a). Francis Galton. https://www.britannica.com/biography/Francis-Galton (Accessed on 11 December 2019).
17. Encyclopaedia Britannica, Inc., (2019b). William Bateson. https://www.britannica.com/biography/William-Bateson (Accessed on 24 December 2019).
18. Eun, H.-M., (1996). RNA Polymerases. In: Enzymology Primer For Recombinant DNA Technology. Academic Press: Massachusetts, U.S.A., pp. 491-565.
19. Genome News Network, (2019). Genetics and Genomics Timeline: 1876, Francis Galton (1822-1911) offers a statistical approach to understanding inheritance. http://www.genomenewsnetwork.org/resources/timeline/1876_Galton.php (Accessed on 11 December 2019).
20. Green, K.D. & Garneau-Tsodikova, S., (2010). Posttranslational Modification of Proteins. Comprehensive Natural Products: Chemistry and Biology, 5, 433-468.
21. Guo, H., Li, J., Gao, F., Li, J., Wu, X. & Liu, Q., (2015) Whole-exome sequencing reveals a novel CHM gene mutation in a family with choroideremia initially diagnosed as retinitis pigmentosa. BMC Ophthalmology, 15, 85. DOI: 10.1186/s12886-015-0081-4 (Accessed on 19 December 2019).
22. Hasan, H., (2005). Mendel and the Laws of Genetics. Rosen Publishing: New York.
23. Kim, B., Nesvizhskii, A.I., Rani, P.G., Hahn, S., Aebersold, R. &Ranish, J.A., (2007). The transcription elongation factor TFIIS is a component of RNA polymerase II preinitiation complexes. Proceedings of the National Academy of Sciences U.S.A., 104(41), 16068-16073.

doi: 10.1073/pnas.0704573104 (Accessed on 26 February 2020).

24. Klug, W.S., Cummings, M.R. & Spencer, C.A., (2006). Concepts of Genetics (8thedn.). Pearson Education, Inc.: New Jersey.
25. Koolman, J. & Roehm, K.H., (2005). Color Atlas of Biochemistry (2ndedn.). Thieme: New York.
26. Korf, B., (2017). Introduction to Human Genetics. In: Clinical and Translational Science: Principles of Human Research (2nd edn.).; Robertson. D.; Williams, G.H., Ed. Academic Press: London; pp.281-311.
27. Lewin, B., (2004). Genes VIII. Pearson Education, Inc.: New Jersey.
28. Liu, Q., Gabriel, S.E., Roinick, K.L., Ward, R.D. & Arndt, K.M., (1999). Analysis of TFIIA function In vivo: evidence for a role in TATA-binding protein recruitment and gene-specific activation. Molecular and Cellular Biology, 19(12), 8673-8685. doi:10.1128/mcb.19.12.8673 (Accessed on 20 February 2020).
29. Lobo, I. & Shaw, K., (2008). Discovery and Types of Genetic Linkage. Nature Education, 1(1), 139.
30. Lodish, H., Berk, A., Matsudaira, P., Kaiser, C.A., Krieger, M., Scott, M.P., Zipursky, L. & Darnell, J., (2003). Molecular Cell Biology (5thedn.). W.H. Freeman: New York.
31. Luo, S., Jin, H., Chen, J. & Zhang, L., (2019). A Novel Variant in Non-coding Region of GJB1 Is Associated With X-Linked Charcot-Marie-Tooth Disease Type 1 and Transient CNS Symptoms. Frontiers in Neurology, 10, 413. doi: 10.3389/fneur.2019.00413 (Accessed on 19 December 2019).
32. Maayan, I., (2011). Theodor Heinrich Boveri (1862-1915). https://embryo.asu.edu/pages/theodor-heinrich-boveri-1862-1915 (Accessed on 8 January 2020).
33. Maderspacher, F., (2008). Theodor Boveri and the natural experiment. Current Biology, 18(7), R279-R286. https://doi.org/10.1016/j.cub.2008.02.061 (Accessed on 8 January 2020).
34. Merchant, J., (2018). Transcription and Epigenetic Regulation. In: Physiology of the Gastrointestinal Tract (6thedn.); Said, H.M. (ed.). Academic Press: Massachusetts, U.S.A. pp.3-30. https://doi.org/10.1016/B978-0-12-809954-4.00001-3 (Accessed on 27 February 2020).
35. Miko, I., (2008a). Gregor Mendel and the Principles of Inheritance.

Nature Education, 1(1), 14.

36. Miko, I., (2008b). Thomas Hunt Morgan and Sex Linkage. Nature Education, 1 (1), 143.

37. Morgan, T.H., (1910). Sex Limited Inheritance in Drosophila. Science, 32, 120-122.

38. Morgan, T.H., (1911). Random Segregation Versus Coupling in Mendelian Inheritance. Science, 34 (873), 384.

39. National Human Genome Research Institute, (2013). 1944: DNA is "Transforming Principle". https://www.genome.gov/25520250/online-education-kit-1944-dna-is-transforming-principle (Accessed on 10 February 2020).

40. O'Connor, C., (2008a). Chromosome Segregation in Mitosis: The Role of Centromeres. Nature Education, 1(1), 28.

41. O'Connor, C, (2008b). Telomeres of Human Chromosomes. Nature Education, 1(1), 166.

42. O'Connor, C & Miko, I., (2008)., Developing the Chromosome Theory. Nature Education, 1(1), 44.

43. Okamoto, T., Yamamoto, S., Watanabe, Y., Ohta, T., Hanaoka, F., Roeder, R.G. & Ohkuma, Y., (1998). Analysis of the Role of TFIIE in Transcriptional Regulation through Structure-Function Studies of the TFIIEβ Subunit. Journal of Biological Chemistry, 273, 19866-19876. doi:10.1074/jbc.273.31.19866 (Accessed on 24 February 2020).

44. Ostrer, H., (1998), Non-mendelian Genetics in Humans. Oxford University Press: New York.

45. Passarge, E., (2001). Color Atlas of Genetics (2ndedn.). Thieme: New York.

46. Paule, M.R. & White, R.J., (2000). Transcription by RNA polymerases I and III. Nucleic Acids Research, 28(6), 1283-1298.

47. Piro, A., Tagarelli, G., Lagonia, P., Quattrone, A. & Tagarelli, A., (2010). Archibald Edward Garrod and alkaptonuria: "Inborn errors of metabolism" revisited. Genetics in Medicine, 12, 475-476.

48. Pray, L.A., (2008). Discovery of DNA Structure and Function: Watson and Crick. Nature Education, 1(1), 100.

49. Raven, P. & Johnson, G., (2002). Biology (6thedn.). McGraw Hill Co.: New York.

50. Rimel, J.K. & Tattjes, D.J., (2018). The essential and multifunctional

TFIIH complex. Protein Science, 27(6), 1018-1037. doi: 10.1002/pro.3424 (Accessed on 26 February 2020).

51. Satzinger, H., (2008). Theodor and Marcella Boveri: chromosomes and cytoplasm in heredity and development. Nature Reviews Genetics 9, 231-238.
52. Sawan, C. & Herceg, Z., (2010). Histone Modifications and Cancer. Advances in Genetics, 70, 57-85.
53. Schleif, R., (1993). Genetics and Molecular Biology (2ndedn.). The Johns Hopkins University Press: Maryland.
54. Sims, R.J. III, Belotserkovskaya, R. & Reinberg, D., (2004). Elongation by RNA polymerase II: the short and long of it. Genes & Development, 18, 2437-2468. doi:10.1101/gad.1235904 (Accessed on 26 February 2020).
55. Soutourina, J., (2018). Transcription regulation by the Mediator complex. Nature Reviews Molecular Cell Biology, 19, 262-274.
56. Sturtevant, A.H., (1913). The Linear Arrangement of Sex-linked Factors in Drosophila, As Shown By Their Mode of Association. Journal of Experimental Zoology, 14, 43-59.
57. Valenzuela, M.S., (2012). Initiation of DNA Replication in the Human Genome. Hereditary Genetics, Suppl. 1(3), 4903.
58. van der Heijiden, G.W., Ejipe, M. & Baarends, W.M., (2011). The X and Y chromosome in meiosis: how and why they keep silent. Asian Journal of Andrology, 13(6), 779-780.
59. Watson, J.D. & Crick, F.H.C., (1953). A structure of deoxyribose nucleic acid. Nature, 171, 737-738.
60. Weinzieri, R.O.J. & Wiesler, S.C., (2011). Revealing the functions of TFIIB. Transcription, 2(6), 254-257. doi:10.4161/trns.2.6.18076 (Accessed on 24 February 2020).

2

Genetics and Diseases

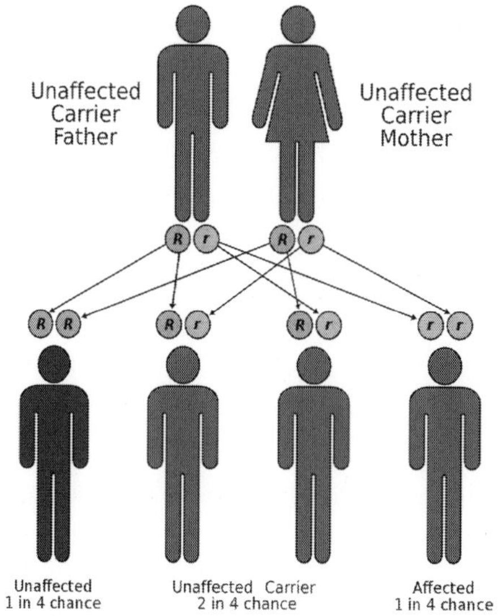

(Source: C. Burnett, Creative Commons License)

The DNA is considered to be the blueprint of life as it stores the molecular instruction for the synthesis of proteins that are necessary for the normal development of an organism's anatomy, physiology, and even behavior. Therefore, significant errors in the DNA that are uncorrected usually result in diseases.

2.1. GENETIC DISEASES DUE TO CHROMOSOMAL ABERRATIONS

Chromosomes are the subject of study of the specialized field of cytogenetics. Most species, especially higher forms of animals as well as humans, have a fixed chromosome number. A basic method used to determine an organism's chromosome number is to make a **karyotype** or **karyogram**, which is a diagram or picture of an organism's chromosomes that are arranged based on chromosome number and with the homologous ones paired together. This is prepared from mitotic cells that were arrested in the prometaphase or metaphase stage, where the chromosomes are at their densest configuration. Cell division is arrested by the addition of mitotic poisons such as colchicine. The nuclei from the sample cells are then isolated and the chromosomes are stained. Then the chromosomes are visualized and may be photographed through the microscope. These are distinguished based on their size and the placement of their centromere. The application of a specialized stain that reveals the chromosome banding patterns can also allow for a more detailed differentiation of chromosomes based on their structure (O'Connor, 2008a).

Humans have 46 chromosomes, composed of 22 homologous pairs of autosomes and a pair of sex chromosomes. According to international standards, the homologous autosomes are numbered from 1 to 22 and are generally lined up in a karyogram in descending order based size (see Figure 2.1). The sex chromosomes are also paired together, whether a pair of X or X and Y and are positioned last in the karyogram. The chromosomes are positioned with their short p arm at the top and the q arm at the bottom. Karyogram analysis can reveal whether a person has the normal chromosome number or not and if the chromosomes have normal morphology. Analyzing the banding patterns on the chromosomes can also help a clinician determine if one or more chromosomes exhibit structural aberration (O'Connor, 2008a). Deviation from the normal number and structure of chromosomes may cause or increase the risk of congenital defects, developmental delays, behavioral disorders, and intellectual disabilities (Sun *et al.*, 2015).

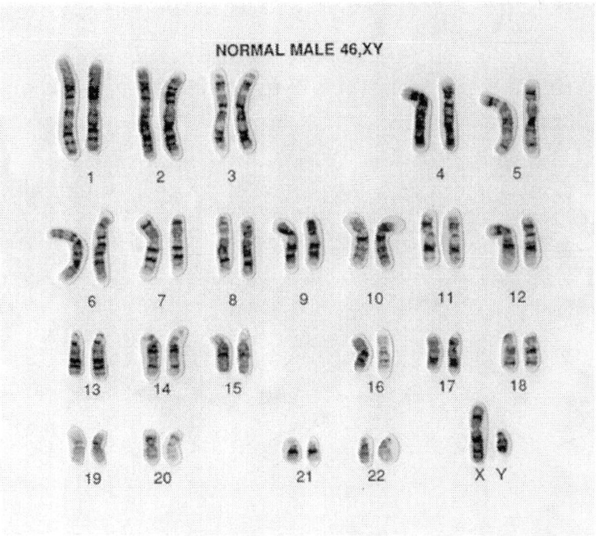

Figure 2.1. The karyogram of a normal human male shows 22 pairs of autosomes as well as the X and Y sex chromosomes. The autosomes are arranged in descending order according to size except for chromosome 21, which is actually smaller than chromosome 22. (Source: Wessex Reg. Genetics Centre, Creative Commons License).

2.1.1. Abnormal Chromosome Number

The normal chromosome number in humans is 2n=46, with the gametes having n=23 chromosomes. In some cases, a chromosome number of 47 or 45 is observed. This condition of having fewer or more chromosomes than the normal count, by one or a small number of chromosomes, is called **aneuploidy**. This is not inherited but is caused by spontaneous nondisjunction during mitosis or meiosis, where the homologous chromosomes or sister chromatids fail to separate from each other and move to opposite poles. This means that two chromosomes or chromatids (that eventually become two chromosomes) go to one pole while there is a missing chromosome in the other pole (Griffiths *et al.*, 2000). Nondisjunction mostly occurs in the oocyte during meiosis (see Figure 2.2).

If this occurs during meiosis I so that a pair of homologous chromosomes do not segregate, half of the formed female gametes will have more chromosomes than normal while the other half will have missing chromosomes. Meanwhile, if there is nondisjunction of sister chromatids during meiosis II, 50% of the eggs are aneuploids while the other 50% are normally diploid. This is somewhat similar to what happens during mitotic nondisjunction. If an aneuploid egg is involved in a fertilization event, then the resulting embryo is an aneuploid (Nagaoka *et al.*, 2012).

Aneuploidy may also be observed in sperms and this may be due to either non-disjunction or anaphase lag. The significantly lower incidence of aneuploid sperms, as opposed to eggs, maybe because of more effective checkpoint mechanisms in spermatogenesis than in oogenesis (Templado *et al.*, 2013).

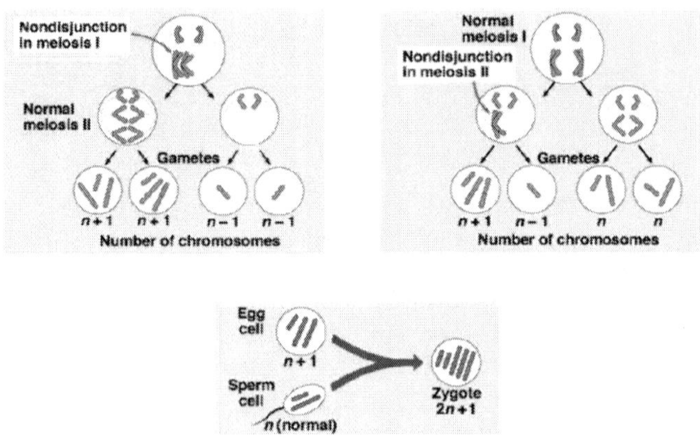

Figure 2.2. Aneuploidy may result from nondisjunction or non-separation of homologous chromosomes during meiosis I (top left) or of sister chromatids during meiosis II (top right). This gives rise to gametes, mostly egg cells, with extra or missing chromosomes. (Bottom) When an aneuploid egg cell is fertilized by a normal sperm, the resulting zygote is an aneuploid. (Source: Ahmad, M.; Silvera-Redondo, C.; Rodriguez, M.H. SaludUninorte 26(1), 117-133, 2010, Creative Commons License).

Aneuploid embryos constitute at least 10% of human pregnancies. This significantly increases to more than 50% for women who are nearing the end of their reproductive period. Therefore, increased maternal age is a significant determinant of the occurrence of aneuploidy. Studies conducted in the 1960s and 1970s revealed that about 0.3% of live births were aneuploids and that these infants show congenital birth defects. In addition, most aneuploid embryos die in the womb so that they make up approximately 35% of spontaneous abortions (Nagaoka *et al.*, 2012). Since each chromosome contains hundreds of genes, the addition or loss of even just one chromosome disturbs the equilibrium of gene products so that in most cases, this makes life non-viable (O'Connor, 2008b).

If chromosome nondisjunction occurs during mitosis, it results in somatic aneuploids. In this case, only the daughter cells resulting from the erroneous mitosis have aneuploid chromosome number while the other cells remain to be diploid. Cancer cells demonstrate significant aneuploidy due to mitotic errors (Herman, 2001; Nicholson and Cimini, 2011).

2.1.1.1. Disorders due to having an extra chromosome

The most common aneuploidy in humans is **trisomy**. This is characterized by having one extra chromosome. Trisomic embryos usually do not survive but some can complete development and may result in a live birth. However, this is restricted to having an extra copy of certain chromosomes.

2.1.1.1.1. Trisomy 21

The most common trisomy is having an extra chromosome 21 or trisomy 21 (see Figure 2.3). This results in **Down syndrome**, which is observed about 1 in every 700-750 births (O'Connor, 2008b; Centers for Disease Control and Prevention, 2019b). The likelihood of giving birth to a child with this condition increases with maternal age. The non-lethality of having an extra copy of chromosome 21 maybe because it is the smallest chromosome with a small number of genes. Therefore, it causes a small change in the gene dose, which is tolerable (Ahmad *et al.*, 2010).

People with Down syndrome usually have mild to moderately low IQ and are delayed in speech development compared to other children. They share common facial features which include, a flattened face and nose bridge, slanting eyes that are almond-shaped with an epicanthal fold or skin fold that comes out from the upper eyelid and covers the inner corner of the eye, tiny white spots on the iris of the eye called Brushfield spots, and a large, wrinkled tongue that stick out of the mouth (see Figure 2.4). They are also characterized by having small ears, short neck, small hands and feet, a single palmar crease, and small pinky fingers that sometimes curve toward the thumb. They also exhibit hypotonia (i.e., poor muscle tone) or loose joints as well as a shorter than average height (Centers for Disease Control and Prevention, 2019a; Griffiths *et al.*, 2000; National Institutes of Health, 2017a). Females may be fertile and thus, may produce normal or trisomic progeny. On the other hand, males have never reproduced (Griffiths *et al.*, 2000).

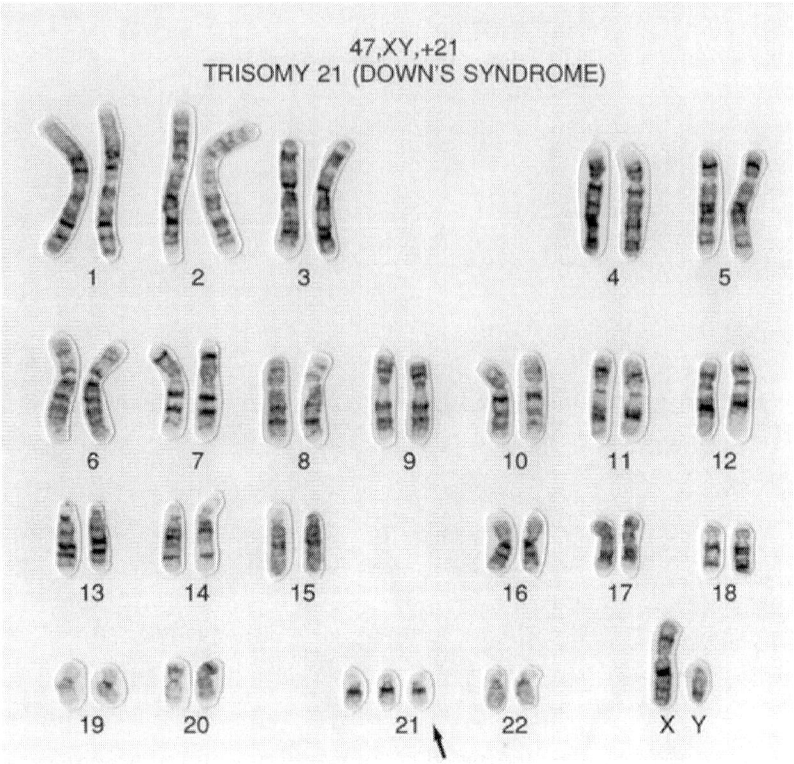

Figure 2.3. The most common viable trisomy in humans is trisomy 21, i.e. having an extra copy of chromosome 21 due to nondisjunction. (Source: Wessex Reg. Genetics Centre, Creative Commons License).

Those with Down syndrome are also at higher risk to have other medical problems such as hearing loss (up to 75%), obstructive sleep apnea (50-75%), ear infections (50-70%), eye diseases (up to 60%), and congenital heart defects (50%). Infants with Down syndrome who have very low birth weight, i.e. 3.3 pounds, are 24 times more likely to die in the first 28 days of life compared to those who have normal birth weight, i.e. 5.5-8.8 pounds. Those with a congenital heart defect (CHD) are five times more likely to die in the first year of life than those with no CHD. In the 1960s, the life expectancy of those with Down syndrome is about 10 years. This significantly increased to 47 years in 2007, with some reaching up to 60 years (Centers for Disease Control and Prevention, 2019a and 2019b). Some studies indicate that those with Down syndrome have a much higher incidence of developing Alzheimer's disease (AD) at an early age and for those who reach the age of 60 or more, 60-75% of them will have AD.

This may be linked to having three copies of chromosome 21, which carries genes for the production of beta-amyloid proteins. Accumulation of these proteins leads to the formation of amyloid plaques in the brain, which can develop into AD (Sara, 2008).

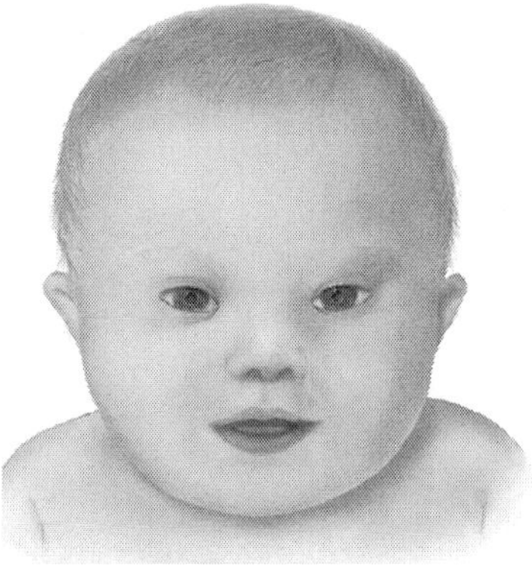

Figure 2.4. Down syndrome is characterized by certain physical features like a flat, broad face and nose bridge, upwardly slanted eyes with epicanthal fold, white Brushfield spots on the iris, small ears, and a large protruding tongue. (Source: Centers for Disease Control and Prevention, Public Domain).

The majority of people with Down syndrome have trisomy 21, i.e. approximately 95%. On the other hand, some have the condition even if they have 46 chromosomes. This is another type of condition called **translocation Down syndrome**, which accounts for about 3% of the cases. This occurs when a part or a whole extra chromosome 21 is translocated or fused to another chromosome instead of being a separate chromosome. The other 2% of cases have **mosaic Down syndrome**. Individuals having this have some of their cells having the normal chromosome number while others have an extra chromosome 21 (Centers for Disease Control and Prevention, 2019a). Mosaicism is a result of chromosomal aberration or mutation that occurs in one somatic cell during the early embryonic stage. Cells that arise from the division of this cell will have the aberration while others will not (Genetics Home Reference, 2020c). Those with mosaic Down syndrome may have fewer features characteristic of Down syndrome because some of their cells are normal (Centers for Disease Control and Prevention, 2019a).

2.1.1.1.2. Trisomy 13

Another human autosomal trisomy is having an extra chromosome 13, a condition resulting in **Patau syndrome** (see Figure 2.5). This occurs in about 1 in every 16,000 newborns. Just like other aneuploidies, the chance of having a child with this condition increases as the mother gets older. A mother who gave birth to an infant with this condition has a 1% chance of having another child with the syndrome. Others may have the mosaic type of Patau syndrome while some have the translocated type. The physical signs and symptoms of these may differ from those of the full trisomy 13, with the former types being milder. These also occur rarely as opposed to full trisomy 13 (National Institutes of Health,2016; Genetics Home Reference, 2013).

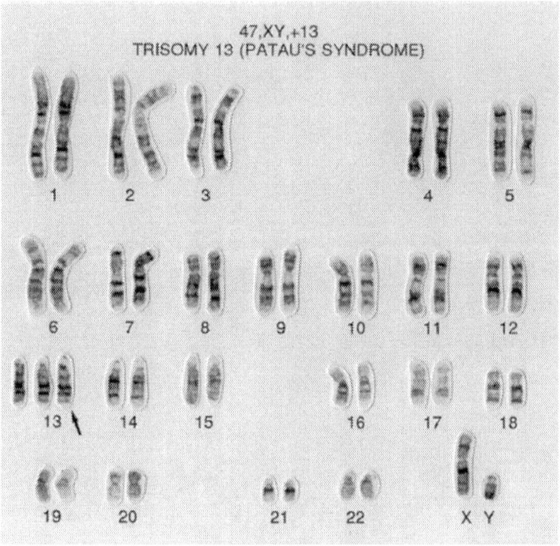

Figure 2.5. Random nondisjunction of chromosome 13 during meiosis may lead to gametes with extra chromosome 13, which when involved in fertilization will result in an infant with Patau's syndrome. (Source: Wessex Reg. Genetics Centre, Creative Commons License).

Infants with trisomy 13 show severe intellectual disability and a host of physical abnormalities. They often have congenital heart defects, brain or spinal cord abnormalities, microphthalmia (i.e., very small or underdeveloped eye), polydactyly (i.e., having extra fingers and/or toes), cleft lip or palate (see Figure 2.6), and hypotonia. Many of them fail to grow at the expected rate, have severe feeding difficulties, and have episodes of apnea (i.e., temporary cessation of breathing). Other features of trisomy 13 include clenched hands with the outer fingers on top of the inner fingers, close-set eyes, hernias, coloboma (i.e., a hole, slit or cleft in the iris of the eye), low set ears, scalp defects, seizures, single palmar crease, skeletal abnormalities, microcephaly (i.e., small head), micrognathia (i.e., small lower jaw), and cryptorchidism (i.e., undescended testicle). Many of these abnormalities are life- threatening so that 80% of infants with trisomy 13 do not survive past their first month. For those that survive infancy, they are at increased risk for cancers apart from the aforementioned complications (National Institutes of Health, 2016).

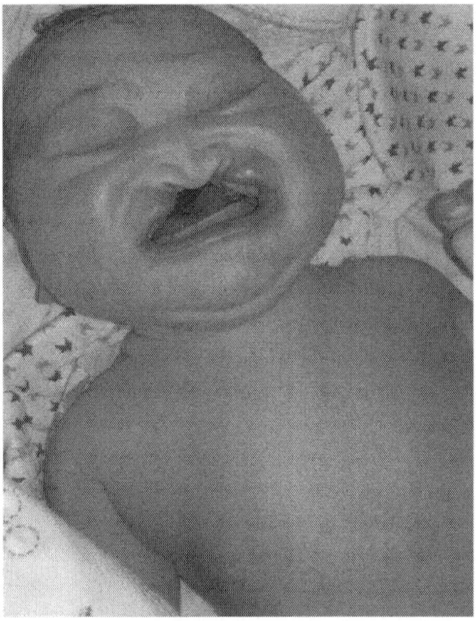

Figure 2.6. Cleft lip and palate are some of the many physical deformities associated with trisomy 13, which is the case for this male infant. (Source: Gashi, A.M.Journal of Pregnancy and Reproduction 1, 2017, Creative Commons License).

While most cases of trisomy 13 are not inherited, the translocation type may be inherited. This may be due to chromosomal rearrangement between chromosome 13 and another chromosome of the unaffected parent. This is called balanced rearrangement because there is no extra or missing material from chromosome 13. A parent with a balanced translocation involving chromosome 13 has an increased likelihood of passing extra material from chromosome 13 to his/her children (National Institutes of Health, 2016).

2.1.1.1.3. Trisomy 18

The second most prevalent autosomal trisomy is having an extra copy of chromosome 18, which leads to **Edwards syndrome** (see Figure 2.7). This syndrome may also result from mosaic trisomy 18 or translocated chromosome 18, although these types occur less frequently. Edwards syndrome occurs in 1 out of every 6,000-8,000 live births. Its prevalence increases with maternal age and the recurrence risk for a mother to have another child with trisomy 18 is 1%. Paternal age has also been shown to have a small association with trisomy 18 (Cereda and Carey, 2012).

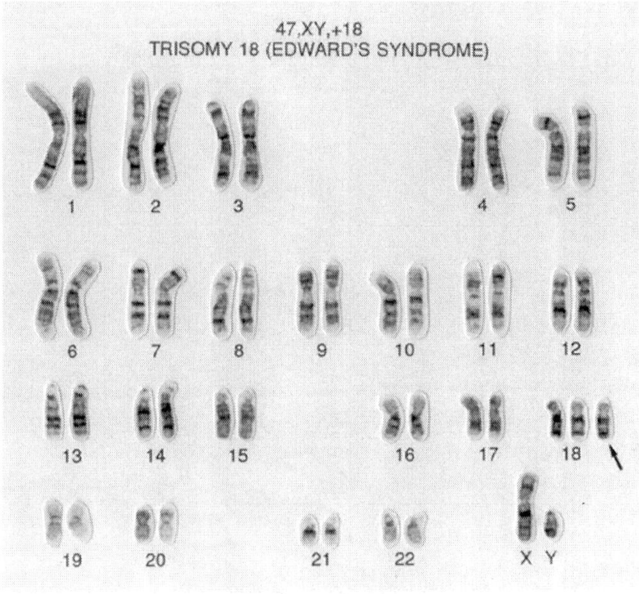

Figure 2.7. Nondisjunction of chromosome 18 during meiosis results in gametes having an extra copy of this chromosome. When these are involved in a fertilization event with a normal gamete, then the resulting individual exhibits trisomy 18. (Source: Wessex Reg. Genetics Centre, Creative Commons License).

Even prior to birth, those with trisomy 18 exhibit growth retardation and when these infants are eventually born, they have low birth weight and exhibit delayed growth rate due to feeding difficulties. They also have severe intellectual disability, heart defects, and other congenital organ (i.e., kidneys, central nervous system, eye, limbs) abnormalities. Other physical features include dolichocephaly (i.e., tall, narrow head), abnormally shaped ears, short palpebral fissures (i.e., the distance between the open eyelid), small jaw and mouth, redundant skin at the back of the neck, clenched fists with overlapping fingers (see Figure 2.8), and club feet. A high percentage of fetuses (38.5%) die during labor. Among the live births, most of them die within their first month while 5-10% may live past their first year (see Figure 2.8). (Genetics Home Reference, 2012a; Cereda and Carey, 2012).

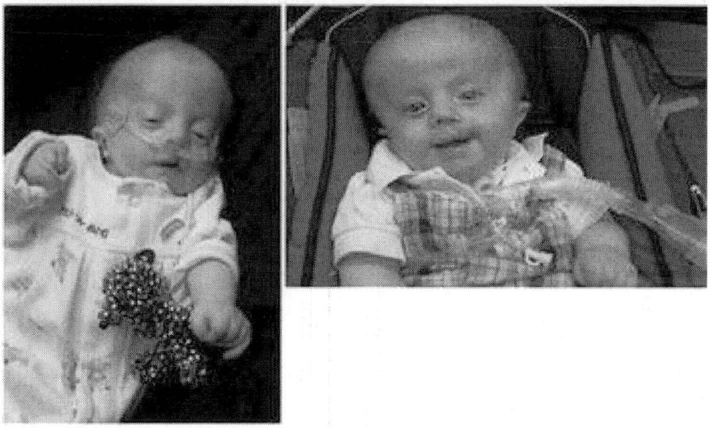

Figure 2.8. (Left) One physical feature of trisomy 18 is the clenched fist with overriding fingers. (Right) While most infants with this condition do not live past their first month, this boy is over 2 years of age and apart from the tracheostomy to aid his breathing, is quite stable medically. (Source: Cereda, A.; Carey, J.C. Orphanet Journal of Rare Diseases 7 (81), 2012, Creative Commons License).

Those with the mosaic type of trisomy 18 exhibit extremely variable phenotype ranging from early mortality to phenotypically normal adults. Those with the translocation type also exhibit variable phenotype depending on the location and extent of the triplicated segment and the possible associated deletion of a chromosome segment (Cereda and Carey, 2012).

Full trisomy 18 is a result of random nondisjunction during meiosis. This is also the case for mosaic trisomy 18, which occurs as a random event during mitotic cell division early in embryonic development. In contrast, the translocated type of trisomy 18 can be inherited. Therefore, people who carry a balanced translocation between chromosome 18 and another chromosome have an increased likelihood of having children with Edwards syndrome (Genetics Home Reference, 2012a).

2.1.1.1.4. Trisomy 10

A very rare case of trisomy is that of mosaic trisomy 10 (see Figure 2.9), of which only seven cases of live births are reported so far. In 2001, a complete trisomy 10 was diagnosed prenatally but the mother went into preterm labor at 35 weeks and 4 days into her pregnancy and the fetus died in the uterus (Gao *et al.*, 2018).

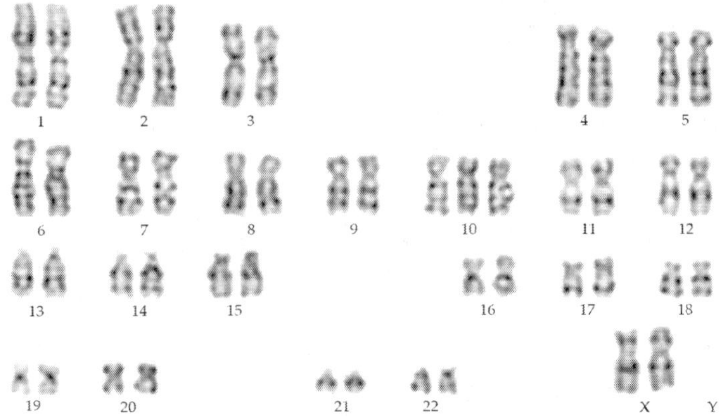

Figure 2.9. A case of mosaic trisomy 10 in a female child showed a karyotype of 47 chromosomes (extra chromosome 10), XX in 42% of metaphases that were observed in her blood cells. (Source: Gao, Y.; Ma, Y.-C.; Ju, Y.-H.; Li, Y.-N. BMC Pediatrics 18 (266), 2018, Creative Commons License).

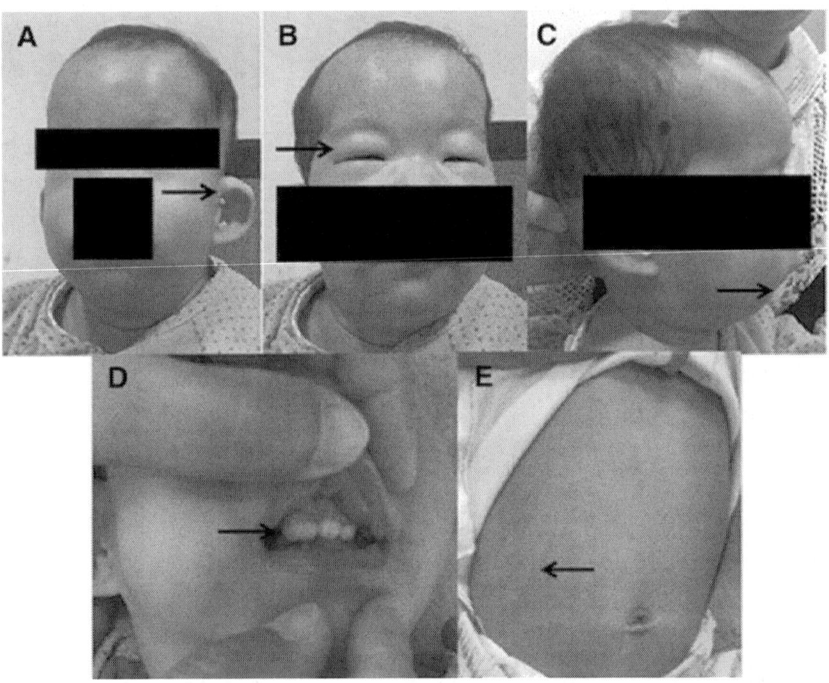

Figure 2.10. A 14-month old girl with mosaic trisomy 10 showing morphological features such as (a) low-set ears, (b) small palpebral fissure, (c) retrognathia,

(d) fused teeth, and (e) patchy and uneven skin pigmentation. (Source: Gao, Y.; Ma, Y.-C.; Ju, Y.-H.; Li, Y.-N. BMC Pediatrics 18 (266), 2018, Creative Commons License).

The infants with mosaic trisomy 10 exhibit features like blepharophimosis (i.e., underdeveloped eyelids that permanently cover part of the eyes), low-set ears, high-arched palate, cleft lip or palate, retrognathia (i.e., the lower jaw is set to further back than the upper jaw), long slender trunk, and marked plantar or palmar furrows (see Figure 2.10). In addition, they also have malformations in the heart, kidneys, central nervous system, and bones. They manifest feeding problems, growth retardation, and failure to thrive. The lifespan of those with this condition ranged from 37 days to 5 years, 4 months (Gao *et al.*, 2018; Orpha.net, 2020).

2.1.1.1.5. Trisomy of sex chromosomes

Humans are much more able to tolerate having an extra copy of sex chromosomes than autosomes. This affects 0.1% of the human population. This is also the most common genetic cause of infertility. This is associated with the reduced sexual development observed among sex chromosome trisomies but otherwise, these individuals often have normal lifespans and many of their symptoms can be treated by hormone supplementation. The tolerance of humans to an extra copy of sex chromosomes may be related to the small number of genes on the Y chromosome as well as the inactivation of the X chromosome in excess of one (O'Connor, 2008b; Hirota *et al.*, 2017).

2.1.1.1.5.1. Extra copy of the X chromosome

Normal females have two copies of the X chromosome while males have one X and one Y chromosome. While the very small Y chromosome contains a small number of genes that specify primary sexual traits in males, the X chromosome has more than 1,000 genes that are needed for normal development and cell viability. However, having a double dose of these X-linked genes in females is potentially toxic. Thus, a dosage compensation mechanism is at work among mammalian and human females, which involves the random and permanent inactivation or **lyonization** of one of the X chromosomes during early embryonic development. This means that one set of X-linked genes is not expressed. The inactivated X chromosome condenses into a compact structure called a **Barr body** (see Figure 2.11). Because X chromosome inactivation occurs randomly, the maternal X chromosome may be active in some cells while the paternal X chromosome is active in other cells (Ahn and Lee, 2008; Genetics Home Reference, 2018a).

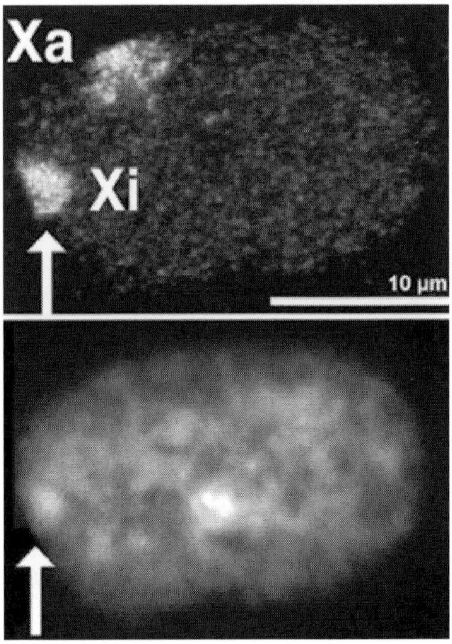

Figure 2.11. One of the two copies of X chromosomes in females is randomly inactivated. The inactivated X chromosome (Xi) condenses into a compact Barr body (indicated by the arrow), appearing as a densely staining structure at the periphery of the nucleus of an amniotic fluid cell. (Top) The inactive and active X chromosome (Xa) were detected by fluorescence *in situ* hybridization (FISH) and visualized with a confocal microscope. (Bottom) The same nucleus was stained with DAPI and recorded with a charged coupled device (CCD) camera. (Source: Eils, R.; Dietzel, S.; Bertin, E.; Schrock, E.; Speicher, M.R.; Ried, T.; Robert-Nicoud, M.; Cremer, C.; Cremer, T. Journal of Cell Biology 135, 1427-1440, 1996, Creative Commons License).

Some genes in the X chromosome escape inactivation. These are located in the terminal region of the chromosome called the pseudoautosomal region. The genes in this region are also present in the Y chromosome so that both males and females have two functional copies of these genes (Genetics Home Reference, 2018a).

Triple X syndrome

The most common chromosomal abnormality among females is having an extra X chromosome or the *triple X syndrome* (see Figure 2.12). It is estimated to occur in 1 out of 1,000 female births. The risk of having a child with trisomy X increases with maternal age. Since some individuals are mildly affected while others are asymptomatic, only about 10% of females with this condition are actually diagnosed. Approximately 10% of females with this condition have the mosaic type with various combinations such as 46,XX/47,XXX, 47,XXX/48;XXXX, 45;X/47,XXX or 45,X/46;XX/47,XXX (Tartaglia *et al.*, 2010; Genetics Home Reference, 2014a).

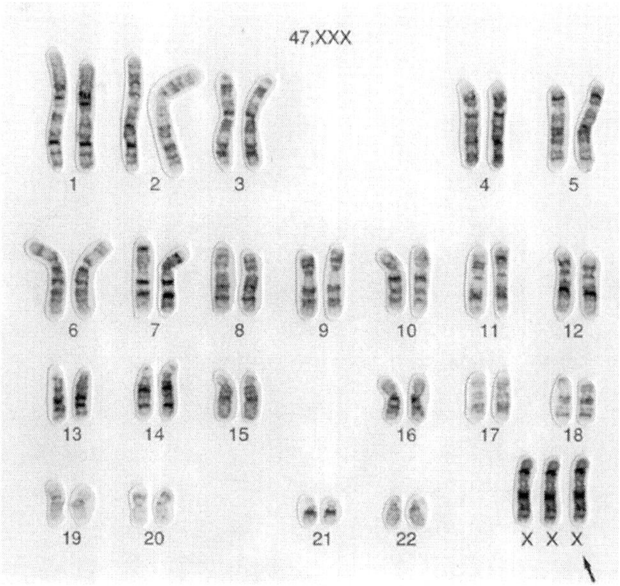

Figure 2.12. A type of aneuploidy specific among females is having three copies of the X chromosome or the triple X syndrome. (Source: Wessex Reg. Genetics Centre, Creative Commons License).

This syndrome causes a variety of phenotypes with no striking facial or physical features, (see Figure 2.13c) but some features observed among the affected females include tall stature, epicanthal folds, hypertelorism (i.e., increased distance between the eyes (see Figure 2.13 a and b)), hypotonia, and clinodactyly (i.e., abnormally bent or curved finger).In about 10% of the affected females, this is associated with seizures, as well as renal and genitourinary abnormalities. Most have normal sexual development and are able to conceive children, but some exhibit oligomenorrhea (i.e., infrequent or very light menstruation) and premature ovarian failure (POV). As they develop, affected children have higher rates of motor and speech delays (with more impaired expressive language than receptive language) as well as increased risk of cognitive deficits and learning disabilities. They also have an increased likelihood of having psychological disorders such as attention deficits and mood disorders (e.g., anxiety and depression) (Tartaglia *et al.*, 2010; Tsai *et al.*, 2009).

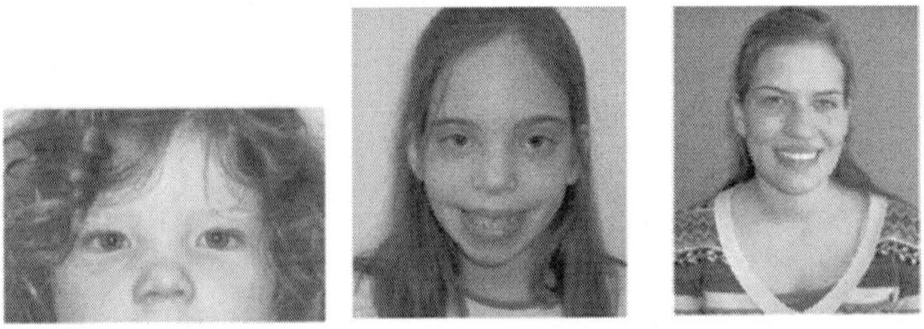

a. b. c.

Figure 2.13. Females with trisomy X exhibit variable facial features such as epicanthal folds[observed in (a) a 2-year old girl] and hypertelorism [observed both in (a) and (b) a 9-year old girl]. (c) On the other hand, a 19-year old girl with trisomy X does not show any dysmorphic features. (Source: Tartaglia, N.R.; Howell, S.; Sutherland, A.; Wilson, R.; Wilson, L. Orphanet Journal of Rare Diseases 5 (8), 2010, Creative Commons License).

In trisomy X, two of the three X chromosomes are inactivated. However, genes in the pseudoautosomal region ,as well as 5-10% of the genes outside this region escape inactivation. It is hypothesized that the abnormalities associated with the triple X syndrome are due to the overexpression of these genes. Most of these genes are not yet identified except for the *SHOX* gene, which is associated with the tall stature of affected females (Tartaglia *et al.*, 2010).

Trisomy X is a result of a nondisjunction event during meiosis when gametes are being formed. The nondisjunction may also occur after conception or post-zygotically. One study suggests that mosaic trisomy X is a result of post-zygotic nondisjunction (Tartaglia *et al.*, 2010). Therefore, both the full and mosaic types of trisomy X *is* not inherited (Genetics Home Reference, 2014a).

There were also recorded cases of tetrasomy X (48, XXXX) and pentasomy X (49, XXXXX) syndromes. These share most features of trisomy X but are usually associated with more severe developmental delays, dysmorphic features, and congenital abnormalities compared to trisomy X. Moreover, females with pentasomy X usually have short stature (Tartaglia *et al.*, 2010).

XXY syndrome

A type of sexual trisomy in males is having an extra copy of the X chromosome, a condition called *Klinefelter syndrome* (see Figure 2.14). This affects about 1 in 650 newborn boys. Less than 10% of affected males have the mosaic type, which may manifest milder symptoms depending on the percentage of trisomic cells. Some of those affected may not show any symptoms or the manifestation of the syndrome may be mild so that it is not diagnosed until puberty or adulthood. Therefore, it is estimated that up to 75% of affected males are not diagnosed (Genetics Home Reference, 2019a). Those with this condition usually have a normal life expectancy(National Institutes of Health, 2018a).

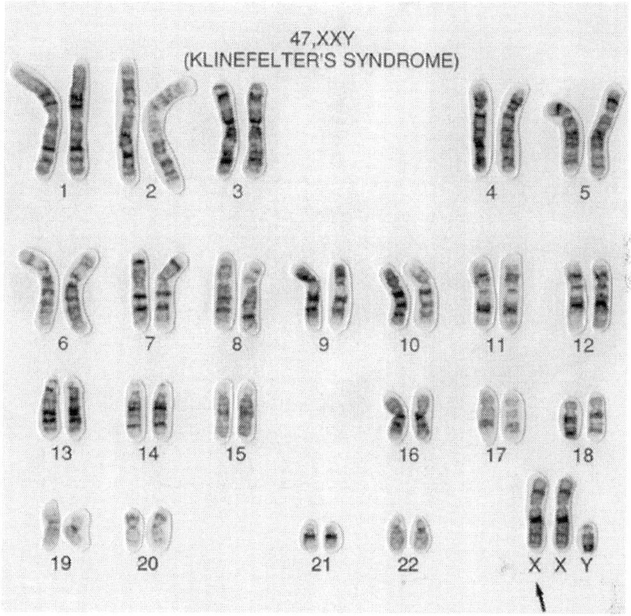

Figure 2.14. Males with Klinefelter syndrome have an extra copy of the X chromosome. (Source: Wessex Reg. Genetics Centre, Creative Commons License).

The most common features of those with Klinefelter syndrome are their tall stature and infertility. Their infertility is primarily due to their small testes that produce a reduced amount of testosterone (see Figure 2.15). If this deficiency is not treated, it can lead to delayed or incomplete puberty with a lack of secondary sexual characteristics such as the reduced amount of facial, body or sexual hair, a high-pitched voice, and more body fat deposited in the hips and buttocks instead of around the chest and abdomen. This can also result in gynecomastia (i.e., breast enlargement), decreased muscle mass, and decreased bone density. Other defects in the genitalia observed in some cases include cryptorchidism, hypospadias (i.e., an opening of the urethra is on the underside instead of the tip of the head of the penis), or having a micropenis (i.e., an unusually small penis). Other physical features associated with the syndrome are radioulnar synostosis (i.e., a fusion of the two bones in the lower forearm), clinodactyly, and pes planus (i.e., flat feet). Affected males may also exhibit hypotonia, which may delay their motor development (Genetics Home Reference, 2019a; National Institutes of Health, 2018a).

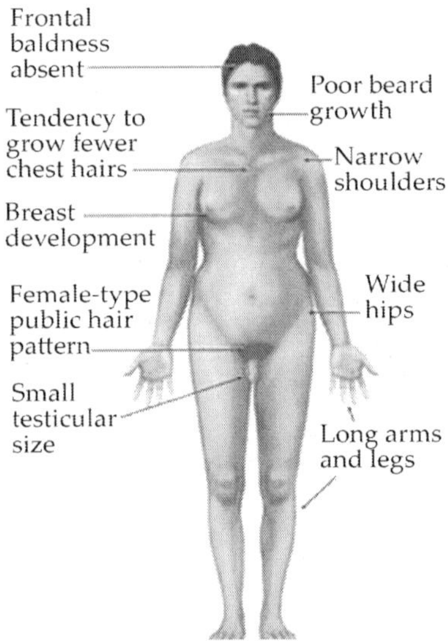

Figure 2.15. A male with Klinefelter syndrome usually exhibits small testicles that produce a reduced amount of testosterone. The lack of the hormone testosterone results in aberrant secondary sexual characteristics (e.g. lack of facial and body hair, female-type pubic hair pattern, wide hips) as well as enlarged breasts. (Source: http://smithperiod6.wikispaces.com/Klinefelter's+Syndrome, Creative Commons License).

Those with Klinefelter syndrome often have learning disabilities with mild delays in speech and language development as well as problems with reading. They also tend to have anxiety, depression, impaired social skills, behavioral problems (e.g., emotional immaturity and impulsivity, attention-deficit/hyperactivity disorder or ADHD), and limited problem-solving skills. Approximately 10% of affected males have autism spectrum disorder (ASD) (Genetics Home Reference, 2019a).

Nearly 50% of males with this condition have metabolic syndrome, which is characterized by type 2 diabetes, hypertension, increased belly fat, and high lipid levels in the blood. Affected individuals also have a slightly increased risk of developing tremors, breast cancer, lung disease, osteoporosis, and autoimmune disorders (e.g., systemic lupus erythematosus and rheumatoid arthritis) (Genetics Home Reference, 2019a; National Institutes of Health, 2018a).

Similar to those with triple X syndrome, males with Klinefelter syndrome have an extra set of certain X-linked genes. The overexpression of these genes may disrupt the development process, especially sexual development, and may be responsible for the anomalies exhibited by affected males. However, the specific genes involved are not yet identified (Genetics Home Reference, 2019a).

A random nondisjunction event may produce either an egg or a sperm with an extra copy of the X chromosome. If the trisomic gamete is fertilized by another gamete, then the resulting child will have Klinefelter syndrome. The mosaic type is also not inherited but is a result of a random error during the division of a cell in early fetal development (Genetics Home Reference, 2019a).

There are variants of Klinefelter syndrome where there are more than two copies of the X chromosome such as 48, XXXY or 49, XXXXY. Affected individuals often exhibit more severe abnormalities in terms of intellectual, skeletal, motor, speech, and behavioral characteristics. They also exhibit serious heart and teeth problems (National Institutes of Health, 2018a).

2.1.1.1.5.2. Extra copy of the Y chromosome

Another type of sexual trisomy in males is having an extra Y chromosome resulting in the **XYY syndrome** or **Jacob syndrome** (see Figure 2.16). This occurs in about 1 in 1, 000 newborn males. It does not cause any unusual physical features or a recognizable pattern of neurodevelopmental or behavioral traits so that many of the affected males are either never diagnosed or diagnosed only later in life. In fact, only about 15% are diagnosed shortly after birth (Genetics Home Reference, 2018b; Gravhold, 2013).

Many males with this condition are taller than average. While they have an increased likelihood of being infertile, affected males usually produce normal levels of testosterone so that they are fertile. They may exhibit severe acne, increased belly fat, macrocephaly (i.e., large head), macrodontia (i.e., unusually large teeth), pes planus, clinodactyly, ocular hypertelorism, and scoliosis (i.e., abnormal side-to-side curvature of the spine). Some of them may also develop tremors, motor tics, seizures, and asthma (Genetics Home

Reference, 2018b; Gravholt, 2013). Increased mortality is observed among affected males with an average loss in the lifespan of 12.4 years. This is due to a variety of causes such as cancer, pulmonary, neurological, and urological diseases (Gravholt, 2013).

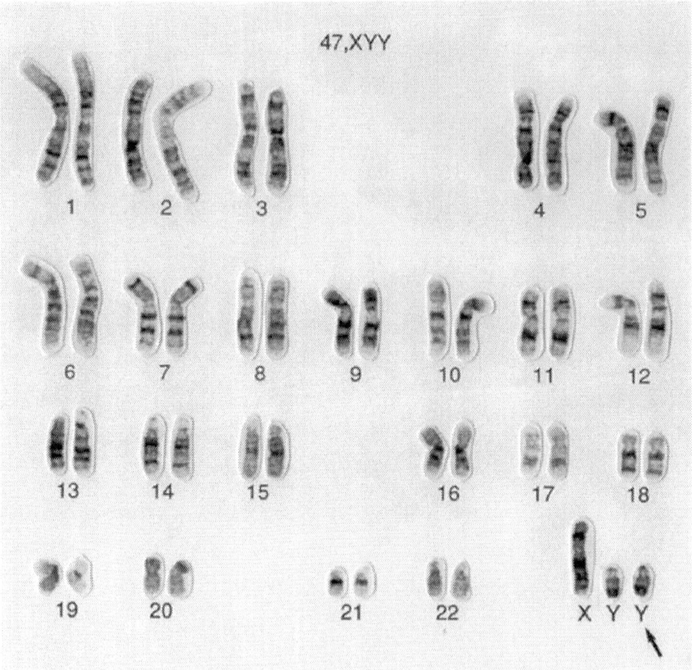

Figure 2.16. Nondisjunction during the formation of sperms may lead to sperms carrying two Y chromosomes so that when they fertilize an egg, a male off-spring with an extra copy of the Y chromosome results. (Source: Wessex Reg. Genetics Centre, Creative Commons License).

Studies indicate that the intellectual ability of affected males is slightly lower than their normal siblings or matched controls. However, it is usually within the normal range. They may have an increased risk of learning disabilities and delayed speech and language skill development. They may also display hypotonia which may cause delayed development of motor skills. Moreover, they have an increased risk of having behavioral (e.g. attention-deficit/hyperactivity disorder), social (e.g., ASD), and emotional disorders especially when they grow up in stressful environments emotional disorders (e.g., depression, anxiety) (Genetics Home Reference, 2018b; Gravholt, 2013). There were early studies associating the XYY condition with increased antisocial or criminal behaviors, but these were eventually considered to be inaccurate. (Genetics Home Reference, 2018b; Hutaff-Lee *et al.*, 2013; Gravholt, 2013). A recent study was performed that investigated the incidence of criminal convictions among 15-70-year old males with Klinefelter syndrome or XYY syndrome in Denmark. Results showed increased convictions for sexual abuse, burglary, and arson among affected males compared to controls. This may be partly or fully due to poor or stressful socioeconomic conditions of the affected males, which may be a result of their cognitive, psychological, emotional, and social maladjustment (Stochholm *et al.*, 2012; Garvholt, 2013).

The XYY condition is not inherited but usually stems from a random nondisjunction that occurs during spermatogenesis so that sperm cells with an extra copy of the Y chromosome are formed. If one of these trisomic sperms fertilize a haploid egg, this results to a child with XYY. The mosaic type of XYY is a result of nondisjunction that occurs during cell division in early embryonic development. It is also not inherited (Genetics Home Reference, 2018b). Moreover, this syndrome is not associated with increased maternal or paternal age (Gravholt, 2013).

2.1.1.2. Disorder due to a missing chromosome

Human chromosomes exist in pairs since one copy is paternally inherited while the other is maternally inherited. If a chromosome exists as a single copy, the condition is called **monosomy**. When there is nondisjunction of a chromosome during meiosis, the recipient of the unseparated chromosomes become trisomic gametes while the other gametes produce monosomic ones (see Figure 2.2).

If a copy of an autosome is missing, this means that the whole set of genes it contains is missing. Therefore, this leads to severe abnormality so that it inevitably causes the embryo to be spontaneously aborted. Monosomy may also be observed in tumor tissue as a result of defective mitosis. On the other hand, some embryos with only one X chromosome can survive (Ferguson-Smith, 2001a).

2.1.1.2.1. XO syndrome

About 1 in every 2,500 newborn girls has only one copy of the X chromosome, a condition called **Turner syndrome** (see Figure 2.17). However, most embryos with this condition (about 99%) are spontaneously aborted during the first trimester. These constitute approximately 10% of spontaneous abortions. Additionally, those that progress to later stages of fetal development are mostly stillborn. There is also a mosaic type of this syndrome. In rare instances, Turner syndrome can also occur in females with a pair of X chromosomes if there is partial deletion or rearrangement of the chromosome (Genetics Home Reference, 2017a; Tsai *et al.*, 2009).

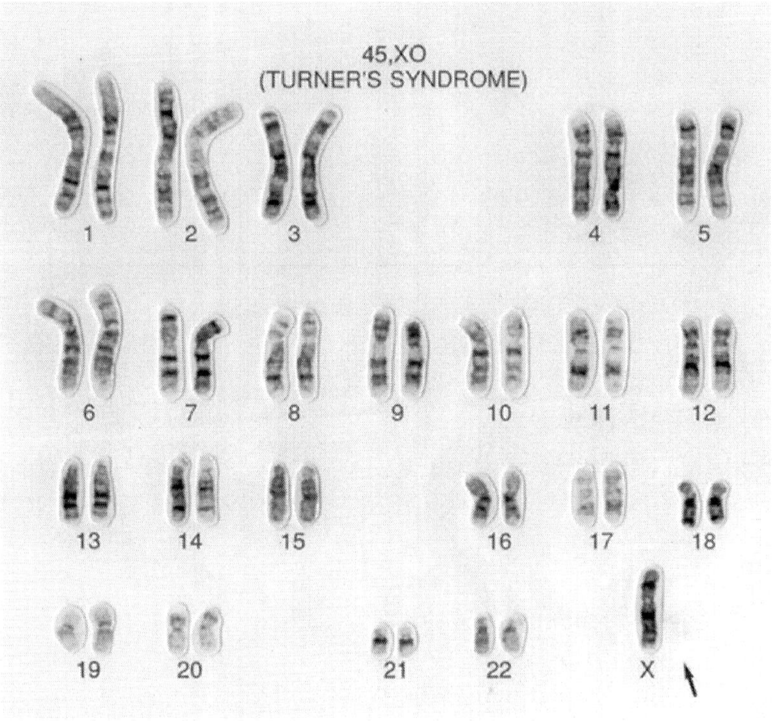

Figure 2.17. A female with Turner syndrome has only one X chromosome as a result of nondisjunction during cell division. (Source: Wessex Reg. Genetics Centre, Creative Commons License).

The signs and symptoms observed among affected females may differ significantly from one individual to another. The most common feature of Turner syndrome is short stature. This may be due to having only a single copy of the *SHOX* gene, which is essential for bone growth and development. Most affected females are infertile due to early loss of ovarian function. Initially, the ovaries may develop normally but egg cells die prematurely and then most of the ovarian tissues degenerate (i.e., gonadal dysgenesis) before birth. This also leads to reduced levels of sex hormones (e.g., estrogen and progesterone) so that the affected females do not undergo puberty and develop secondary sexual characteristics (e.g., development of breasts and normal body form, growth of bones, and menstruation) unless they receive hormone replacement therapy. One third to one half of those with Turner syndrome are born with a heart defect (e.g., narrowing of the aorta or abnormalities of the aortic valves), which can be life-threatening. About a third of affected females have a webbed neck (i.e., extra folds of skin on the neck), a low hairline at the back of the neck (see Figure 2.18), lymphedema (i.e., puffiness or swelling) of the hands and feet, skeletal abnormalities, or kidney problems (Genetics Home Reference, 2017a; National Organization of Rare Disorders, 2020).

A B

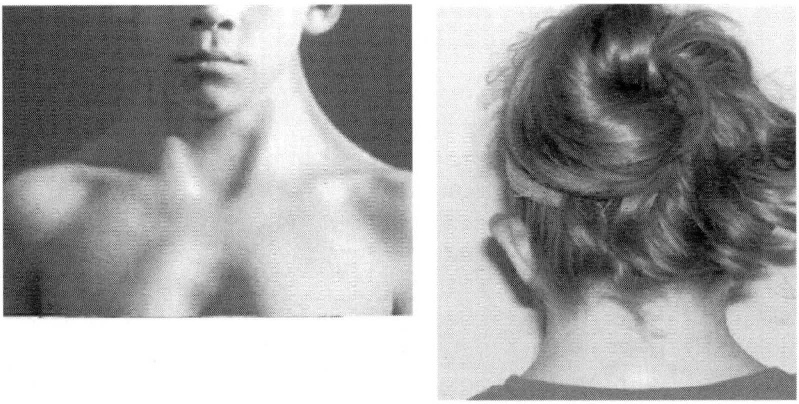

Figure 2.18. Some of the physical features exhibited by females with Turner syndrome include (A) neck webbing and (B) a low hairline at the; (Source: National Human Genome Research Institute, Public Domain).

Most females with Turner syndrome have normal intelligence. Some may show varying degrees of developmental delays, nonverbal learning difficulties (e.g., math, sense of direction, visual-spatial relationships) as well as behavioral and social problems (Genetics Home Reference, 2017a; National Organization of Rare Disorders, 2020).

Turner syndrome due to absence of one X chromosome is not inherited, but a result of a random nondisjunction of the X chromosome during gametogenesis or anaphase lag of the Y chromosome during spermatogenesis. If a sperm or an egg with no copy of a sex chromosome is involved during fertilization, the resulting embryo will have Turner syndrome. The mosaic type of the syndrome is also not inherited but a product of nondisjunction during mitosis in early fetal development. On the other hand, Turner syndrome due to a partially deleted segment of the X chromosome is heritable (Genetics Home Reference, 2017a; Tsai *et al.*, 2009).

2.1.1.3. Diagnosis of aneuploidy

There are several methods employed to detect if an individual has an incorrect number of chromosomes. These may be used prenatally or postnatally (from birth to old age). Nevertheless, the earlier that a diagnosis is made, it is more advantageous because it allows for early intervention in order to support the pregnancy as well as it guides treatment and management decisions to be made by family members. However, certain signs and symptoms appear later in life such as during puberty that necessitate a referral to a genetic specialist (Genetic Alliance, 2010).

2.1.1.3.1. Physical and clinical examination

Certain genetic disorders, like Down syndrome, are characterized by distinctive facial and other physical features, which can aid in their diagnosis. A physician or a geneticist will perform a thorough physical examination that may include measuring the head circumference, the distance between eyes, and the length of the limbs. Congenital abnormalities are sometimes present in aneuploidies as well as other genetic disorders so that specialized examination of the nervous system, the eye, and other organs may also be performed. To see the structures inside the body, imaging techniques may also be used like x-rays, computerized tomography (CT) scans, or magnetic resonance imaging (MRI). Developmental delays, mental retardation as well as social, behavioral, and emotional maladjustments are also noted down especially if these occur along with the other clinical features (Genetics Home Reference, 2020a; Genetic Alliance, 2010).

2.1.1.3.2. Analysis of medical history

Obtaining information about the individual's health, as early as from birth, can help in making a correct diagnosis. A person's medical history includes past health issues, hospitalizations and surgeries, allergies, medications, and the result of any medical or genetic testing that has been previously done (Genetics Home Reference, 2020a).

Information about the health of family members (i.e., parents, siblings, children, and even more distant relatives) is also important in diagnosing genetic disorders since these may be heritable. The data obtained can be used to make a pedigree in order to determine the inheritance pattern of a genetic condition in a family (Genetics Home Reference, 2020a). Pedigree construction and analysis are discussed in Chapter 1, Lesson 1.1.4. of this book. Additionally, the age of the mother and the father during conception is taken into account since most types of aneuploidy are associated with increased parental age.

2.1.1.3.3. Genetic testing

Genetic testing employs laboratory methods in order to determine the genetic make-up of an individual. These can be done prenatally where a tissue sample from the fetus is obtained and then used to determine if there are changes in the fetus' genes or chromosomes. Samples are taken either via amniocentesis or chorionic villus sampling (Genetic Alliance, 2010). In instances where conception is via *in vitro* fertilization, the embryo may undergo genetic testing before it is implanted in the uterus.

Genetic testing can also be employed at any time after birth to confirm a diagnosis in a symptomatic individual or to monitor the response of the individual to treatment (Genetic Alliance, 2010). A variety of tissues can be used as samples such as tumor biopsies or bone marrow for cancer diagnosis or from peripheral blood specimens or a skin biopsy for other diagnoses (O'Connor, 2008a).

After obtaining cell samples, the genetic composition of the cells is then determined. There are various methods of genetic testing but those that examine a person's chromosomal makeup are critical in order to diagnose aneuploidy. This includes the cytogenetic method of karyotyping as well as molecular methods like polymerase chain reaction and DNA sequencing.

2.1.1.3.3.1. Preimplantation genetic diagnosis

For couples who conceive via *in vitro* fertilization and are at risk of having an offspring with a genetic disorder (i.e., advanced maternal age, recurrent miscarriage, family history of a certain genetic disease), **preimplantation genetic diagnosis (PGD)** may be done (see Figure 2.19). Cells from the embryo such as blastomeres at the three-day cleavage stage or trophectoderm cells at the five to six-day blastocyst stage are removed and undergo genetic testing after which only those embryos with no genetic defect are transferred to the uterus (de Rycke and Sermon, 2010).

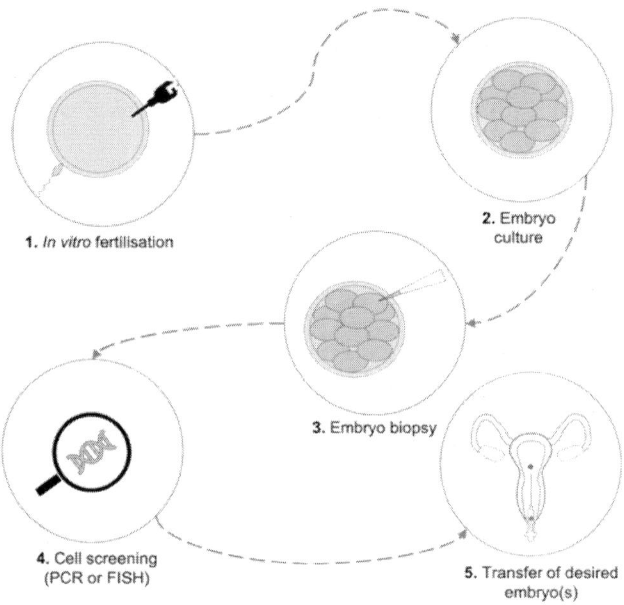

Figure 2.19. Preimplantation genetic diagnosis (PGD) is performed after (1) *in vitro* fertilization. After (2) the embryo has been cultured for 3-6 days, (3) a cell is obtained from the embryo. This cell undergoes (4) genetic screening or testing in order to determine the genetic soundness of the embryo. Based on the testing results, only the embryos that are free from genetic aberrations are (5) implanted into the uterus. (Source: Bananapancake 212, Creative Commons License).

More than half of the cleavage-stage embryos produced *in vitro* have chromosomal abnormalities and this may increase to 80% in women who are more than 42 years old. In the case of embryos at the blastocyst stage, more than 40% are abnormal and this proportion increases with advanced maternal age. Much like what is observed in normally conceived pregnancies, embryos with chromosomal anomalies have very low chances of implanting successfully in the uterus. Therefore, selecting normal embryos increases the likelihood that an assisted pregnancy will be successful (Munné, 2012).

2.1.1.3.3.2. Amniocentesis

Amniocentesis is a procedure where a little amount of amniotic fluid is obtained from the uterus (see Figure 2.20). This fluid contains fetal cells, which are processed for genetic testing. Amniocentesis is usually performed between the 15th and 20th weeks of pregnancy. There is a risk of 0.1-0.3% miscarriage when amniocentesis is performed. The risk is higher if it is done before the 15th week of pregnancy. Other risks, although minimal, include amniotic fluid leaks through the vagina, needle injury, infection or infection transmission. Another associated risk is **Rh sensitization**, which happens if the mother has Rh- negative blood type and the baby is Rh- positive. Rarely, amniocentesis may cause fetal blood cells to enter the mother's bloodstream inducing the production of Rh antibodies. These antibodies can cross the placenta and damage the baby's red blood cells. This is prevented by administering **Rh immunoglobulin** to the mother after amniocentesis (Mayo Clinic, 2019a).

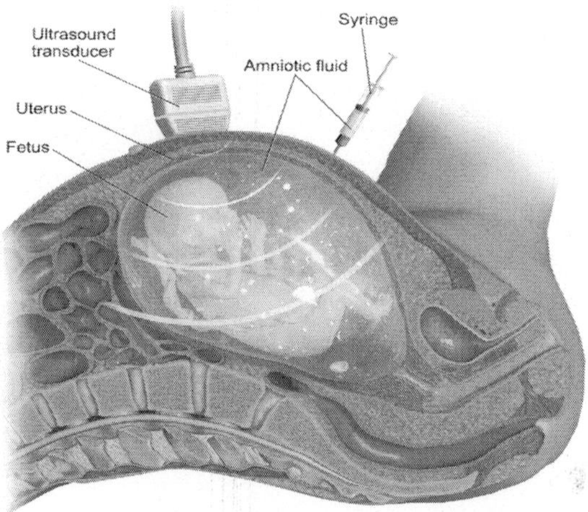

Figure 2.20. During amniocentesis, an ultrasound transducer is used to show on a monitor the position of the fetus in the uterus. Then a sample of the amniotic fluid is withdrawn via a syringe. (Bruce Blaus, Creative Commons License).

Amniocentesis is recommended if the mother is 35 years or older, has a previous pregnancy with a chromosomal anomaly or a neural tube defect or has abnormal ultrasound findings such as increased nuchal translucency (Mayo Clinic, 2019a). **Nuchal translucency (NT)** is the sonographic appearance of a fluid-filled subcutaneous space identified at the back of the fetal neck during the late first trimester and early second trimester. Increased NT has been associated with chromosomal aneuploidies, structural fetal abnormalities, cardiac malformations, a high risk of miscarriage, and intrauterine death (Socolov *et al.*, 2017). Amniocentesis if also offered is the mother or father has a family history of a genetic condition or is a known carrier of a genetic disease (Mayo Clinic, 2019a).

2.1.1.3.3.3. Chorionic villus sampling

Chorionic villus sampling (CVS) is another method of obtaining fetal cells for prenatal testing. In this method, a sample of chorionic villi, which are part of the placenta and are made up of cells with the baby's genetic makeup (see Figure 2.21), is removed from the placenta. The sample tissue is taken either though a catheter inserted through the cervix (transcervical) or a syringe injected through the abdominal wall (transabdominal). It is performed between the 11th and 14th week of pregnancy (Mayo Clinic, 2019b).

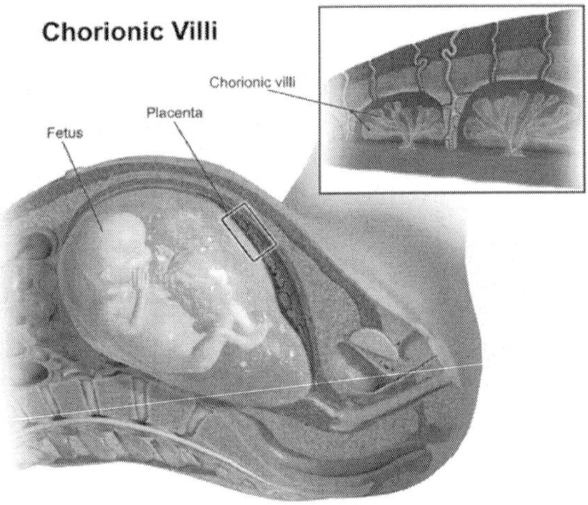

Figure 2.21. The chorionic villi are tissues that project from the placenta. These are formed from embryonic cells so that they have the same genetic composition as the fetus. (Source: Bruce Blaus, Creative Commons License).

The risk of miscarriage due to CVS is approximately 0.22%. Other risks include infection or Rh sensitization. Some studies also indicate that if the procedure is done earlier than the 10th week of pregnancy, there is a risk for the baby to have defects in the fingers or toes (Mayo Clinic, 2019b).

2.1.1.3.3.4. Cell-free DNA Screening

Due to the risks posed to the pregnancy by obtaining fetal cells via amniocentesis and CVS, these are recommended only to women whose pregnancy is at high risk of having a genetic abnormality. One way of determining whether these invasive procedure are necessary is by performing non-invasive pre-natal testing called **cell-free DNA screening**. In this procedure, a blood sample is taken from the woman after 10 weeks of pregnancy. A pregnant woman's blood contains small fragments of DNA that are free- floating instead of inside a cell's nucleus. These **cell-free DNAs (cfDNAs)** are a mixture of maternal and fetal DNA. The cfDNA can be used for genetic testing such as sequencing and then computing for the percentage of DNA from chromosome 21, 18, 13, X or Y. If the percentage of cfDNA fragments from a certain chromosome is higher than expected, the fetus has an increased likelihood of having a trisomy condition. For example, if a fetus has Down syndrome, there will be slightly more DNA fragments belonging to chromosome 21 in the sample. This method has been validated to be 99% sensitive and specific in determining the risk for trisomy 21, 97% for trisomy 18, and 87% for trisomy 13 (UCSF Health, 2020; Genetics Home Reference, 2018c; Rafi *et al.*, 2017).

False negatives may be observed due to low levels of fetal DNA(i.e., 4% or lower) in the cfDNA or it might be detecting the genetic condition of the mother instead of the child. For these reasons, a positive screening result requires further diagnostic testing (e.g., karyotyping) to confirm the result. In this case, amniocentesis or CVS may be suggested to obtain the needed fetal tissue samples (UCSF Health, 2020; Genetics Home Reference, 2018c; Rafi *et al.*, 2017).

2.1.1.3.3.5. Construction and analysis of karyotype

Karyotyping involves taking an image of the chromosome set in a cell as seen through the microscope and then pairing and ordering them. The resulting karyotype or karyogram can reveal any changes in the chromosome number or structure of the chromosomes. For this reason, it can provide diagnostic information for genetic diseases and even cancers (O'Connor, 2008a).

The tissue samples obtained are used to perform short-term cell culture, in which the cells undergo cell growth and cell division. Then the cells are treated with a mitotic poison like colchicine or colcemid, which disrupts the spindle apparatus that facilitates chromosome segregation. This causes the cells to be arrested, preferably at the prometaphase or metaphase stage, where the chromosomes assume their densest form. Then the cells are treated with a hypotonic solution that makes their nuclei swell and causes the cells to burst. The nuclei are then treated with a chemical fixative, which aims to preserve the cellular structures in their natural or life-like state. Afterward, the samples are then dropped on a glass slide and treated with certain stains to reveal the form and structural features of the chromosomes (see Figure 2.22) (O'Connor, 2008a).

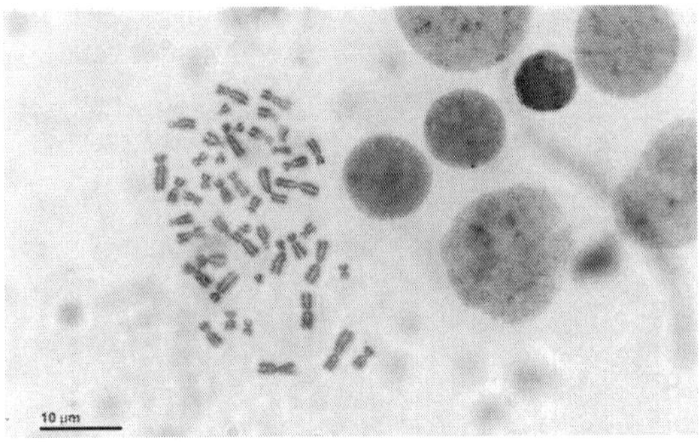

Figure 2.22. The 46 human chromosomes arrested at the metaphase stage are stained and the resulting image of chromosomes, magnified 3000x by a microscope, are then arranged to make a karyotype or a karyogram. (Source: Josef Reischig, Creative Commons License).

The stained images of the paired chromosomes are then arranged from chromosome 1 to 22 following international conventions. The sex chromosomes are placed at the end of the karyogram (O'Connor, 2008a). Analyzing the karyogram can reveal if a chromosome is missing or an extra copy is present.

Karyotyping during the prenatal stage (i.e., using samples taken by amniocentesis or CVS) is recommended if the mother had two or more consecutive spontaneous abortions or even after just one spontaneous abortion if the mother is advanced in age. It is also performed if severe retardation in intrauterine growth is observed or if ultrasound images suggest one or more malformations or ambiguous genitalia after 13 weeks (Lester, 2010; Horii *et al.*, 2018). Karyotyping after birth can also be performed to confirm a diagnosis of aneuploidy based on the observed signs and symptoms, especially among individuals that show mild forms of the condition. Karyotype analysis is considered the gold standard in terms of diagnosing aneuploidies. However, it presents challenges since it requires cell culture, considerable time, and viable cells and it has low sensitivity. Therefore, in order to save time and cost, without compromising accuracy, newer methods are utilized such as quantitative fluorescence polymerase chain reaction, fluorescence *in situ* hybridization, and comparative genomic hybridization (Huo *et al.*, 2019).

2.1.1.3.3.6. Fluorescence in situ hybridization

Fluorescence *in situ* **hybridization (FISH)** depends on the principle of specific base pairing. A single-stranded nucleotide molecule anneals via hydrogen bonding to another molecule containing its complementary nucleotides. Thus, a fluorescently labeled DNA or RNA sequence can be used as probe to determine the position of its complementary DNA sequence *in situ* (i.e., their natural location in the chromosome) (O'Connor, 2008c). The fluorochromes used to label the probe can emit either red, green or blue fluorescence.It is essential that the probe and the target sequences are denatured into single- stranded forms (see Figure 2.23). This is accomplished either with chemicals or heat. The chromosomes used as a target can either be at the interphase or metaphase configuration (Huber *et al.*, 2018). The target DNA sequences are affixed onto a microscope slide (Bishop, 2010). The probe is then added to the target. It is expected to specifically hybridize with its complementary sequence on the target chromosome. The resulting fluorescent hybrid can be detected directly or indirectly using a fluorescence microscope (O'Connor, 2008c).

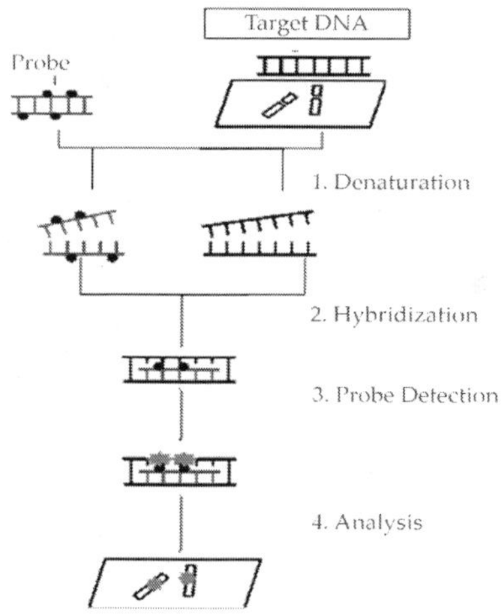

Figure 2.23. In fluorescence *in situ* hybridization (FISH),, the probe and target sequences should be denatured so that they become single-stranded. Then these are incubated together to allow the probe to hybridize with its complementary target sequence. The fluorescence emitted by the probe is detected by a fluorescence microscope and the fluorescence data is then analyzed. (Source: Ratan, Z.A.; Zaman, S.B.; Mehta, V.; Haidere, M.F.; Runa, N.J.;Akter, N. Cureus 9(6), e1325, 2017, Creative Commons License).

In the direct FISH procedure, the fluorochrome is incorporated directly into the probe (see Figure 2.24A). Therefore, the probe-target hybrid can be directly detected by fluorescence microscopy without additional steps. This is faster and simpler to perform than the indirect method, but its main disadvantage is its reduced sensitivity (Downie *et al.*, 1997).

Indirect FISH uses a probe with modified nucleotides carrying a **hapten** (i.e., a small reporter molecule that can bind to an antibody or a certain protein) like digoxigenin (DIG) or biotin (see Figure 2.24B). After the probe hybridized to its target, the hapten is detected by binding to a fluorochrome-conjugated protein such as an antibody (for DIG) or avidin (for biotin) or by a colored histochemical reaction that is visible under a light microscope. Indirect FISH is highly sensitive, and its signal can be intensified by using antibodies. On the other hand, it is expensive, is takes a long time to do the staining, and it requires higher background labeling (Downie *et al.*, 1997; Bishop, 2010).

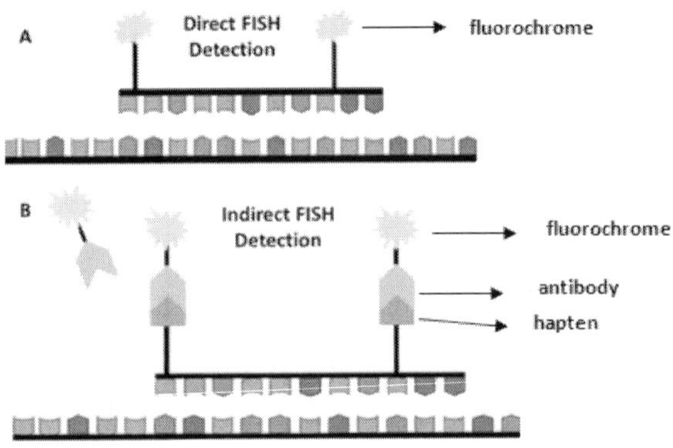

Figure 2.24. (A) Direct fluorescence *in situ* hybridization (FISH) utilizes a probe with fluorochrome that is incorporated directly into the nucleotides. (B) Indirect FISH, on the other hand, employs a probe with a hapten which is then detected by a fluorochrome-labeled antibody. (Source: Adapted from Escott16, Creative Commons License).

Using multiple probes at once can be performed in order to simultaneously detect several chromosomes. This may be done by using three different probes that target three different chromosomes. These probes may be labeled with red, green or blue fluorochromes, respectively. The main limitation of this is that only two or three chromosomes. Another approach is to label a single probe with up to three different haptens or fluorochromes so that it produces up to three different signals *in situ*. Seven different probes had been visualized simultaneously on metaphase chromosomes using a combination of three single-labeled probes, three double-labeled probes, and one triple-labeled probe. A third method is by ratio labeling. Aliquots of the probe are directly labeled with different fluorochromes and then these are mixed in different ratios before incubating with the target sequences. After hybridization, composite signals with up to 12 different colors from the three primary colors can be produced (Downie *et al.*, 1997). Further discussion on multicolor FISH is provided in section 2.1.2.5.2.

There are three types of DNA probes for FISH. One of these is a **centromeric probe**, which recognizes repetitive DNA sequences in the centromeric region (see Figure 2.25). These are commonly used to detect aneuploidy. It can also detect Robertsonian translocation that changes chromosome structure. A **sequence-specific probe** can detect specific regions in the chromosome such as unique single-copy genes or satellite DNA (see Figure 2.25). If it is unique to a certain chromosome, then it can also be used to detect aneuploidy. The third type is the **whole chromosome painting (WCP)** probe, which labels entire chromosomes and can be used to detect structural anomalies in chromosomes (Downie *et al.*, 1997).

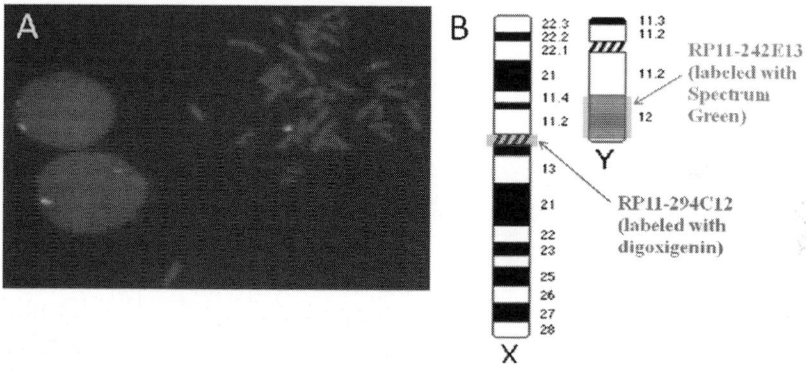

Figure 2.25. (A)The result of fluorescence *in situ* hybridization (FISH) shows a red signal, which is specific for the X chromosome, and a green one for the Y chromosome. (B) The probe used to detect the X chromosome is labeled with digoxigenin, which is detected by incubating it with rhodamine (i.e., red in color)-conjugated antibodies. The probe targets the centromeric region. The probe for the Y chromosome is directly labeled with a green fluorochrome and it targets a satellite DNA in the long arm of the chromosome. (Source: Hsu, J.H.; Zeng, H..; Lemke, K.H.; Polyzos, A.A.; Weier, J.F.; Wang, M.; Lawin-O'Brien,

A.R.; Weier, H.-U.G.; O'Brien, B. International Journal of Molecular Science 14(1), 57-71, 2013, Creative Commons License).

Fluorescent signals can be viewed directly in the fluorescence microscope or via charged coupled devices (CCD) camera connected to a microscope and a computer. Images from each filter set are captured and subjected to image analysis using certain software.

If the signal to noise ratio is too low, either because of high background noise or low intensity of signals, the software can improve the ratio. The actual color of the images is gray, but these are overlaid and given artificial colors (Kjeldsen and Kølvraa, 2002). After obtaining images of the sample, the standard approach of analysis is by manually counting the FISH signals per cell or nucleus. A large number of high-resolution images may be obtained and so several algorithms for the automatic detection and counting of FISH signals have been developed (Huber *et al.*, 2018).

2.1.1.3.3.7. Quantitative fluorescence- polymerase chain reaction

Quantitative fluorescence-polymerase chain reaction (QF-PCR) is a method used to amplify (i.e., increase the copy number) certain DNA segments, after which the amount of DNA present in those regions are determined. Specifically, these segments are called short tandem repeats (STRs) which are found along the length of chromosomes. Just like other PCR-based methods, this requires extraction of DNA from the samples. Primers are then designed to amplify the STRs and these primers are fluorescently labeled. PCR increases the number of copies of certain DNA regions by following the principle of DNA replication. Hence, apart from the DNA sample serving as a template and the primers, a polymerase enzyme as well as deoxyribonucleotides are added into the mix. The amplification process occurs with the aid of a machine called the thermocycler, which can be programmed to adjust its temperature to allow DNA template denaturation, annealing of primer to the template, and extension of the primer by the DNA polymerase (see Figure 2.26). This temperature cycle is repeated 20 to 40 times. The PCR products of each cycle then undergo capillary electrophoresis, where they are injected into capillaries containing polymers. High voltage is applied so that the DNA fragments are separated by size. The fluorescence emitted by the DNA is activated and detected by a laser/camera system. The fluorescence and the size of the DNA that has been copied are measured. The number of peaks and the height or area of each peak show the number of copies of STRs at that chromosome region

(Elucigene, 2014; ThermoFisher Scientific, 2020).

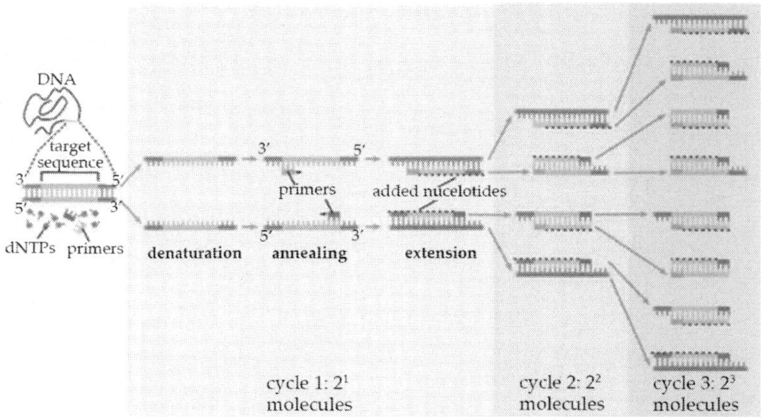

Figure 2.26. Polymerase chain reaction (PCR) is an *in vitro* technique that consists of a series of temperature cycles composed of template DNA denaturation, annealing to the template of a pair of primers that flank the target region, and the elongation of the primers by a DNA polymerase using deoxyribonucleotide triphosphates (dNTPs) as substrates. This cycle is repeated 20-40 times in order to amplify the target DNA segment and produce millions of copies from a small quantity of DNA template. (Source: Quan, P.-L.; Sauzade, M.; Brouzes, E. Sensors 18(4), 1271, 2018, Creative Commons License).

QF-PCR is a rapid method for detecting chromosome copy number, with results available within 24-48 hours or after 5 days at the latest, while karyotyping takes longer (i.e. 15-30 days). It is also reliable exhibiting100% correspondence with karyotyping results. Thus, this can be used to detect prenatally various aneuploidies using amniotic fluids and CVS samples. Primers that amplify 3-5 STRs across chromosomes 13, 18, 21, X, and Y are utilized in order to check for all possible aneuploidies. The different primers are labeled with fluorescent tags with different colors. These also yield amplification products with different sizes (Elucigene, 2014). The peak area of measured fluorescence for each STR is used to compute the allele ratios (shorter allele/longer allele). An allele ratio between 0.8-1.4 is considered normal. On the other hand, ratios of >1.8 or <0.65 or three alleles with equal peak areas indicate trisomy. At least two informative markers for the chromosome of interest must be detected for a reliable diagnosis (Huo*et al.*, 2019).

In one study that used QF-PCR to detect sex chromosome aneuploidies, primers that target 10 STR markers distributed across the chromosomes were designed and utilized (see Figure 2.27). These include five X-linked STRs such as DXS981, DXS6809, DXS1187, DXS1053, and DXS8377. Two sex-determining genes were also employed, namely, the amelogenin gene with variable alleles found in the X and Y chromosomes (i.e., AMX and AMY, respectively) as well as the sex-determining region Y (SRY) gene found only in the Y chromosome (Xie and Liang, 2014). In the case of the amelogenin gene, the X-linked allele (AMX) is longer with ~1.5 kilobases (kb) than the Y-linked one (AMY) with ~1.3 kb (Dutta *et al.*, 2017). Another marker, X22, is found in both the X and Y chromosomes. Finally, two autosomal STRs, D13S305 and D21S11, were used as internal controls (Xie and Liang, 2014).

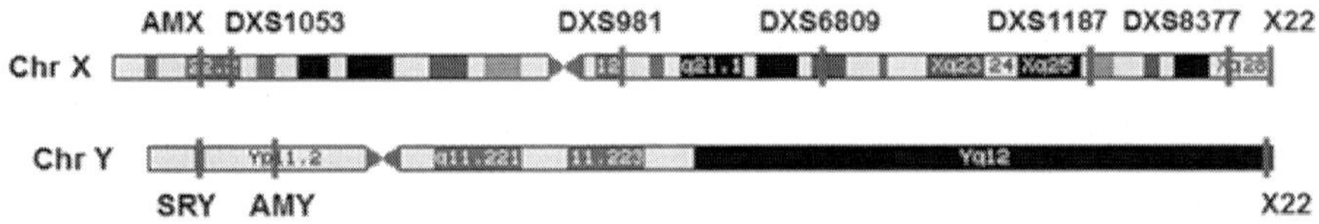

Figure 2.27. Detection of sex chromosome aneuploidies via quantitative *fluorescent polymerase* chain reaction (QF-PCR) requires primers that target STRs located in both X and Y chromosomes such as X22 as well as those exclusively located in the X chromosome like DXS1053, DXS981, DXS6809, DXS1187, and DXS8377. Two known sex-determining genes can also be included such as the SRY gene found only in the Y chromosome and the two alleles of the amelogenin gene, one found in the X chromosome (AMX) and the other in the Y chromosome (AMY). (Source: Xie, X. and Liang, Q. PLoS ONE 2014, 9(9), e106307, Creative Commons License).

After performing capillary electrophoresis of the PCR products, the profile of STR fluorescence peaks exhibited by known normal and aneuploidy samples were then analyzed and compared. Normal female samples (46, XX) showed a 1:1 (or 1 which means the two peaks have the same area) allele ratio for all X-linked STRs, which means that the alleles of these STR are present in the two X chromosomes at the same dosage (see Figure 2.28A). Moreover, a single peak for the AMX allele (or AMXY) and the X22 STR was observed, which is to be expected since the Y-linked equivalents of these regions are absent. Expectedly, no SRY signal is detected from these samples. For the normal male samples (46, XY), these showed single peaks for the X-linked STRs since only a single copy of these are present in the samples (see Figure 2.28B). The results also showed a 1:1 allele ratio for the AMX/AMY gene and X22 STR indicating the presence of alleles of these in both X and Y chromosomes at equal dosage, and a single peak for the SRY gene (Xie and Liang, 2014).

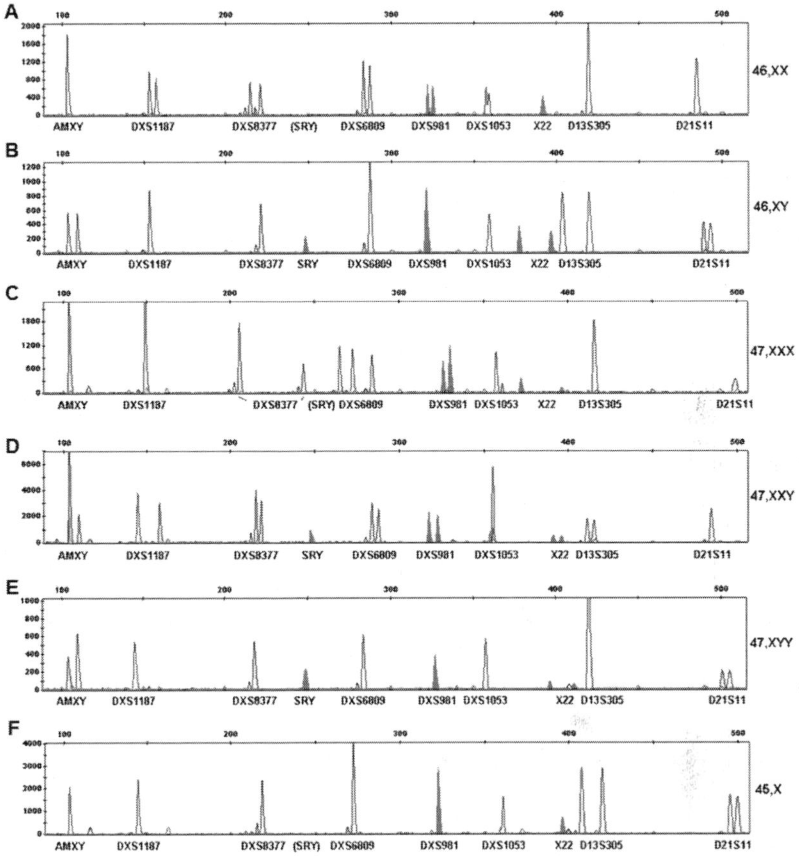

Figure 2.28. After capillary electrophoresis, analyzing the resulting fluorescence peaks of the products of quantitative fluorescence-polymerase chain reaction (QF-PCR) reveals significant differences between (A) normal XX and (B) XY karyotypes and the sex chromosome aneuploidies such as (C) triple X syndrome, (D) Klinefelter syndrome, (E) Jacob syndrome, and (F) Turner syndrome. (Source: Xie, X. and Liang, Q. PLoS ONE 2014, 9(9), e106307, Creative Commons License).

For samples with the triple X syndrome (47, XXX), the X-linked STRs showed either abnormal peaks with allele ratio of 2:1 (or 1:2) such as those of DXS1187, DXS8377, DXS981, and DXS1053 or three peaks like that of

DXS6809 (see Figure 2.28C). In the case of those with Klinefelter syndrome (47, XXY), some aspects of the sample profile are similar to those of the XY karyotype such as a single SRY peak and a single peak for some X-linked STRs (see Figure 4.28D). However, certain deviations from the normal male were observed such as other X-linked STRs exhibiting two peaks with 1:1 allele ratio, which is a feature of the XX karyotype; AMX and AMY showing an allele ratio of 2:1; and the X/Y-linked STR, X22, having a 2:1 allele ratio or having three peaks (Xie and Liang, 2014).

XYY samples are similar to the XY profile in terms of having a single peak for SRY and for all X-linked STRs (see Figure 2.28E). It differs from the normal male profile in terms of the AMX and AMY alleles having a 1:2 ratio as well as the X22 STR showing either a single peak, two peaks with 1:2 allele ratio or three peaks. Those with Turner syndrome (45, XO) exhibited single peaks for all X-linked STRs including X22 and the AMX gene (see Figure 2.28F) (Xie& Liang, 2014).

2.1.1.3.3.8. Comparative genomic hybridization

Comparative genomic hybridization (CGH) combines cytogenetic and molecular techniques to screen the entire genome for **copy number variations (CNVs)**. CNVs are relatively large segments of the genome that have been deleted or duplicated. These may increase the risk of having a genetic disorder but do not by themselves cause the disorder (Sun *et al.*, 2015). CGH was initially used to detect changes in the number of chromosomes or chromosome regions in tumor cells and eventually extended to detect aneuploidies or structural changes in the chromosome. Two genomes are used as probes in this method, a test genome (i.e., the tumor DNA or the DNA extracted from the sample and the control, which is a known normal genome (see Figure 2.29). These two are fluorescently labeled with differently colored fluorochromes (e.g. test genome is labeled green while the control is labeled red).

The presence of repetitive sequences in the probe may result in non-specific hybridization with the target. In order to prevent this, the probes are denatured together with an excess of unlabeled COT-I DNA. These are 50-300 bp DNA fragments obtained from male human placental DNA and they consist mainly of rapidly annealing repetitive elements. These are expected to hybridize with the repetitive sequences in the target DNA so that the probes will hybridize with the unique or single-copy sequence (Interchim, 2020). The probe mixture is then allowed to simultaneously hybridize with normal human metaphase chromosomes. A fluorescence microscope is needed to visualize the resulting fluorescence. Software that performs image analysis computes the red and green fluorescence ratio for each chromosome. The equal ratio of the red and green means the test genome has the normal amount of DNA in the chromosome or region of interest. On the other hand, increased intensity of the test genome-specific fluorochrome may indicate a net gain in DNA while decreased intensity indicates net loss (Theisen, 2008; Nikiforova and Nikiforov, 2011; Weiss *et al.*, 1999).

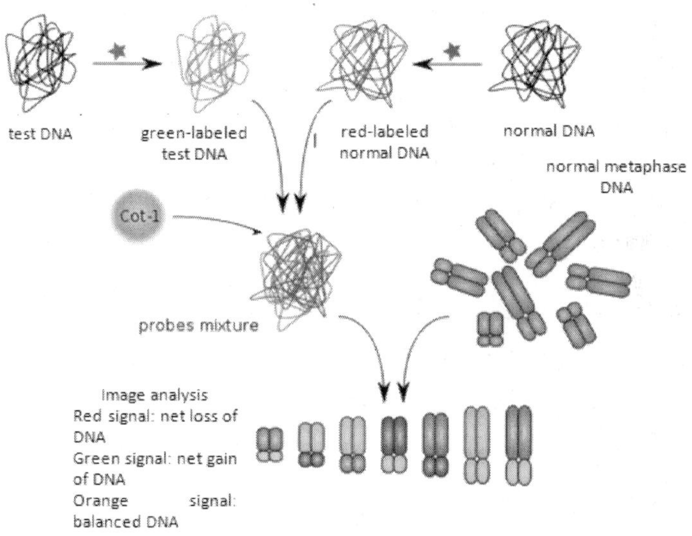

Figure 2.29. Comparative genome hybridization (CGH) allows for the screening of the whole genome to determine changes in the amount of DNA in either the chromosome or region of interest. There are two probes used, the sample or test DNA and a known normal DNA that are labeled with different fluorochromes. Cot-1 DNA is added into the probe mixture to block non-specific hybridization with repetitive DNA. These are then incubated with the target, which consists of normal metaphase DNA. The resulting fluorescence intensity is analyzed to determine if there is no change in DNA amount at the region of interest and if there is, whether it is a net loss or net gain. (Source: Gbdivers, Creative Commons License).

CGH, unlike traditional cytogenetic methods, does not require cell samples that are undergoing division. In combination with PCR, it does not require a large amount of cell samples since DNA extracted from a single cell or a small number of cells can be amplified and used as a probe (Voullaire *et al.*, 1999). Moreover, it allows the whole genome to be scanned instead of certain chromosomes or regions of DNA. Nevertheless, similar to the earlier methods, it can only detect changes of about 5-10 Mb.

In order to improve its resolution, a newer version of CGH has been developed. This combines the principles of CGH with the use of microarrays instead of metaphase chromosomes.

This is called the **array CGH (aCGH)**. A **microarray** is created by depositing and immobilizing thousands of single-stranded DNA fragments, called probes, on a solid substrate such as a glass, plastic or silicon slide (also called a chip). The identity of the probes is known, and these are arranged in rows and columns. Their size varies from oligonucleotides with 25-85 bp to bacterial artificial chromosomes with 80,000-200,000 bp. Nevertheless, the largest probes are still significantly smaller than metaphase chromosomes, which theoretically increases the resolution of aCGH compared to traditional CGH. However, the actual resolution depends not only on the probe size but also on the genomic distance between the probes. For instance, a microarray with probes representing regions of the genome that are 1 Mb apart will not detect changes in the intervening sequence (Theisen, 2008). Still, aCGH has been proven to detect chromosomal aberrations that were missed by conventional karyotyping (Kan *et al.*, 2014).

The steps of aCGH are basically similar to conventional CGH, but the probes are hybridized with a microarray instead of metaphase chromosomes (see Figure 2.30). The fluorescently labeled test and control or reference DNA will then hybridize with their complements among the single-stranded probes on the array slide. A digital imaging system then captures and quantifies the fluorescence ratio of the test and control hybridization signals (Theisen, 2008). Specifically, the log2 ratio or logR ratio (LRR) of the Cy5/Cy3 fluorescence is computed.

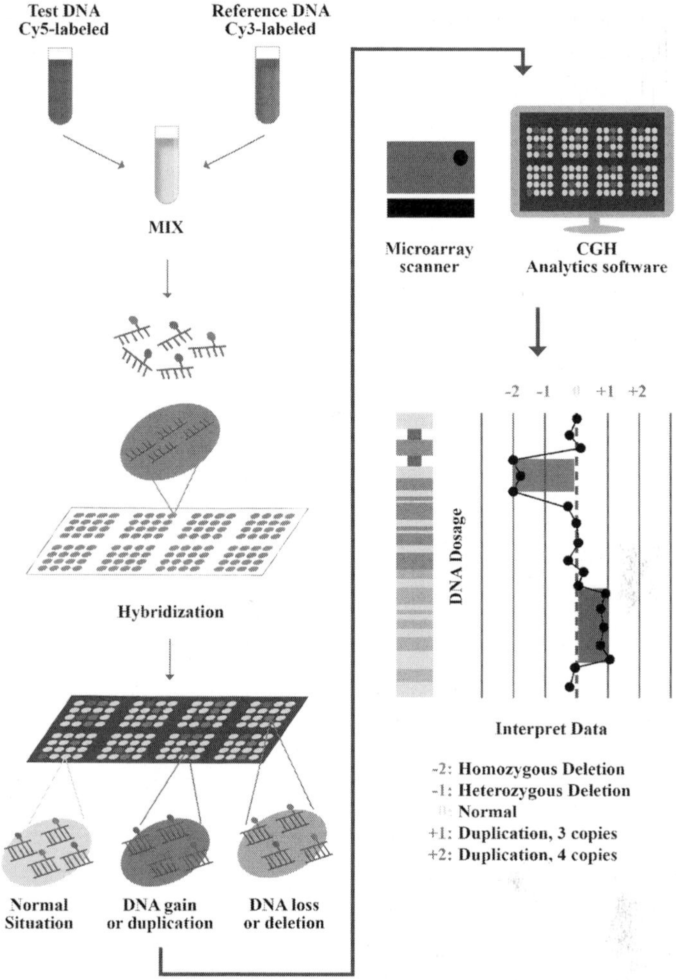

Figure 2.30. Array comparative genomic hybridization (aCGH) employs two sets of DNA samples, one that is extracted from the patient (test DNA) and a normal control (reference DNA). These are labeled with different fluorochromes, respectively (e.g. red cyanine-5 for the test DNA and green cyanine-3 for the reference). These are then allowed to simultaneously hybridize with thousands of known DNA sequences, called probes, arranged on specific spots on an array chip. These probes represent the entire genome. The labeled DNA that successfully annealed with a probe emits a fluorescent signal that is detected by a microarray scanner. A certain software computes the log2 ratio of the Cy5/Cy3 fluorescence. A higher intensity of the test sample color in a specific region of the genome relative to the control indicates the gain of DNA (or excess DNA) in that region while a higher intensity of the control sample color

indicates DNA loss in that region. When the resulting color is intermediate of the two fluorochromes used (e.g., yellow in the case of red and green), then that indicates normal DNA amount in that region. (Source: Colaianni, V.; Mazzei, R.; Cavallaro, S. Neurological Sciences 37(12), 2016, Creative Commons License).

While aCGH can detect alterations involving smaller DNA regions that can be missed by conventional CGH (see Figure 2.31F), it has also been used successfully to screen for aneuploidies such as in embryonic cells during preimplantation genetic diagnosis (see Figure 2.31A and B). This method can simultaneously screen more than one chromosomal aberration in a sample (see Figure 2.31C and D). This can also be used to screen for genomic alterations in newborn infants.

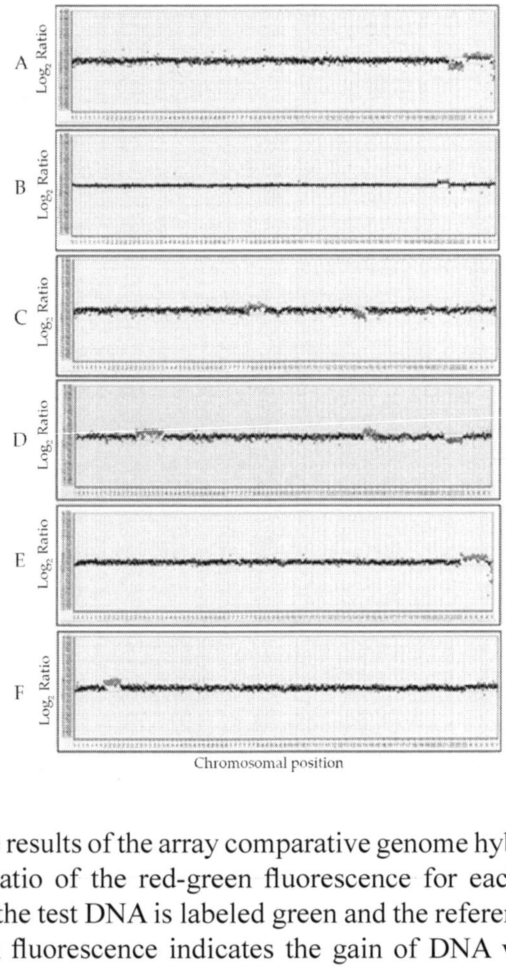

Figure 2.31. The results of the array comparative genome hybridization (aCGH) show the log2 ratio of the red-green fluorescence for each chromosome. In these examples, the test DNA is labeled green and the reference DNA is labeled red. Thus, green fluorescence indicates the gain of DNA while red indicates

a loss. Examples are: (A) A sample from a female embryo showing loss of chromosome 22; (B) a cell from a male embryo with trisomy 21; (C) a sample from a male embryo showing trisomy 18 and loss of chromosome 14; (D) a cell taken from another male embryo exhibits multiple aneuploidies, an extra copy of chromosome 3 and 15 as well as a missing copy of chromosome 22; (E)a normal female embryo shows an intense fluorescence for the X chromosome; and (F) a cultured cell with a duplicated long arm of chromosome 22 that served as a positive control. (Source: Chung, M.K.; Jeong, H.J.; Lee, J.; Park, S.-J. Molecular Cytogenetics 6(1), 22, 2013, Creative Commons License).

2.1.2. Abnormalities in Chromosome Structure

Some types of genetic disorders stem not from having abnormal chromosomal number but due to aberration in the structure of a chromosome. Mutagens such as ionizing radiation may cause the chromosome to break. Recombination errors that occur spontaneously may also result in structural rearrangements in both somatic and germ cells (Ferguson-Smith, 2001b). During meiosis, homologous chromosomes undergo synapsis. This involves the recognition by one homolog of its complementary sequences in the other homolog. Mismatching, or nonallelic homologous recombination (NAHR), may occur at random especially at regions containing tandem repeats, i.e. repetitions of identical or similar sequences. As a result, recombination may occur between sequences of different, non-homologous chromosomes, sequences at different positions of homologous chromosomes, or even two sequences on the same chromosome. This may result in deletion or duplication of the DNA sequences in such regions. If the recombination occurs between repeat sequences and the inverted form of such sequences, it leads to inversions or translocations (Ferguson-Smith, 2001b; Clancy and Shaw, 2008). Embryos with these aberrations that survive and develop to term are mostly heterozygous for such condition, i.e. only one of the homologous pairs has structural abnormality, while the complement remains normal.

2.1.2.1. Deletion of chromosome region

As previously mentioned, deletion of a chromosome segment occurs due to spontaneous errors in the pairing of homologous chromosomes particularly at DNA repeats. Chromosomal deletion is also called **partial monosomy**. It can occur in the **terminal** region or within the chromosome, in which case it is called an **interstitial deletion** (see Figure 2.32). There are also microdeletions, which are too small to be detected by karyotype, but still involve several genes. These mostly occur *de novo* and after fertilization. Therefore, they are not inherited (National Institutes of Health, 2017b).

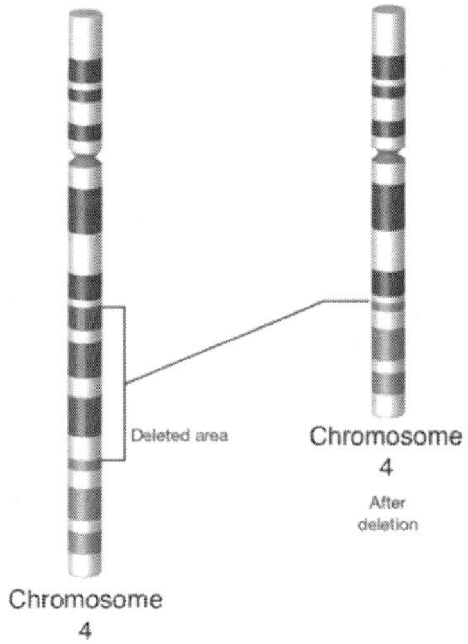

Figure 2.32. If the lost segment is located within the chromosome, then it is classified as interstitial deletion. (Source: National Human Genome Research Institute, Public Domain).

2.1.2.1.1. Disorders due to deleted chromosome region

Since several genes may be included in the deleted region, certain abnormalities are observed in affected individuals. Much like the disorders due to aneuploidies, these abnormalities have no cure. There are several disorders that are due to the deletion of a portion of specific chromosomes like chromosomes 1, 4, 5, 7, 11, 15, 16, 17, 18, or 22. This section discusses some of these.

2.1.2.1.1.1. Chromosome 4p deletion syndrome

Chromosome 4p deletion syndrome is also called **Wolf-Hirschhorn syndrome** or **monosomy 4p syndrome**. Another disorder called Pitt-Rogers-Danks syndrome has features that overlap with Wolf-Hirschhorn syndrome and eventually, these two are now recognized to be part of a single syndrome with variable signs and symptoms. This is caused by the loss of a portion of the short arm of chromosome 4. The signs and symptoms, as well as their severity, vary depending on the size and location of the deletion, particularly the genes involved. The deletion of more than 3 Mb appears to be correlated with a higher risk of heart defects and cleft palate. Its frequency is about 1 in 50,000 births, but this may not include those that are affected but are never diagnosed. There are approximately twice as many females that are affected than males, although the reason for this is yet unknown (National Institutes of Health, 2015b; Genetics Home Reference, 2012b; Orpha.net, 2016).

Affected individuals have the so-called "Greek warrior helmet" appearance, which results from having a broad, flat nasal bridge and a high forehead (see Figure 2.33). Other distinct facial features are widely spaced, protruding eyes, a short philtrum (i.e., shortened distance between the nose and the upper lip), a downturned mouth, micrognathia, malformed ears with pits (i.e., small holes) or tags (i.e., flaps of skin), asymmetrical facial features, and microcephaly. Most of those affected are also short in stature (National Institutes of Health, 2015b; Genetics Home Reference, 2012b).

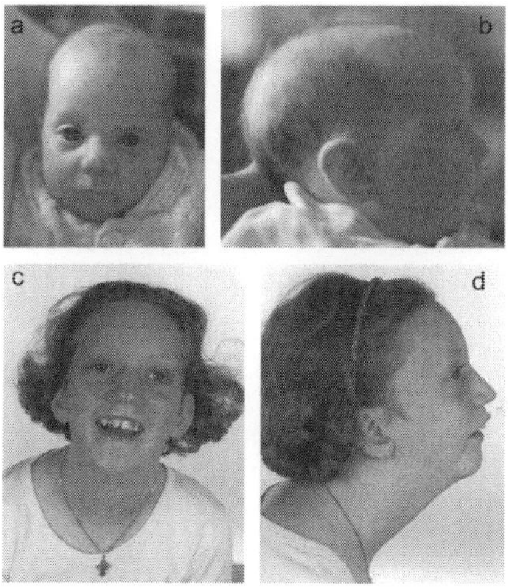

Figure 2.33. A female with chromosome 4p deletion syndrome, specifically a microdeletion, at age (a, b) 3 months and (c, d) 13 years. She exhibits microcephaly, micrognathia, protruding eyeballs, broad nasal bridge, and a high forehead, which are characteristic facial features of the syndrome. (Source: Sogaard, M.; Tümer, Z.; Hjalgrim, H.; Hahnemann, J.; Friis, B.; Ledaal, P.; Pedersen, V.F.; Baekgaard, P.; Tommerup, N.; Cingöz, S.; Duno, M.; Brondum-Nielsen, K. BMC Medical Genetics 6, 21, 2005, Creative Commons License).

Even in the uterus, fetuses with this syndrome exhibit very slow growth and after birth, they usually have problems feeding and gaining weight. As a result, they experience delayed growth and development. They also have hypotonia and underdeveloped muscles, which contribute to their feeding problem and significantly slow down the development of their motor skills. In some instances, they may not even achieve sphincter control as well as self-feeding or dressing skills. Most of them also have seizures, but these normally dissipate as they grow older. Moreover, they have abnormalities in their spine (e.g., scoliosis and kyphosis), teeth, mouth and/or lip (e.g., cleft palate, cleft lip), eyes, heart, genitourinary tract, and brain (Genetics Home Reference, 2012b; Orpha.net, 2015).

Those affected exhibit intellectual disability ranging from mild to severe. Additionally, they have weak verbal communication and language skills, which are usually limited to guttural or disyllabic sounds. Despite all these, they have strong social skills (Genetics Home Reference, 2012b; Orpha.net, 2015).

The signs and symptoms of these disorders are due to the loss of several genes on the short arm of chromosome 4 like *NSD2, LETM1*, and *MSX1*. These genes have significant roles in early development. Specifically, the deletion of *NSD2* is probably responsible for the distinctive facial features and the developmental delay observed among those who are affected. The loss of *LEYM1* may be correlated with seizures or other abnormal electrical activity in the brain. Meanwhile, the deletion of *MSX1* seems to be associated with the observed dental abnormalities, cleft lip, and cleft palate (Genetics Home Reference, 2012b). Translocation is discussed in more detail in a succeeding section of this chapter.

In most cases, this syndrome is not inherited. Therefore, it can occur in people with no family history of the disorder. A small proportion of individuals have the syndrome because of a ring chromosome 4 (see Figure 2.34). This formed because the chromosome broke in two places and the ends of the chromosome arms fused to form a closed structure. The genes in the regions that broke off are then lost (Genetics Home Reference, 2012b).

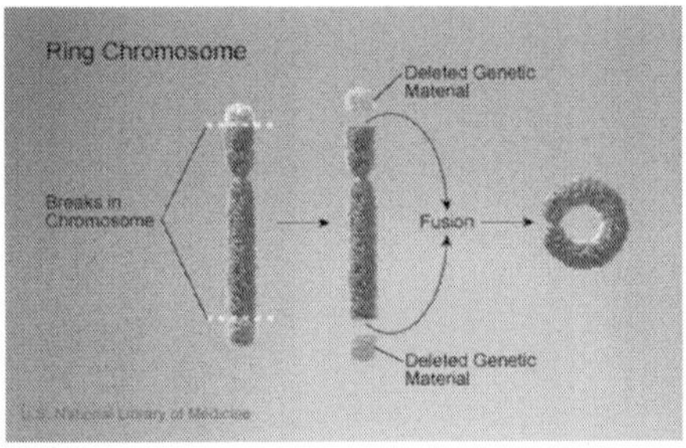

Figure 2.34. A ring chromosome is formed when the two termini of a chromo-

some break off and the remaining chromosome arms fuse to form a closed, circular structure. (Source: U.S. National Library of Medicine, Public Domain).

The remaining cases of the syndrome are a result of inheriting an unbalanced translocation from a parent that carries a balanced translocation involving chromosome 4 and another chromosome. The unbalanced translocation leads to the deletion near the end of chromosome 4 (Genetics Home Reference, 2012b).

2.1.2.1.1.2. Cri-du-chat syndrome

Cri-du-chat is a French term for cat cry, so the syndrome is also called cat cry syndrome. It is so named because infants born with this condition often have a high-pitched cry that sounds like that of a cat. This may be because of abnormalities in the structure of the larynx and the epiglottis (Mainardi, 2016). The disorder is also called **5p minus syndrome** because this is due to the deletion of a portion of the p arm of chromosome 5. The size of the deleted segment varies among affected individuals. Certain bands or regions in the 5p are associated with phenotypes that are observed in affected individuals (See Figure 2.35). A larger deletion may result in more severe signs and symptoms than a smaller one. This syndrome occurs in approximately 1 out of 20,000-50,000 births (National Institutes of Health, 2015a; Genetics Home Reference, 2020b).

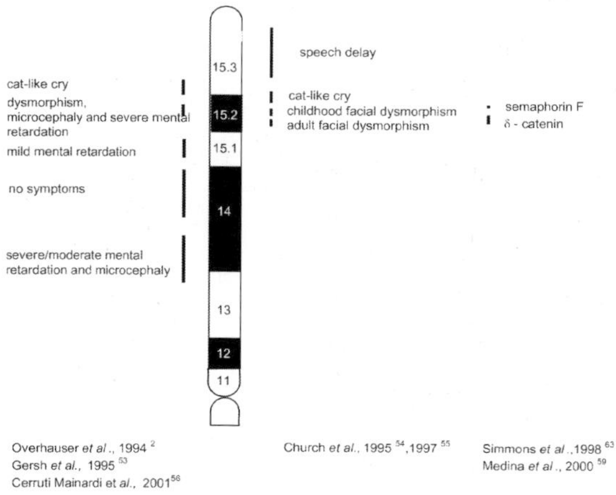

Figure 2.35. The deletion of the terminal band p15.3 of chromosome 5 is associated with speech delay and cat-like cry observed in cri-du-chat syndrome. Removal of band p15.2 is also associated with cat-like cry along with facial dysmorphism and mental retardation. Mental retardation is also correlated with

the deletion of band p15.1 and p14. (Source: Mainardi, P.C. Orphanet Journal of Rare Diseases 1, 33, 2016, Creative Commons License).

This syndrome is characterized by intellectual disability, delayed development, microcephaly, low birth weight, and hypotonia. The loss of the *CTNND2* gene may be associated with the severe intellectual disability observed in the affected individuals. Certain facial features are also observed among affected individuals such as hypertelorism, low-set ears, a small jaw, and a rounded face (see Figure 2.36). The characteristic cat's cry usually becomes less apparent with time. Most of those with this syndrome have a normal life expectancy. However, some affected infants who are born with a heart defect, abnormalities in other organs or other life-threatening conditions may have a shorter life span (National Institutes of Health, 2015a; Genetics Home Reference, 2020b).

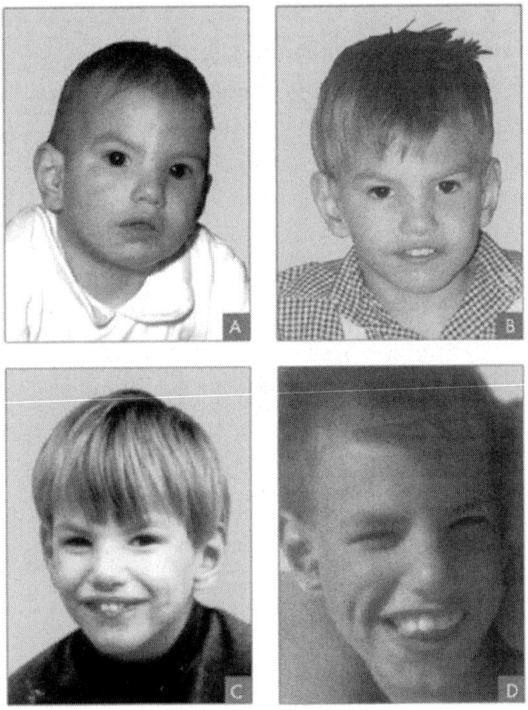

Figure 2.36. A patient with cri-du-chat syndrome at (A) 8 months and (B) 2 years showing facial features associated with the disorder such as microcephaly, round face, large nasal bridge, hypertelorism, epicanthal folds, and downturned corners of the mouth. Certain characteristics develop with age as observed in the patient at (C) 4 and (D) 9.5 years old. These include the face

becoming long and narrow, short philtrum, and dental malocclusion or open bite. (Source: Mainardi, P.C. Orphanet Journal of Rare Diseases 1, 33, 2016, Creative Commons License).

This is usually not inherited since it is due to a random deletion that occurs during meiosis when gametes are formed or during early fetal development. Approximately 10% of affected persons inherit the condition from an unaffected parent that carries a chromosomal rearrangement called balanced translocation. A person who has chromosomes with balanced translocations is phenotypically normal. However, these chromosomes become unbalanced when they are passed on to the offspring. This will then result in either extra or missing genetic material (National Institutes of Health, 2015a; Genetics Home Reference, 2020b).

2.1.2.1.1.3. Jacobsen syndrome

Jacobsen syndrome is due to the partial deletion at the end of the q or long arm of chromosome 11. As such, it is also known as **11q terminal deletion disorder**. The deleted segment ranges from 7-20 Mb. The breakpoints usually occur at or before band 11q23.3 and then it extends to the telomere. Loss of larger segments that include regions beyond band 11q23.3 likely results in death of the embryo in utero. A large terminal deletion occurring at band 11q21 has been reported, but it is in the mosaic form. Nevertheless, the person has a severe complex phenotype. Milder phenotypes are observed among those showing smaller terminal deletions as well as those with rare interstitial deletions. Interstitial deletions commonly occur in the long arm centromeric to the Jacobsen region and these result in a distinct clinical phenotype. Jacobsen syndrome is estimated to occur in 1 out of 100,000 births, with twice the number of affected females than males. About 20% of individuals with this syndrome die during their first two years, mostly because of complications from congenital heart disease and sometimes due to bleeding. More than 90% of the cases have a bleeding disorder called Paris-Trousseau syndrome, a disorder of the platelets. One patient has been recorded to have reached 45 years old, but the life expectancy is generally unknown (Mattina *et al.*, 2009; Genetics Home Reference, 2015a).

The most common features of this syndrome are retarded growth before and after birth, delayed psychomotor development, and facial dysmorphism. The characteristic facial features include skull deformities, hypertelorism, ptosis, coloboma, downslanting palpebral fissures, epicanthal folds, broad nasal bridge, short nose, long and flat philtrum, v-shaped mouth, thin upper lip, and small, low-set, posteriorly rotated ears (see Figure 2.37). Additionally, those with the syndrome have a short neck, thin fingers, abnormal palmar creases, stubby and flat feet, as well as brachydactyly and syndactyly of the toes. At birth, blood deficiencies such as thrombocytopenia (i.e., deficiency in blood platelets), pancytopenia (i.e., deficiency in red blood cells, white blood cells, and platelets), and abnormal platelet function are usually present. Those who are affected also have abnormalities in their heart, kidney, gastrointestinal tract, genitalia, central nervous system, and skeleton. Some may also have issues in their vision, hearing, immune system, and hormones (Mattina *et al.*, 2009; Genetics Home Reference, 2015a).

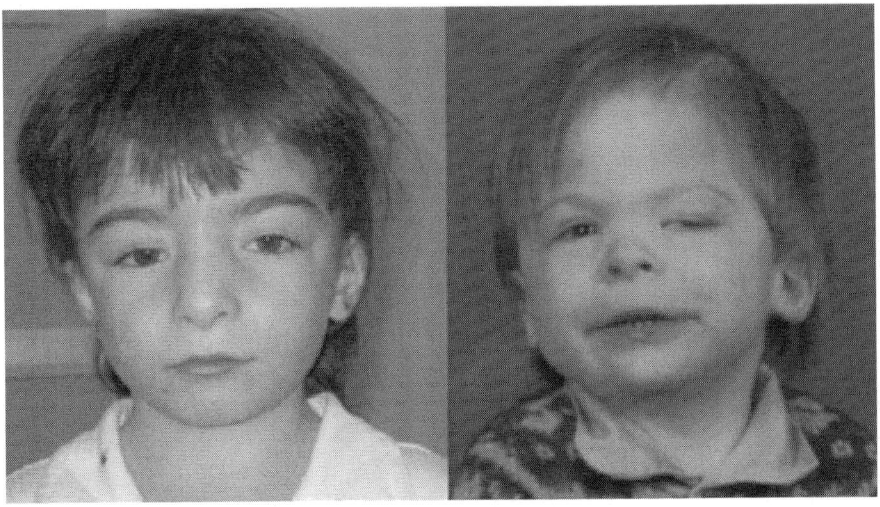

Figure 2.37. Jacobsen syndrome patients have facial dysmorphism, which includes skull deformities such as high prominent forehead, ocular hypertelorism, ptosis, downslanting palpebral fissures, epicanthal fold, ptosis, prominent nasal bridge, flat philtrum, and thin upper lip. (Source: Mattina, T.; Perrotta, C.S.; Grossfeld, P. Orphanet Journal of Rare Diseases 4, 9, 2009, Creative Commons License).

Most affected individuals have delayed development of speech and motor skills as well as cognitive problems and learning difficulties. Many are also diagnosed with ADHD and have an increased likelihood of having ASD (Genetics Home Reference, 2015a).

Depending on its size, the deleted region in Jacobsen syndrome contains 170-342 genes. Some of these genes are vital for the normal development of the brain, facial features, and the heart (Mattina *et al.*, 2009; Genetics Home Reference, 2020d). One of these genes is the *FLI-1*, which is involved in vasculogenesis, hematopoiesis, and intercellular adhesion. It has an essential role in megakaryocyte differentiation so that its heterozygous loss is associated with thrombocytopenia. Another gene is *BARX2*, which encodes a homeobox protein. In mice, this gene is involved in the development of neural and craniofacial structures. Therefore, it is possible that the deletion of this gene results to the facial dysmorphism observed in Jacobsen syndrome. Certain genes are also identified to possibly be related to the CHD of patients. The *JAM3* gene, a member of the junction adhesion molecules family, is considered one of the candidate genes for the observed hypoplastic left heart (i.e., a congenital defect of the left side of the heart) and trigonocephaly (i.e., premature fusion of the metopic suture leading to a triangular forehead) observed among affected persons. The loss of a gene involved in angiogenesis, *ADAMTS8*, is suspected to be involved in conotruncal heart defects (Mattina *et al.*, 2009).

Genes that are involved in brain development are also among those included in the deleted region. Among these is *B3GAT1*, which codes for beta 1-2-glycoronyltransferase. Mice who were knocked out for this gene showed impaired synaptic plasticity and learning disabilities. This might be involved in bipolar disorder observed in some Jacobsen syndrome patients. Another homeobox gene, *BSX*, is expressed during early brain development and is possibly involved in global cognitive development. The loss of *NRGN* (neurogranin), a gene involved in synapse plasticity and long-term potentiation, is suspected to contribute to auditory attention deficit. Abnormalities of the brain white matter are proposed to be due to the loss of hypothetical candidate genes, which include *FEZ1*, a gene for axonal outgrowth, and *RICS*, which is highly expressed in the developing brain (Mattina *et al.*, 2009).

Other genes in the deleted region include *KCJN 1*, which encodes a potassium channel, and *ADAMTS15*, which encodes zinc-dependent proteases. These are expressed in fetal liver and kidney and their loss may be related to kidney abnormalities in affected individuals. *TECTA* is a gene that codes for a non-collagenous component of the tectorial membrane of the inner ear and its deletion may be involved in deafness (Mattina *et al.*, 2009).

In 85% of the cases, the deletion occurred *de novo* and so it is not inherited. In the remaining 15% of the cases, it was a result of the inheritance of balanced translocation or other chromosomal rearrangements from an unaffected parent (Mattina *et al.*, 2009). This translocation becomes unbalanced in the child and results in loss of segments in chromosome 11 and extra genetic material in another chromosome (Genetics Home Reference, 2015a).

2.1.2.1.1.4. Chromosome 18p deletion syndrome

Chromosome 18p deletion syndrome is also known as **de Grouchy syndrome**. This is due to the deletion of the p or short arm of chromosome 18. The signs and symptoms vary depending on the size and location of the deleted region (National Institutes of Health, 2018b). So far, the recorded cases are due to terminal deletion (Turleau, 2008). It occurs in approximately 1 out of 50,000 live births, with more females affected than males at a ratio of 3:2. Affected individuals with heart and brain malformations die prematurely, some living only for a few months. Otherwise, they can live well into their 60's (Goyal *et al.*, 2017; Turleau, 2008).

Generally, most of the affected individuals are developmentally deficient. During early childhood, the observed signs and symptoms are similar to those of Turner syndrome and trisomy 21 (Turleau, 2008). They have certain facial features such as round face, short philtrum, protruding ear with abnormal antihelix, and hypodontia (i.e., 1-6 teeth fail to develop) (see Figure 2.38). They are also usually short in stature with brachydactyly (i.e., short fingers or toes). Their speech and language development are delayed, and they also exhibit mild to severe intellectual disability. Some also show skull, bone, and other dental abnormalities (e.g., microcephaly, brachycephaly or short and broad skull, micrognathia, kyphoscoliosis, misaligned teeth, dental cavities, wide rib cage, funnel chest). They may also have ptosis or drooping eyelids, drooping corners of the mouth, cleft palate, short and webbed neck, muscular hypotonia, cardiovascular problems, autoimmunity as well as behavioral issues (National Institutes of Health, 2018b; Goyal *et al.*, 2017; Turleau, 2008).

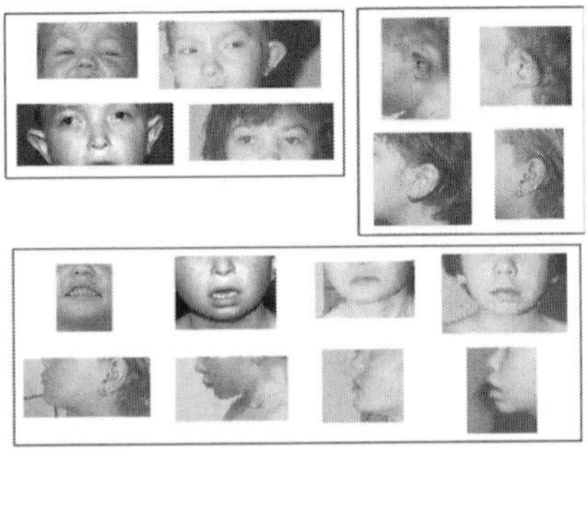

Figure 2.38. People with chromosome 18p deletion syndrome exhibit certain

facial features like flat midface, ptosis of the eyelid, flat and broad nose bridge, short philtrum, large and low-set ears with detached pinnae, misaligned teeth, small chin in children, and a wide mouth with the short protruding upper lip. (Source: Turleau, C. Orphanet Journal of Rare Diseases 3(4), 2008, Creative Commons License).

The syndrome arises from *de novo* deletions in 85% of the known cases. The remaining cases result from the inheritance of balanced translocation between the long arm of chromosome 18 and an acrocentric chromosome. In the offspring, this results in loss of a region of chromosome 18 p arm, inversion, or the formation of a ring chromosome (Goyal *et al.*, 2017; Turleau, 2008). In some cases, those with the deletion may pass it on to their offspring (see Figure 2.39). Most cases of inherited chromosome 18p deletion were maternally inherited (Maranda *et al.*, 2006).

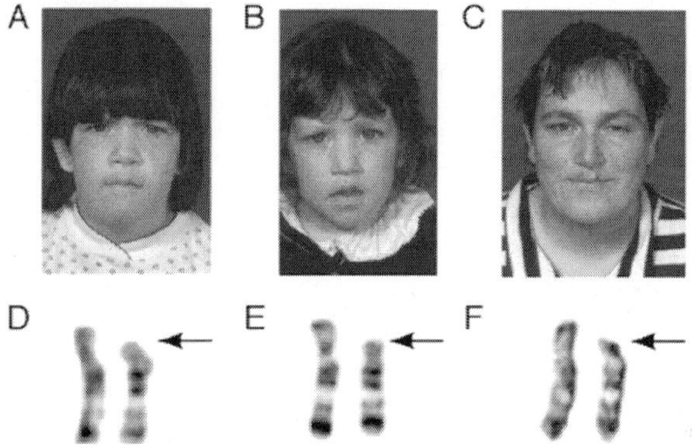

Figure 2.39. (D,E,F) Deletion of the p arm of chromosome 18 may be inherited from a parent. Example, (A,B) two daughters inherited it from their (C) mother. (Source: Maranda, B.; Lemieux, N.; Lemyre, E. BMC Medical Genetics 7, 60, 2006, Creative Commons License).

2.1.2.1.1.5. Chromosome 18q deletion syndrome

Chromosome 18q deletion syndrome results from the loss of a portion of the long or q arm of chromosome 18. The signs and symptoms differ among affected individuals, depending on the size of the deleted region. This occurs in an estimated 1 in 55,000 newborns. This has two types, the distal and the proximal 18q deletion syndromes. Most of the affected individuals have the distal type and are females with a 3:2 ratio relative to affected males (Genetics Home Reference, 2018d; National Organization of Rare Disorders, 2009).

The distal type of this syndrome occurs when the missing chromosomal segment is near the end of the chromosome, specifically between band 18q21 and the terminus of the long arm. The features commonly associated with this disorder include microcephaly (see Figure 2.40A), short stature (mostly because of growth hormone deficiency), hypotonia, hearing loss due to either aural stenosis (i.e., narrow ear canals) or aural atresia (i.e., absent ear canals), and foot abnormalities (e.g., clubfoot, rocker-bottom feet) (see Figure 2.40B). In some cases, certain facial features are observed like deep-set eyes (see Figure 2.40C), epicanthal folds, midface hypoplasia (i.e., flat or sunken appearance of the middle face) (see Figure 2.40D), broad nasal bridge, wide or "carp-shaped" mouth, mandibular prognathism (i.e., protruding lower jaw), cleft palate, cleft lip, and prominent ears (Genetics Home Reference, 2018d; National Organization of Rare Disorders, 2009).

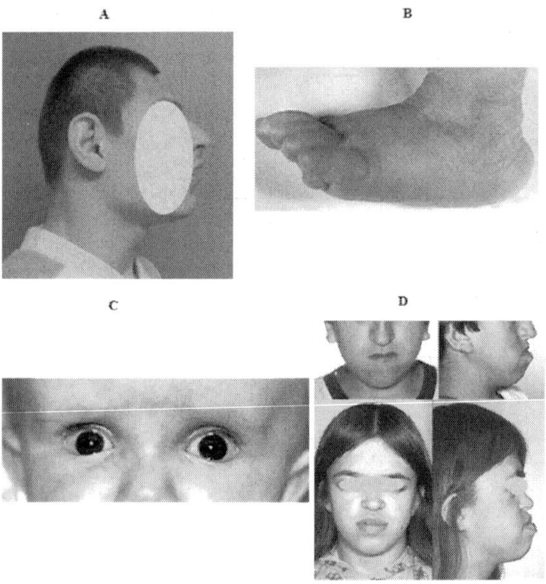

Figure 2.40. Distal chromosome 18q deletion syndrome is commonly characterized by certain features such as (A) microcephaly, (B) rocker-bottom feet, (C) deep-set eyes, and (D) midface hypoplasia or the sunken appearance of the middle part of the face. (Source: National Human Genome Research Institute, Public Domain).

Distal 18q deletion syndrome may also result to abnormalities in the eye (e.g., eye movement disorders, vision problems, microphthalmia, iris coloboma, opaque cornea, retinal and optic disk defect), thyroid gland (i.e., hypothyroidism), bones (e.g., rib and hip deformities), heart (e.g., atrial or ventricular septal defects, pulmonary stenosis, patent ductus arteriosus), kidney, genitals (e.g., hypoplastic labia, cryptorchidism, micropenis, hypospadias), and skin. It also impacts the nervous system causing impaired production of the myelin sheath, a fatty substance around nerve cells that promotes rapid impulse transmission. People with this syndrome exhibit neurological issues, but it is not yet certain if these are due to abnormal myelination. Some of their problems include delayed development, learning and intellectual disability, seizures, hyperactivity, mood disorders, and features of ASD that affect communication and social interaction. The deletion of the gene *TCF4* in this condition is associated with severe intellectual disability and breathing problems (Genetics Home Reference, 2018d; National Organization of Rare Disorders, 2009).

Most of the individuals with this condition did not inherit it from a parent. In some cases, it is inherited from a parent with a form of the disorder. It can also be inherited from a parent with a balanced translocation, which becomes unbalanced in the offspring that results in distal 18q deletion. This condition is considered to be autosomal dominant (Genetics Home Reference, 2018d).

Proximal 18q deletion syndrome occurs when the lost segment is near the center of the chromosome. In particular, this involves the region between 18q11.2 and 18q21.2. Very few individuals are on record to have this type and so it is difficult to determine the characteristic features of this condition. Most of them experience seizures and show muscle hypotonia, which results in delayed development of motor skills. They are also delayed in speaking with mild to severe intellectual disability. It is also common for them to have behavioral problems like hyperactivity, aggression, and features of ASD that affect their communication and social interaction (Genetics Home Reference, 2018e).

This type of 18q deletion syndrome is autosomal dominant. In most cases, it arises *de novo* during meiosis or in early fetal development so that it is not inherited (Genetics Home Reference, 2018e).

2.1.2.1.1.6. 22q11.2 deletion syndrome

The **chromosome 22q11.2 deletion syndrome** is so named because it is due to the removal of band q11.2 of chromosome 22, which is located near the middle of the chromosome (see Figure 2.42). It is also called **velocardiofacial syndrome (VCFS), DiGeorge syndrome**, and **Sphrintzen syndrome**. It is the most common syndrome due to a chromosomal deletion with approximately 1 affected every 4,000-6, 395 newborns. However, due to the variable degree of severity of the symptoms, it is suspected that some individuals with mild forms of the disorder remain undiagnosed or it may be mistaken for other disorders with similar features. This wide range of severity of symptoms may also influence the life span of affected individuals. Those with a congenital heart defect as well as severe problems with the immune system such as the complete absence of thymus gland and T cells have a shortened life span (National Institutes of Health, 2017c; Genetics Home Reference, 2019b).

The most common facial features observed among affected individuals include an elongated face, almond-shaped eyes, wide nose, small and low-set ears (see Figure 2.41). They may also have submucosal cleft palate (i.e., the muscles of the roof of the mouth are not completely closed although the tissue covering them does), which causes air to come out of the nose during speech so that its nasal sounding. Moreover, they may suffer from hearing impairment, pharynx abnormality, low levels of calcium in the blood, and hypotonia. As previously mentioned, many of them also have heart defects (e.g., abnormalities in the aortic arch, pulmonary valve, atrial septum, ventricular septum, tetralogy of Fallot, truncus arteriosus) and problems with their immune system leading to recurrent infections. It is also common for these individuals to manifest developmental delay, intellectual disability, and learning difficulties. Some have a psychiatric illness, attention deficit disorder, anxiety, repetitive behaviors, and difficulty with social interactions (National Institutes of Health, 2017c; Genetics Home Reference, 2019).

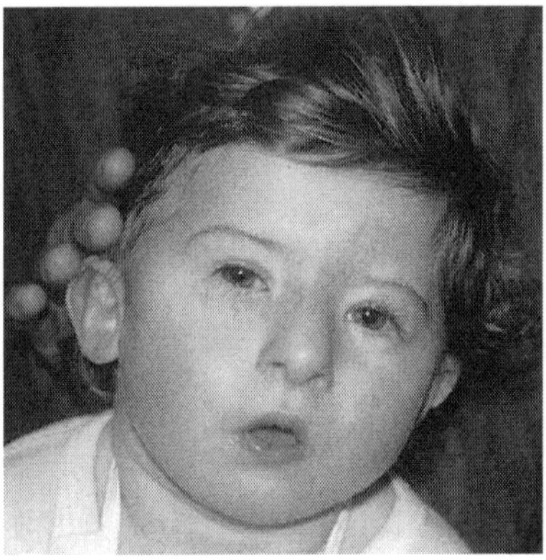

Figure 2.41. A patient with 22q11.2 deletion syndrome showing the characteristic facial features, which include narrow upslanted palpebral fissures, bulbous nose with prominent bridge, narrow mouth, and small, low-set ears. (Source:

Grech, V.; Digilio, M.C.; Capolino, R.; Dallapiccola, B. Images in Paediatric Cardiology 7(2), 23-34, 2005, Creative Commons License).

The deleted segment usually contains about 30-40 genes but in some instances, the deleted region is smaller. One of the genes is *TBX1*, the deletion of which is suspected to be responsible for many of the characteristic signs and symptoms of the syndrome (e.g., the facial features, cleft palate, heart defects, hearing loss) as well as the associated behavioral problems (see Figure 2.42). The deletion of another gene, *COMT*, may also lead to an increased risk of behavioral problems and mental disorders (National Institutes of Health, 2017c; Genetics Home Reference, 2019b).

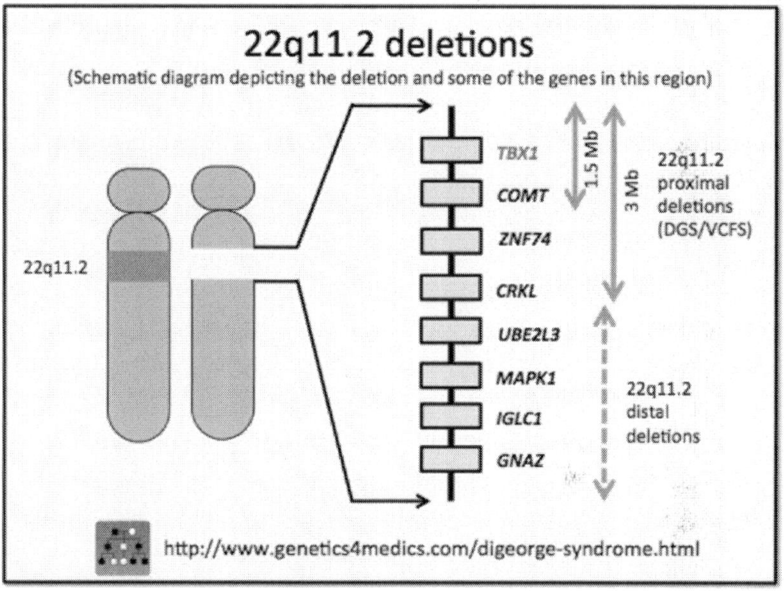

Figure 2.42. Band q11.2 is located in the middle portion of chromosome 22, near the centromere. It contains a number of genes such as *TBX1* and *COMT*, which when lost are associated with the facial features, as well as behavioral and mental abnormalities exhibited by those carrying a deletion in this segment. (Source: genetics4medics, Creative Commons License).

Most cases of VCFS are not inherited but a product of a random error during meiosis or during early fetal development. In approximately 10% of the cases, the deletion is inherited from a parent bearing the deletion. It is inherited in an autosomal dominant manner so that having the deletion in a single copy of chromosome 22 is sufficient to cause the associated phenotypes. This means that a parent having this deletion has a 50%

probability of having a child with the same deletion (National Institutes of Health, 2017c; Genetics Home Reference, 2019b).

2.1.2.1.1.7. Prader-Willi syndrome

Prader-Willi syndrome is an interesting disorder that is due to the loss of function of genes in a certain region of chromosome 15. The loss of function results from both gene inactivation and the deletion of a small segment of the chromosome. Although there are two copies of every chromosome in the human genome, some of the genes are turned on only in one copy, while those in the other copy are inactivated. The gene activation is parent-specific. This phenomenon is called **genomic imprinting**. About 70% of cases of Prader-Willi syndrome occur when the 15q11.2-q13 segment of the paternal copy of chromosome 15, which contains the genes that are supposed to be expressed, is deleted (see Figure 2.44). On the other hand, the maternal copies are present, but these are silenced by epigenetic mechanisms. In 25% of the cases, the affected persons have two copies of chromosome 15 that are both maternally inherited. This phenomenon is called **maternal uniparental disomy**. The rest of the cases may be caused by translocation, mutation, or imprinting defect that abnormally turns off the paternal genes of chromosome 15. An estimated 1 out of 10,000-30,000 people have this syndrome (Genetics Home Reference, 2014c; Cassidy *et al.*, 2011).

Infants with this syndrome exhibit severe muscle hypotonia which might contribute to their feeding difficulties and poor growth. When they reach childhood, they develop an insatiable appetite, which leads to hyperphagia (i.e., chronic overeating) and obesity (see Figure 2.43A). These obese patients are likely to develop type 2 diabetes, heart problems, and respiratory difficulties. They may also have sleep abnormalities. Respiratory illness is one of the most frequent causes of death in children, along with other febrile illnesses, while obesity-related cardiovascular and gastric problems or sleep apnea are the most common causes of death in adults. Their distinctive facial features include a narrow forehead (see Figure 2.43B), almond-shaped eyes, strabismus, and triangular mouth. Most of them have scoliosis and growth hormone insufficiency so that they are also short with small hands and feet (see Figure 2.43C, D). Some have unusually fair skin and light-colored hair. Affected individuals also have hypogonadism (i.e., underdeveloped genitals) so that they have delayed or incomplete puberty and most of them are infertile. They have mild to moderate intellectual impairment and learning disabilities. It is also common for them to have behavioral problems like temper tantrums, stubbornness, compulsive behavior (e.g., picking on skin which can result to wounds, scarring, and infection), and manipulative behavior (Genetics Home Reference, 2014c; National Health Service, 2018; Cassidy *et al.*, 2011).

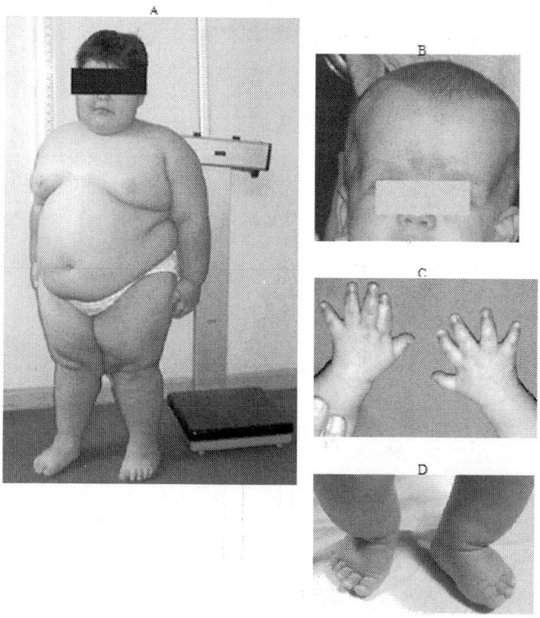

Figure 2.43. (A) An individual with Prader-Willi syndrome may become obese due to excessive eating, and also exhibits distinct facial features such as (B) narrow forehead, (C, D) small hands and feet. (Source: A. Cortés, F.M.; Alliende, M.A.R.; Barrios, A.R.; Curotto, B.L.; Lorena, S.M.V.; Barraza, X.O.; Troncoso, L.A.; Mellado, C.S.; Pardo, R.V. Chilean Medical Journal 133, 33-41, 2005, Creative Commons License. B,C,D. National Human Genome Research Institute, Public Domain).

The deleted 15q11.2-q13 region in Prader-Willi syndrome spans about 6 megabases (Mb), in which 2.5 Mb comprise of genes (see Figure 2.44). This segment can be divided into four regions that are delineated by three common breakpoints. Between the two proximal breakpoints (BP1 and BP2) is a proximal nonimprinted region with four genes that are expressed in both paternal and maternal copies. These are *NIPA1, NIPA2, CYF1P1*, and *TUBGCP5*. Next is a region containing paternally expressed genes which include *MKRN3, MAGEL2, NDN*, and the bicistronic *SNURF-SNRPN*. Included too, are six genes for small nucleolar RNAs (snoRNAs), which regulate the activity of other types of RNA, and a number of antisense transcripts (e.g., IPW, UBE3A). *SNURF-SNRPN* codes for two different proteins. Exons 1-3 produce a polypeptide of unknown function while exons 4-10 encode the SmN protein, which is involved in mRNA splicing. SNURF-SNRPN also regulates the expression of the snoRNA genes. A specific group of snoRNA genes, called the *SNORD116* cluster, is suggested to be critical in causing the signs and symptoms of Prader-Willi syndrome. However, their specific roles in causing such are still unexplained (National Health Service, 2018; Cassidy *et al.*, 2011).

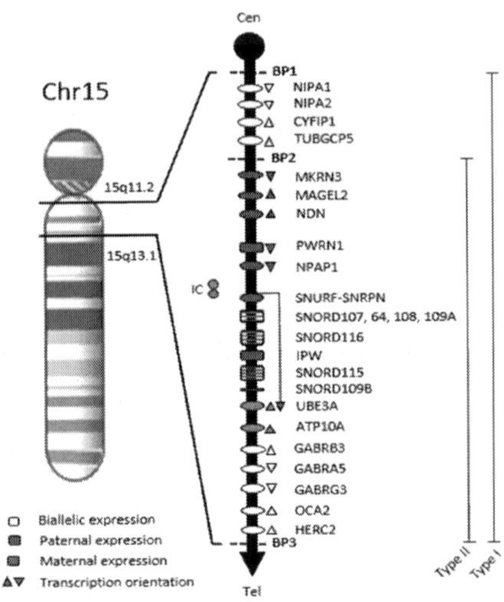

Figure 2.44. Prader-Willi syndrome is due to the deletion of a portion of the long arm of chromosome 15, which may span 15q11.2 (either from breakpoint/ BP1 or 2) to 15q13.1(BP3). The deleted segment may be divided into four segments: (1) a nonimprinted region containing four genes, with both paternal and maternal alleles expressed (biallelic expression), (2) a region with only the paternal copies of the genes expressed, which include protein-coding genes and short nucleolar RNA (snoRNA) genes, (3) a region with maternally-expressed genes, and (4) a distal nonimprinted region with genes that are biallelically expressed. (Source: Costa, R.A.; Ferreira, I.R.; Cintra, H.A.; Gomes, L.H.F.; Guida, L.d.C. Frontiers in Endocrinology 10, 864, 2019, Creative Commons License).

The third region contains maternally expressed genes like *UBE3A* and *ATP10A* (see Figure 2.44). The fourth region is a distal nonimprinted region containing a cluster of three GABA receptor genes, (i.e., *GABRB3*, *GABRA5, GABRG3*), the oculocutaneous albinism type 2 gene (*OCA2*), *HERC2*, and the common distal breakpoint (BP3). These genes all show biallelic expression. The loss of *OCA2* is related to the unusually fair skin and light-colored hair observed among the affected persons. This gene encodes a protein that helps determine the coloring of the skin, eyes, and hair (Genetics Home Reference, 2014c). It is also suspected that the deletion of certain genes in chromosome 15 negatively affects the hypothalamus, which might explain the poor growth and excessive appetite of those with the syndrome (National Health Service, 2018; Cassidy *et al.*, 2011).

Prader-Willi syndrome caused by the deletion of a segment of paternal chromosome 15 or by maternal uniparental disomy is not inherited. In rare cases, it can be inherited if, for example, it is caused by a heritable mutation that abnormally inactivates the paternal genes in chromosome 15 (Genetics Home Reference, 2014c).

2.1.2.2. Duplication of chromosome region

A certain region of the chromosome may be duplicated, a condition is known as **partial trisomy** since there is an extra copy of a certain segment. This results in three copies of a chromosomal region. This can happen at the terminus (**terminal duplication**) or within the chromosome (**interstitial duplication**) (National Institutes of Health, 2017b) (see Figure 2.45). If the duplicated segments are adjacent to the original, it is called **tandem duplication**. On the other hand, if the duplicate is separated from the original by unduplicated regions, it is considered displaced (Clancy and Shaw, 2008).

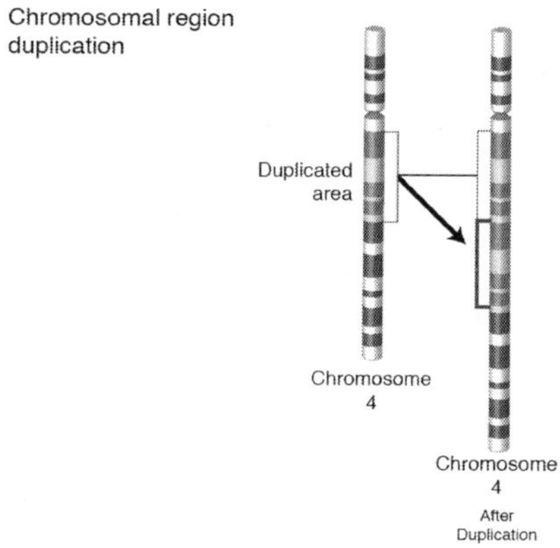

Figure 2.45. If the duplicated region is located within the chromosome, then it is classified as interstitial duplication. Moreover, if it is positioned next to the original segment, then it is considered a tandem duplication. (Source: National Human Genome Research Institute, Public Domain).

2.1.2.2.1. Disorders due to duplicated chromosome region

The duplicated region may contain several genes so that there are extra copies of genes. This may result in having an excess amount of proteins, disrupting normal dosage of proteins that is necessary for normal embryonic development. Therefore, this type of alteration in the chromosomal structure often leads to developmental defects (Clancy and Shaw, 2008). There are several disorders due to duplicated segments in certain chromosomes. Some of these are discussed in this section.

2.1.2.2.1.1. Chromosome 1q21.1 duplication syndrome

Chromosome 1q21.1 duplication syndrome is due to a microduplication on the long arm of chromosome 1, specifically at band q21.1. The length of the duplicated segment varies, with the most common spanning 1.35 Mb. This affects approximately 3 out of 10,000 individuals. However, it is possible that many people with this condition are never diagnosed because its features can be attributed to several causes while some do not show any abnormalities (Genetics Home Reference, 2014b). There are no distinct physical abnormalities that characterize this syndrome. Some affected individuals have physical features such as wide-set eyes, low-set ears, large head, or tall stature and less commonly, clubfeet or hip dysplasia (i.e., misalignment of the hip joint) (see Figure 2.46A). Some are also born with heart defects such as tetralogy of Fallot (i.e., a composite of four defects namely, narrowing of the pulmonary tract, displacement of the aorta, ventricular septal defect, and thickening of the wall of the right ventricle), and in some rare cases, hypospadias or epilepsy. Those with the condition are more likely to develop cysts, varicose veins or carpal tunnel syndrome (i.e., numbness, tingling, and weakness in the hands and fingers). Most of those who are affected exhibit developmental delay and mild to moderate intellectual disability (see Figure 2.46B C,D). They may also show features of ASD. Children with this syndrome have an increased risk of having ADHD and other behavioral problems. Some affected adults may also have psychiatric disorders such as schizophrenia or mood disorders such as anxiety or depression (Genetics Home Reference, 2014b).

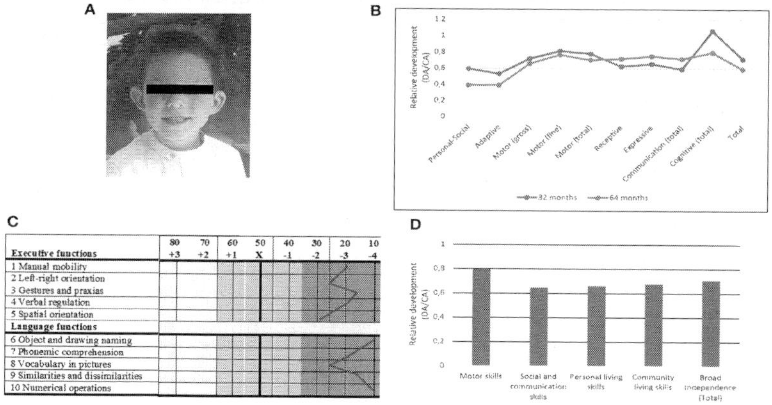

Figure 2.46. (A) A boy with chromosome 1q.21.1 duplication syndrome shows some facial dysmorphisms such as mild hypertelorism, small nose, and protruding ears. (B) Developmental profiles of the patient were determined according to Battelle Developmental Inventories. At 2 years and 8 months (blue), he showed mild developmental delay, mostly affecting his language, social and adaptive abilities. At 5 years and 4 months (orange), his cognitive skills seemed more impaired. (C) The Luria-Nebraska Neuropsychological Battery for Children (LNNB-C) suggests that at 5 years and 6 months, the boy has significant impairment of language and executive functions. (D) The Inventory for Client and Agency Planning (ICAP), which was administered in the same period, indicates an almost 2-year delay in the child's global development, especially his social and communication skills. (Source: Benítez-Burraco, A., Barcos-Martínez, M., Espejo-Portero, I., Fernández-Urquiza, M., Torres-Ruiz, R., Rodríguez-Perales, S. & Jiménez-Romero, M.S. Frontiers in Pediatrics, 6, 163, 2018. Creative Commons License).

Within the duplicated segment, the genes *BCL9, GJA8, PDZK1*,and *PRKAB2* are thought to be involved in the neuropsychiatric symptoms observed in the disorder. *BCL9* has a significant role in embryonic body axis development and it is highly expressed in the brain, particularly in the cerebellum. It is one of the candidate genes for schizophrenia. In one study involving a person with 1q21.1 duplication syndrome, the duplicated genes *CDH1L* and *ROBO1* showed higher than normal expression. *CDH1L* encodes a DNA helicase protein involved in DNA repair, and its overexpression has been linked to several types of cancer. It is also a candidate gene for ASD and ADHD. It is highly expressed in the cerebellum, a part of the brain that is involved in speech and language processing, as well as in social interaction and behavior. *ROBO1* codes for an axon guidance receptor involved in interneuron migration and axon tract development in the forebrain. Specifically, it performs a critical function in the development of thalamocortical axons that are involved in various cognitive functions, consciousness, and alertness. It also regulates interaural interaction in auditory pathways and is a possible candidate gene for speech development. Therefore, *ROBO1* may be involved in speech problems like dyslexia. On the other hand, the duplicated genes *ACP6* and *TLE3* were downregulated. *ACP6* codes for a histidine acid phosphatase involved in G protein-coupled receptor signaling and balancing the lipid composition in the cell. Its differences in methylation status have been associated with hearing impairment. Also, its pathogenic variants are observed in children with cerebral visual impairment, which usually co-occurs with intellectual disability. *TLE3*, on the other hand, encodes a repressor of gene expression and plays an integral role in brain development (Benítez-Burraco *et al.*, 2018).

This syndrome can occur by chance due to a *de novo* mutation during gametogenesis or early embryonic development. It may be inherited in an autosomal dominant manner from a parent having the duplicated segment (National Institutes of Health, 2013).

2.1.2.2.1.2. Chromosome 7q11.23 duplication syndrome

Chromosome 7q11.23 duplication syndrome is due to a microduplication in the q or long arm of chromosome 7, particularly at band q11.23. This region is called the Williams-Beuren syndrome critical region (WBSCR) because its deletion causes Williams-Beuren syndrome. The signs and symptoms associated with this syndrome may be mild and not recognized in some affected individuals. It is estimated that 1 out of 7,500-20,000 people have the 7q11.23 duplication syndrome (Genetics Home Reference, 2017b).

Facial features observed among people with this condition include a large head with flattened back, broad forehead, straight eyebrows, deep-set eyes with long eyelashes and nose broad at the tip with low insertion of the columella (i.e., the area separating the nostrils attaching lower than usual on the face) resulting to shortened philtrum. Some may have a high-arched palate and ear abnormalities. Delayed development of speech and motor skills are typically observed. Speech problems and abnormalities in walking and standing may persist for life. Affected individuals may also have hypotonia, abnormal movements like synkinesis (i.e., involuntary movements of one side of the body that mirrors the intentional movements of the other side), or seizures. About 50% of the affected individuals have enlargement of the aorta that worsens over time. This can be life-threatening if the wall of the aorta separates into layers (see Figure 2.47) or ruptures. Approximately one-third of those with the syndrome have an intellectual disability or borderline intellectual disability. Most of them have below-average to average intelligence. There are behavioral problems associated with the syndrome such as anxiety disorders (e.g., social phobias, selective mutism), ADHD, physical disorder, oppositional disorder (i.e., excessively defiant behavior), and autistic behaviors (Genetics Home Reference, 2017b).

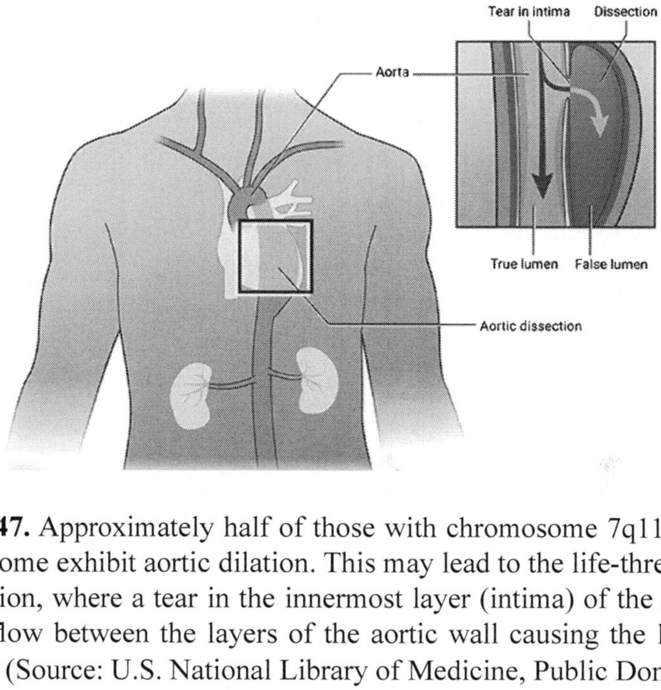

Figure 2.47. Approximately half of those with chromosome 7q11.23 duplication syndrome exhibit aortic dilation. This may lead to the life-threatening aortic dissection, where a tear in the innermost layer (intima) of the aorta causes blood to flow between the layers of the aortic wall causing the layers to get separated. (Source: U.S. National Library of Medicine, Public Domain).

The duplicated region is 1.5-1.8 Mb in length, which encompasses 26-28 genes. Extra copies of several genes such as *ELN* and *GTF21* possibly contribute to the characteristic features of the syndrome. *ELN* gene codes for tropoelastin, which attaches to one another to form the elastin. Elastin is the major component of elastic fibers that provide strength and flexibility to connective tissues. Elastic fibers are also present in the extracellular matrix, which gives structural support to organs and tissues (Genetics Home Reference, 2012c). Having an extra copy of the *ELN* gene in each cell is suspected to be associated with an increased risk of aortic dilatation. On the other hand, *GTF21* encodes two proteins, BAP-135 and TFII-I. BAP-135 is active in B cells. It is activated to transmit chemical signals that allow B cells to fight infection. TFII-I is a transcription factor that is active in the brain and other tissues. It is believed to be involved in coordinating cell growth and division (Genetics Home Reference, 2008). Having an extra copy of *GTF21* may be associated with some of the behavioral issues observed in the 7q11.23 duplication syndrome (Genetics Home Reference, 2017b).

Most cases of this syndrome arose due to duplication that occurred at random during meiosis. In approximately 25% of the cases, the duplicated segment was inherited from a parent, who may have subtle signs and symptoms of the syndrome. The syndrome itself is an autosomal dominant condition (Genetics Home Reference, 2017b).

2.1.2.2.1.3. Chromosome 12q duplication

Chromosome 12q duplication syndrome is a result of duplication on the q or long arm of chromosome 12. The severity of the condition depends on the size and location of the duplication and the genes that are involved. Affected individuals often show growth and developmental delay, intellectual disability, behavioral problems, and certain facial features such as ptosis. The condition is mostly a result of a *de novo* duplication and so it is not inherited. In some cases, a person bearing the duplication may pass it on to his child (National Institutes of Health, 2015c).

In a case of two brothers with this syndrome, they showed inverse Marcos Gunn phenomenon, which is a rare form of congenital facial synkinesis. This is manifested by temporary ptosis, either unilateral or bilateral, due to jaw movement, such as when a child sucks on his bottle. This may be due to dysinnervation of facial muscles during the development of the nervous system. A previous case of a patient with classic Marcus Gunn had duplicated 12q24.1-q24.2 region. Karyotype and array CGH performed on samples from the two brothers indicated the presence of the same duplicated region, which contains an estimated 44 protein-coding genes (Myers *et al.*, 2016).

A rare case of this syndrome involves the microduplication of the 12q13.13 region, which spans 802 kb. Affected individuals commonly exhibit dysmorphic facial features, abnormal nails, deformed digits or limbs, and intellectual disability. Those with a microdeletion in this region also show similar phenotypes although they have a more severe deformity of their digits or limbs. A 14-year, 8-month old male with this condition had dysmorphic facial features such as a long and narrow face, hypertelorism, broad nose bridge and tip, short philtrum (see Figure 2.48a), receding anterior hairline (see Figure 2.48b), and low-set ears (see Figure 2.48d). He also had bifid or cleft uvula and high arched palate (see Figure 2.48c). He also had hyperconvexed nails, stubby thumbs, fallen or collapsed foot arches (see Figure 2.48d), prominent elbows, and slight scoliosis. He had a history of developmental delay and was diagnosed with ADHD. At 14 years old, he was in special education classes working at a second-grade level (Hu *et al.*, 2017).

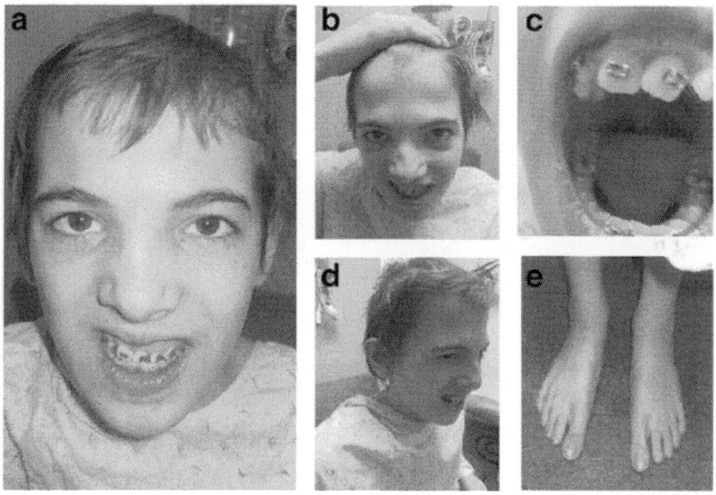

Figure 2.48. A nearly 15-year old boy with duplicated 12q13.13 region showing certain physical features such as (a) long and narrow face, high arched heavy eyebrows with medial flare, hypertelorism, broad nasal bridge and tip, and short philtrum; (b) receding anterior hairline with widow's peak; (c) bifid uvula and abnormal teeth; (d) low-set ear; and (e) abnormal toenails. (Source: Hu, J., Ou, Z., Infante, E., Kochmar, S.J., Madan-Khetarpal, S., Hoffner, L., Parsazad, S., Surti, U. Molecular Cytogenetics 10, 24, 2017, Public Domain).

There are at least 16 genes in the 12q13.13 region. One gene in this region, the *SP1* gene, is most likely dosage sensitive. It codes for a transcription factor that binds the GC-rich motifs of many promoters and thus, it is involved in a variety of cellular processes such as cell growth, apoptosis, differentiation, and immune responses, DNA damage response, and chromatin remodeling. It also influences the activity of another gene in the region, *SP7*. This gene encodes a bone-specific transcription factor, which regulates osteogenesis during embryonic development. Patients with either a deleted or duplicated 12q13.13 region and those with homozygous *SP7* mutation have certain clinical features, namely short stature, high-arched palate, mild bone deformities, mild scoliosis, and delayed motor development. It is still unclear whether concurrent dosage changes in the *SP1* and *SP7* genes cause these abnormal features (Hu *et al.*, 2017).

Some of the genes in the duplicated region are expressed in the central nervous system (CNS), such as *NPFF, SP1, MAP3K12*, and *PCBP2*. Just like *SP1, PCBP2* is also predicted to be dosage sensitive. It codes for the heterogeneous nuclear riboprotein E2. The changes in the dosage of these CNS genes may contribute to the developmental delay and intellectual disability of the affected individuals (Hu *et al.*, 2017).

2.1.2.2.1.4. Chromosome 17p11.2 duplication

The duplication of about 1.5 Mb of DNA on segment p11.2 of chromosome 17 results in a certain type of *Charcot-Marie-Tooth (CMT) disease*. This is one of the most common inherited neurological disorders that affect an estimated 1 person for every 5,000-10,000 people. Specifically, it is a type of peripheral neuropathy affecting the transmission of impulses between the CNS and the rest of the body. It has many forms namely, CMT1, CMT, 2, CMT3, CMT4, and CMTX. In turn, CMT1 has three main types, i.e., CMT1A, hereditary neuropathy with predisposition to pressure palsy (HNPP), and CMT1B. It also has less common types, namely CMT1C, CMT1D, and CMT1E. It is CMT1A that is caused by the duplication of chromosome 17p11.2 and it accounts for 70% of CMT cases. Most CMT1 patients have a normal lifespan and exhibit phenotypes of varying severity (National Institutes of Health, 2007; Orpha.net, 2016; National Institutes of Health, 2018c).

The onset of CMT1 signs and symptoms ranges from infancy to age 40 or even later. In most cases, pes cavus (i.e., high-arched foot) or pes planus with hammertoes is present since childhood (see Figure 2.49). For those with CMT1A, they typically start experiencing weakness and atrophy of the lower leg muscles in their adolescence. They may also develop scoliosis at this stage. Inability to sense position, vibration, pain, and temperature commonly occurs in the feet and later in the hands. CMT, in general, is slowly progressive, and among those with CMT1, less than 5% become wheelchair dependent (National Institutes of Health, 2007; Orpha.net, 2016; National Institutes of Health, 2018c).

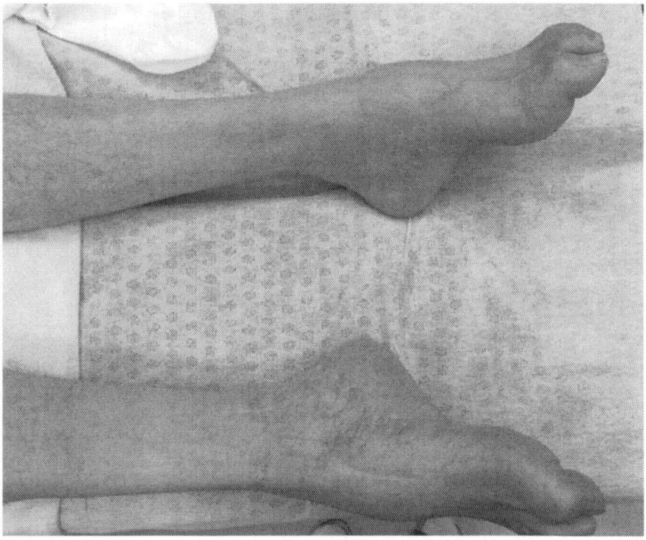

Figure 2.49. One feature observed commonly among those with Charcot-Marie-Tooth type 1 is pes cavus (high-arched foot) with hammertoes. (Source: Itoh, N.; Hoshijima, H.; Takeuchi, R.; Nakamura, C.; Iwase, Y.; Nagasaka, H.; Sato, T.; Yoda, T. Stomatological Disease and Science 2,2, 2018, Creative Commons License).

A study conducted among families affected with CMT1A showed that in about 10% of the cases, the duplicated region arose in the parent due to *de novo* mutation. It is thought to be a result of unequal crossing over between parental chromosome 17 homologues during both paternal and maternal meiosis (Blair *et al.*, 1996). The duplication is then inherited in an autosomal dominant manner (National Institutes of Health, 2018c). The duplicated region contains the *PMP-22* gene, which encodes the peripheral myelin protein-22 (PMP-22). This protein is a vital component of the myelin sheath, and its overexpression causes abnormalities in myelin sheath structure and function. Repeated episodes of segmental demyelination and remyelination in CMT1A lead to the accumulation of Schwann cell (i.e., these are the cells producing myelin sheath) around axons. Therefore, the cross-section of CMT1A nerves usually shows numerous, multilayered onion bulbs composed of Schwann cell processes and axons encircled with abnormally thin or relatively thick myelin sheaths (see Figure 2.50) (Vallat, 2003). This, in turn, affects the function of peripheral nerves so that they are unable to activate target muscles or relay information from the limbs back to the brain. As a result, peripheral motor nerves slowly degenerate, which results in muscle weakness and atrophy in the arms, legs, hands or feet. In some cases, peripheral sensory nerves degenerate leading to reduced ability to feel the heat, cold, pain, and vibration (National Institutes of Health, 2007; Orpha.net, 2016; National Institutes of Health, 2018c).

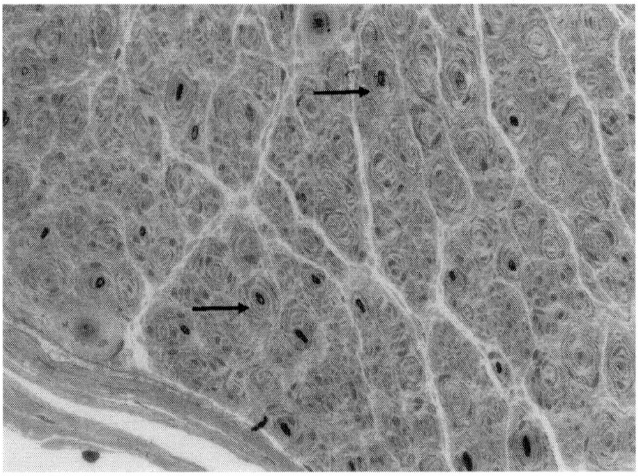

Figure 2.50. Onion bulbs (black arrows), which form concentric layers around nerve cells, are usually observed in Charcot-Marie-Tooth type 1A disease. These are made up of accumulated Schwann cells, axon, and myelin sheath. (Source: Jensflorian, Creative Commons License).

2.1.2.2.1.5. Chromosome Xq duplication

Chromosome Xq duplication syndrome results from the presence of an extra copy of the genetic material located on the q or long arm of the X chromosome. The degree of severity of the condition depends on the size and location of the duplicated region as well as the sex of the affected person. Generally, affected males exhibit more severe signs and symptoms than their female counterparts (National Institutes of Health, 2015d). The most frequently reported Xq duplication involves the Xq28 segment resulting in MECP2 duplication syndrome or trisomyXq28. More than 200 individuals are reported to have this specific syndrome. It is estimated to be responsible for 1-2% of cases of intellectual disability caused by abnormalities in the X chromosome (Genetics Home Reference, 2017c; Sanlaville *et al.*, 2009).

Among affected males, they show common features such as microcephaly, craniofacial abnormalities, hypotonia, feeding difficulties during infancy, growth and developmental delay, muscle stiffness, and short height. They exhibit distinctive facial features namely, brachycephaly, midface hypoplasia, deep-set eyes, large ears with abnormalities, pointed nose with a flat bridge, and prominent chin (see Figure 2.51) (National Organization for Rare Disorders, 2017). They also have brain abnormalities, abnormally developed reproductive organs, as well as dysfunctional endocrine system and immune system, the latter causing frequent infections. Recurrent respiratory infections are a major cause of death among those with MECP2 duplication syndrome, with only 50% surviving past the age of 25 (Genetics Home Reference, 2017c). Those with Xq duplication syndrome also show intellectual disability, absence or delay in speech and language development, and they experience seizures. On the other hand, most affected females merely have short height and do not have any symptoms. However, some may be as severely affected just like the males (National Institutes of Health, 2015d).

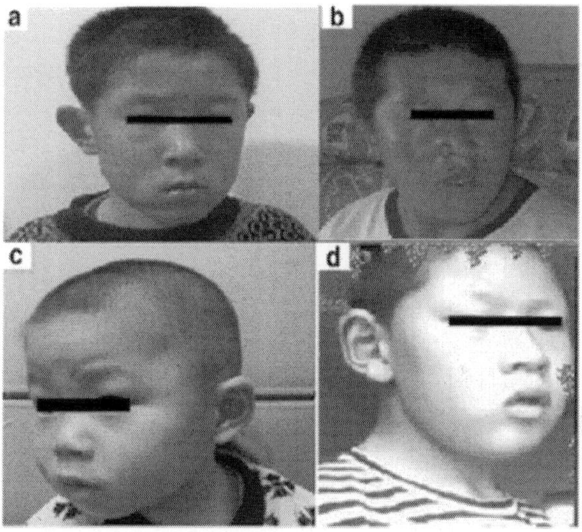

Figure 2.51. Members of the same family (a and b are first cousins, c and d are siblings) with MECP2 duplication syndrome exhibiting typical facial features of the condition such as midface hypoplasia, low-set ears with large lobe and crease, small mouth, and protruding jaw. (Source: Li, X.; Xie, H.; Chen, Q.; Yu, X.; Yi, Z.; Li, E.; Zhang, T.; Wang, J.; Zhong, J.; Chen, X. BMC Medical Genetics 18(1), 131, 2017, Creative Commons License).

The features of the condition may be a result of the duplication or disruption of certain genes in the duplicated region. Although the effects of the overexpression of several genes in the region are not yet elucidated, there are three genes that are well-characterized. One of these is the *MECP2* gene, which is suspected to be responsible for many of the signs and symptoms in affected people. It codes for the methyl-CpG binding protein 2 that binds specifically to methylated DNA. The gene is expressed in many tissues, but it its highest expression is in the brain. Therefore, the protein product must be essential for normal brain function. Specifically, it is believed that this protein regulates the expression of other genes in the brain. An extra copy of the MECP2 gene leads to an increased quantity and function of its protein, which may lead to abnormal neuronal function. Mild overexpression of this gene in mice resulted in neurodevelopmental abnormalities. Duplicated MECP2 among males is associated with moderate to severe intellectual disability, developmental delay, hypotonia during infancy, feeding difficulties, and frequent respiratory infections. In females, an extra copy of this gene is associated with behavioral and psychiatric symptoms like depression, anxiety, and features of ASD (Sanlaville *et al.*, 2009; Genetics Home Reference, 2017c). An extra copy of another gene, *PLP1*,is associated with brain abnormalities that are often observed in a group of disorders called leukodystrophies. The third gene is *OPHN1* and it is reported to be duplicated in people with severe intellectual disability, microcephaly, and certain facial features (National Institutes of Health, 2015d).

Chromosome Xq duplication is inherited in an X-linked manner. In most cases, it is inherited from a mother who does not show any signs or symptoms of the duplication. As previously discussed, there is the random inactivation of an X chromosome in the female cells (see Figure 2.11). However, when a female has an X chromosome with a duplicated copy of its q arm, the abnormal chromosome is preferentially inactivated in many or all cells. This is called *skewed X-inactivation*. This is why some women with the syndrome do not develop the associated abnormal features. As described previously however, some females with an extra copy of the *MECP2* gene exhibit behavioral and psychiatric symptoms despite skewed X-inactivation. Although there is no clear explanation for this yet, it is hypothesized that in these females, some cells in their brain may have a different pattern of X-inactivation so that the X chromosome with the duplicated *MECP2* gene is active (Genetics Home Reference, 2017c).

Some cases of the chromosome Xq duplication syndrome occur because of *de novo* duplication during meiosis of germ cells or in early fetal development. This is a random event and thus, not inherited (National Institutes of Health, 2015d).

2.1.2.3. Inversion of chromosome region

Inversions occur when a chromosome segment breaks off and then is re-inserted into the chromosome, but in a reversed orientation. These are often generated by non-allelic homologous recombination (NAHR) between inverted repeats. These can also result from double-strand repair mechanisms such as non-homologous end joining or from replication-based mechanisms like fork stalling and template stalling (Puig *et al.*, 2015). When the inverted segment includes the centromere, it is called **pericentric inversion**; if the centromere is not involved, it is called **paracentric inversion** (see Figure 2.52). Between the two inversion types, pericentric inversion is more common (Mennuti, 2019). In the case of heterozygous inversion, one of the two chromosomes rotate 180° to form an inversion loop while the other molds around it so that homologous segments are correctly paired together during synapsis (see Figure 2.53). Still, there may still be regions of incorrect pairing.

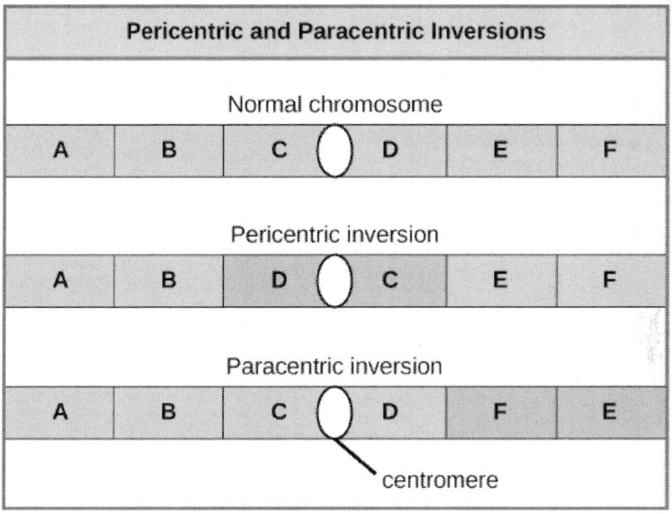

Figure 2.52. Pericentric inversion occurs if the inverted chromosome segment includes the centromere, but if it does not, it is called paracentric inversion. (Source: Boundless.com, Creative Commons License).

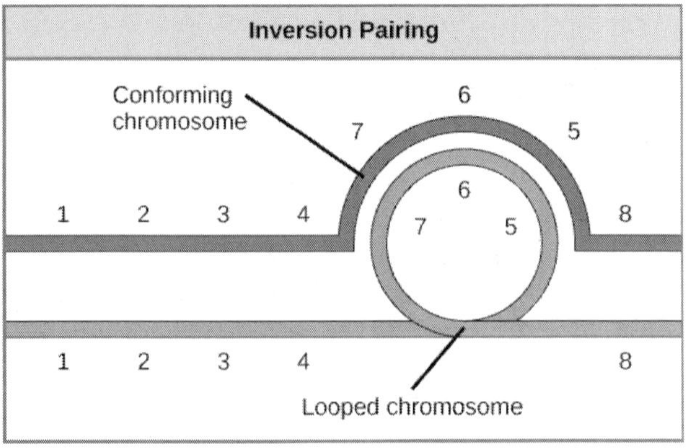

Figure 2.53. When one of the two homologues has an inverted segment, an inverted loop is formed by one chromosome in order to ensure that homologous regions interact during synapsis. (Source: Boundless.com, Creative Commons License).

As described previously, heterozygous inversions will affect the way that a chromosome will pair with its homologue during meiosis. Depending on the size of the inverted segment, it is difficult for the chromosome with inversion to align and recombine with the normal chromosome during meiosis. If there is an even number of crossovers in the inverted segment, there is a 50% chance of producing gametes with either the balanced inversion or the normal homolog with no inversion. On the other hand, if an odd number of crossovers occur in the paracentric inversion loop, gametes will either contain an acentric (i.e., without centromere) chromosome fragment that will be lost in subsequent cell divisions or a large dicentric chromosome with the deleted segment. When either type of gamete participates in fertilization, the resulting embryo fails to develop, or the fetus eventually dies *in utero*. In the case of pericentric inversion, an odd number of crossovers produce gametes with chromosomes having duplicated and deleted regions. This combination of trisomic and monosomic regions may not allow the resulting embryo from these gametes to develop unless the monosomic segment is small. This is possible if the inverted segment is large (Mennuti, 2019).

2.1.2.3.1. Disorders due to inverted chromosome region

Unlike deletions and duplications, inversions may not change the overall dosage of genetic material, a condition called **balanced inversion**. Therefore, affected individuals do not show any phenotype of clinical significance. This is the case for pericentric inversions of heterochromatic regions in chromosomes 1,2,3,5,9,10, and 16. However, if the breakpoint is within a gene, then it may lead to a lethal gene mutation, which may directly affect gene function or regulation. An inversion may also predispose the offspring of carriers to have secondary chromosomal aberrations. This is the case for disorders like Williams-Beuren syndrome, Angelman syndrome, or Sotos syndrome, which are caused by the deletion of segments of certain chromosomes. In addition, the transmitting parents exhibit inversions in the same chromosome region (Griffiths *et al.*, 2000; Puig *et al.*, 2015). Balanced inversion may be inherited, but it can decrease the reproductive potential of the carrier (Mennuti, 2019).

2.1.2.3.1.1. Chromosome 9 inversion

Chromosome 9 inversion is one of the most common balanced chromosomal variants. The pericentric inversion in this chromosome has variable frequencies in different human populations ranging from 0.26% in Asians to 3.57% in individuals of African origin (National Institutes of Health, 2017d; Puig *et al.*, 2015).

It is still unclear if this inversion has any clinical significance. In some cases, it has been associated with congenital anomalies, growth delay, infertility, recurrent pregnancy loss, and cancer (National Institutes of Health, 2017d). In one study conducted involving newborns with congenital anomalies, those with an inverted p11q13 segment of chromosome 9 exhibited intrauterine growth restriction, polydactyly, clubfoot, microtia (i.e., underdeveloped external ear), deafness, and an asymmetric face. They also have various organ abnormalities namely, giant Meckel's diverticulum, duodenal diaphragm, small bowel malrotation, pulmonary stenosis, cardiomyopathy, and arrhythmia. Cytogenetic analysis of the parents of the affected babies showed that the inversion was not inherited but arose *de novo*. However, congenital anomalies among those with inversions were not significantly different from those of the normal population. This indicates that the cause of the observed anomalies can not be attributed to the inversion (Jeong *et al.*, 2010).

2.1.2.3.1.2. Chromosome 17q21.31 inversion

A well-studied case of inversion is that of **chromosome 17q21.31**, which is a common inversion polymorphism in populations with European ancestry. An approximately 970-kb segment is inverted, which contains a number of genes. There are two different haplotypes of this inversion namely, H1 and H2. H1 haplotype is associated with neurodegenerative diseases like progressive supranuclear palsy, corticobasal degeneration, Alzheimer's disease, and Parkinson's disease. The H2 haplotype, on the other hand, is associated with higher mean recombination rates in the female carriers. These women tend to have more children. It is also associated with recurrent deletion events associated with the 17q21.31 microdeletion syndrome, characterized by developmental delay and learning disability. It is relatively common in European populations (approximately 20%) and almost absent in African and Asian populations (Puig *et al.*, 2015; de Jong *et al.*, 2012).

One of the genes within the inverted segment includes the *CRHR1* gene for corticotrophin releasing hormone receptor 1. Its expression is significantly decreased in the H1/H1 haplotype, particularly in the cerebellum. This gene has a role in mediating stress response and has been implicated in stress-related psychiatric disorders, such as depression and alcoholism. On the other hand, this gene shows increased expression in the H2 haplotype and its altered expression may contribute to the H2 phenotypes. Another gene in the inverted segment is *MAPT*, which encodes microtubule-associated protein tau. Increased expression of this gene is associated with the number of alleles in the frontal cortex and cerebellum. The accumulation of hyperphosphorylated tau protein in neuronal cell bodies is associated with neurodegenerative diseases (Puig *et al.*, 2015; de Jong *et al.*, 2012).

2.1.2.3.1.3. Hemophilia A

Hemophilia A or **factor VIII (FVIII) deficiency** is a heritable disorder in which the blood does not clot normally so that affected persons bleed more than normal after an injury, surgery, or dental procedure. In severe cases, bleeding may also occur spontaneously or even when there is no injury. It is inherited in an X-linked recessive pattern since the mutated gene involved is located in the X chromosome. This means that more males are affected since they just need one copy of the mutant allele to have the disorder while females need to have two copies. In rare cases, the mutation causing the disorder occurred *de novo* (National Institutes of Health, 2018d; Puig *et al.*, 2015).

This bleeding disorder is caused by a mutation in the X-linked *F8* gene, which codes for the FVIII protein that is required to initiate blood clot formation. A homologous recombination between an intragenic 9.5 kb repeat in intron 22, referred to as int22h-1, of the gene and one of two other extragenic copies (i.e., int22h-2 and int22h-3) located about 565 kb farther in the X chromosome results to an inversion (see Figure 2.54). This disrupts the *F8* gene and is observed in approximately 45% of hemophiliacs with severe symptoms. In 4.8% of the cases, the disruption is caused by an inversion mediated by repeats in intron 1. Similar to intron 22 inversion, intron 1 inversion occurs via homologous recombination between an intragenic repeat (int1h-1) and an extragenic one (int1h-2) in the promoter region (see Figure 2.55) (Puig *et al.*, 2015; Hwang *et al.*, 2012).

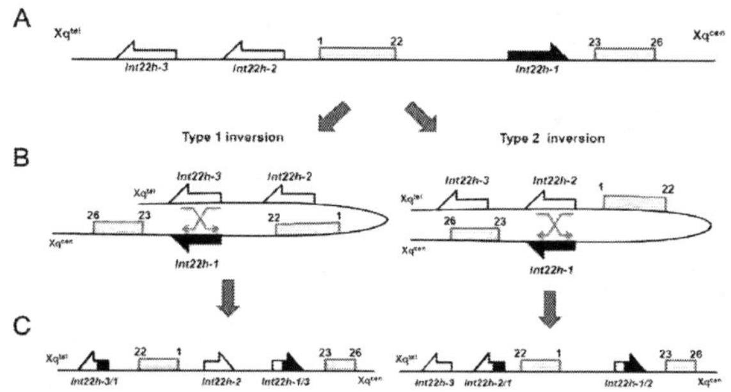

Figure 2. 54. (A) The normal structure of the *F8* gene with 26 exons (light blue boxes). Intron 22 is its largest intron with int22h1 repeat that is homologous with telomeric repeats, int22h-2 and int22h-3. (B) Homologous recombination may occur between int22h-1 and int22h-3 (producing a type 1 inversion) or between int22h-1 and int22h-2 (type 2 inversion). (C) The recombination disrupts *F8* gene so that exons 1 to 22 are displaced toward the telomere and are inverted in orientation (Xqtel: X chromosome q arm telomere. Xqcen: X chromosome q arm centromere). (Source: Hwang, S.H.; Kim, H.-J.; Kim, H.S. Hemophilia, 2012, Creative Commons License).

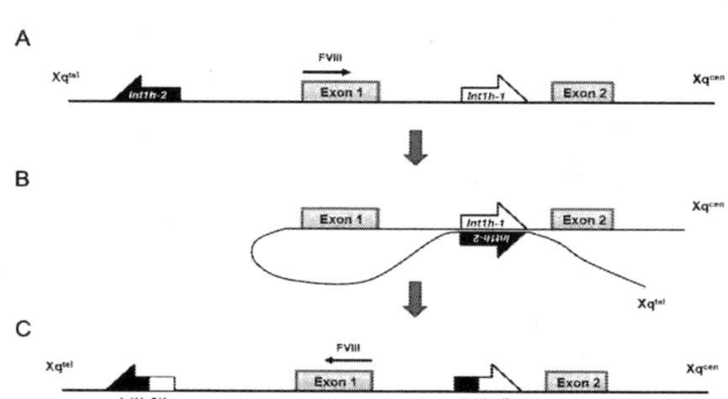

Figure 2.55. (A) A repeat sequence within intron 1 (int1h-1) is homologous with an extragenic repeat (int1h-2) near the promoter region. (B) Homologous recombination may occur between the two repeats. (C) This inverts the direction of transcription of exon 1, opposite the location of the promoter, so that

a complete Factor VIII protein cannot be synthesized (Xqtel: X chromosome q arm telomere. Xqcen: X chromosome q arm centromere). (Source: Hwang, S.H.; Kim, H.-J.; Kim, H.S. Hemophilia, 2012, Creative Commons License).

2.1.2.4. Translocation of chromosome regions

Translocation occurs when a chromosome region breaks off and then fuses with another chromosome, whether a nonhomologous chromosome or to a new site on the same chromosome. Translocation may be **nonreciprocal**, in which a chromosomal segment is transferred to a nonhomologous chromosome. On the other hand, **reciprocal translocation** involves the exchange of segments between nonhomologous chromosomes. If there is no genetic material lost during the exchange, then it is considered to be a balanced translocation (see Figure 2.56). This does not usually cause any abnormal phenotype. Certain chromosome regions are vulnerable to balanced translocation such as 11q23, 17q11, and 22q11. It is an unbalanced translocation when some genetic material has been lost or gained. Another type is **Robertsonian translocation**, in which the q arms of two acrocentric chromosomes are fused at a single centromere producing a novel chromosome (see Figure 2.57). On the other hand, the p arms also fuse but are mostly lost after the first cell divisions. Since these are very short regions with repetitive sequences, their loss does not have any phenotypic effects (National Institutes of Health, 2017b; O'Connor, 2008d; Dolskiy *et al.*, 2018).

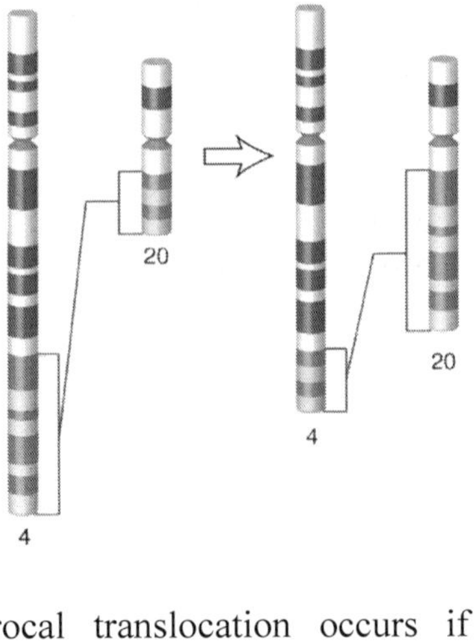

Figure 2.56. A reciprocal translocation occurs if two nonhomologous chromosomes exchange segments (in this case, chromosome 4 and chromosome

20). It is a balanced type of translocation if DNA is neither gained nor lost during the exchange. The rearranged chromosomes are called derivative (der) chromosomes and the particular translocation is designated as "t". After translocation, the chromosomes in this example are referred to as der(4) t(4; 20) for the derivative chromosome with a centromere derived from chromosome 4 and der(20); t(20;4) for the other derivative chromosome with its centromere from chromosome 20. (Source: National Human Genome Research Institute, Public Domain).

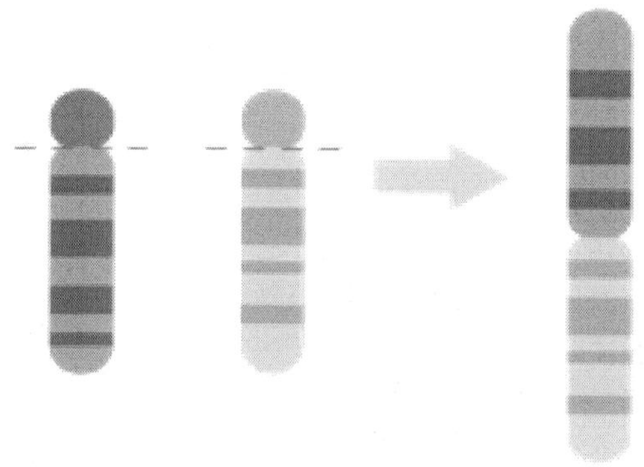

Figure 2.57. Robertsonian translocation happens when the long arms of two acrocentric chromosomes are joined at a single centromere and the two short arms are lost. (Source: yourgenome, Creative Commons License).

Translocations may occur as *de novo* events but there are also cases of recurrent, non-random translocations. The mechanisms that cause these non-random translocations are not yet well understood. There are potential contributory factors such as illegitimate V(D) J recombination in the lymphocyte genome, class switch recombination in B lymphocytes, homologous recombination, non-homologous end joining, and genome fragile sites (Aplan, 2006). Exposure to mutagens that cause double-strand breaks (DSBs) in the DNA, such as ionizing radiation, can also result in translocation (O'Connor, 2008d).

2.1.2.4.1. Disorders due to translocated chromosome region

The previous discussion on trisomy disorders indicated that these can also be caused by translocation, particularly Robertsonian translocation. The most common fusion occurs between the q arms of acrocentric chromosomes 13 and 14. The q arm of chromosome 21 can also fuse with that of chromosome 13, 14, 15, or 22 (Healthline, 2018). While a carrier of a balanced or Robertsonian translocation may not present any clinical phenotypes, his translocated chromosomes may exhibit irregular synapsis and segregation during meiosis. In the case of reciprocal translocation, a quadrivalent is formed by the derivative chromosomes and their original homologues. There are various ways by which this quadrivalent may separate, but only the alternate type will produce gametes with normal, balanced chromosomal content. Adjacent segregation or non-disjunction, on the other hand, produces aneuploid gametes, with missing and extra whole chromosomes or chromosome regions (see Figure 2.58a). In the case of Robertsonian translocation, a trivalent is formed by the hybrid chromosome and the original homologues. Irregular segregation may occur producing aneuploid gametes (see Figure 2.58b) (Tulay, 2017). This may be correlated with fertility problems and recurrent spontaneous abortions observed among carriers. Also, he may pass on to his offspring unbalanced translocation, i.e. chromosome with missing or extra DNA. This may result in *translocation monosomy*, in which case, the affected embryo is not viable. It may also result in *translocation trisomy* leading to disorders such as Down syndrome, Patau syndrome, or trisomy 14 (Healthline, 2018; Ozen *et al.*, 2014; O'Connor, 2008d).

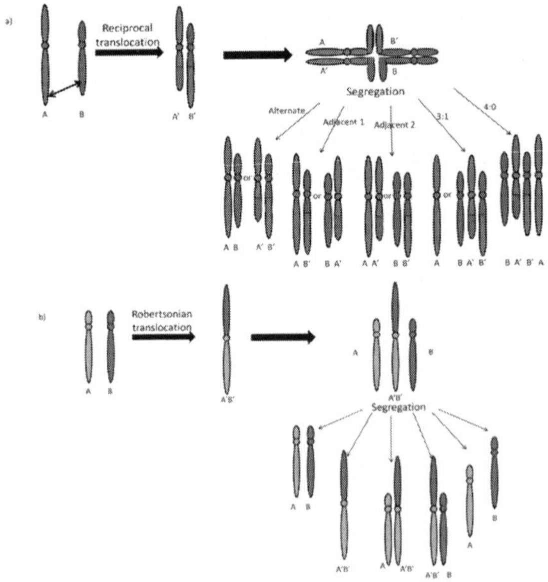

Figure 2.58. Translocation carriers are usually phenotypically normal but may produce abnormal gametes. (a) In reciprocal translocation, the derivative

chromosomes are partially homologous with each other and with their original homologue. This results in the formation of a quadrivalent. This can segregate in an alternate manner producing either normal gametes or gametes with a balanced translocation. If segregation is in an adjacent manner, gametes with unbalanced translocations are formed. If there is non-disjunction, aneuploid gametes are formed. (b) In Robertsonian translocation, synapsis among the hybrid chromosome and the original homologues results to a trivalent. Normal gametes are produced only if two normal chromosomes segregate together. Otherwise, gametes with unbalanced translocations or aneuploid gametes are formed. (Source: Tulay, P. IntechOpen, 2017, Creative Commons License).

Chromosomal translocations are also frequently associated with various cancers, especially leukemias, lymphomas and childhood sarcomas. Some translocations disrupt normal gene expression and regulation producing oncogenes, which may be responsible for malignant tumor formation. There are two possible ways that this can happen, one is by the translocation of a gene near a strong regulatory region. This usually results in a significant overexpression of the oncogene. Another possibility is that the translocation event fuses two genes from the participating chromosomes, and the resulting hybrid protein may induce malignancy. However, the same chromosomal translocations may be observed among healthy individuals indicating that translocation alone is not sufficient to induce malignant transformations, but it has to be complemented by other mutations (Aplan, 2006; O'Connor, 2008d; Gatt and Izraeli, 2019).

2.1.2.4.1.1. Robertsonian translocation 13/14

Robertsonian translocation between chromosomes 13 and 14 accounts for 75% of all cases of this type of chromosomal aberration. The carriers are usually normal and healthy because they have the balanced type (see Figure 2.59), but when their offspring inherit the translocation, pathological phenotypes may be observed. Translocation trisomy 13 or translocation trisomy 14 (see Figure 2.60) may be observed among the offspring. Between the two, an embryo with the former (refer to section 2.1.1.1.2. Trisomy 13 for a detailed discussion of the phenotypes) can possibly develop to term while that with the latter is usually lost in the first trimester. In one instance, a healthy mother with this chromosomal rearrangement has a son with intellectual disability, speech impairment, ataxia (i.e., lack of voluntary coordination of muscle movements), scoliosis, and hydrocephaly. These clinical manifestations may be attributed to uniparental disomy, *de novo* rearrangements, or changes in methylation patterns and gene expression (Scriven *et al.*, 2001; Dolskiy *et al.*, 2018).

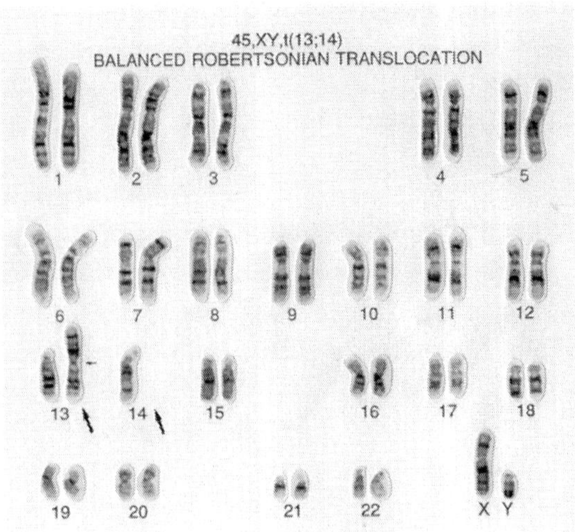

Figure 2.59. Balanced Robertsonian translocation between chromosomes 13 and 14 is the most common type of this chromosomal rearrangement, in which a novel chromosome made up of fused long arms of these two chromosomes is formed. Since there is no net gain or loss of genetic material, carriers of this aberration do not manifest clinically significant phenotypes. (Source: Wessex Reg. Genetics Centre, Creative Commons License).

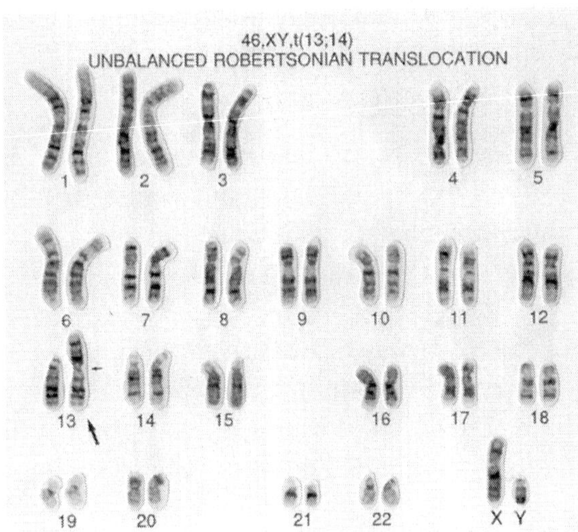

Figure 2.60. Robertsonian translocation of chromosomes 13 and 14 causes irregular pairing and segregation of these chromosomes during gamete formation. As a result, the resulting offspring may inherit the unbalanced type of

Robertsonian translocation. One possible outcome is translocation trisomy 14, in which there are three copies of chromosome 14 long arms. This does not produce a viable embryo. (Source: Wessex Reg. Genetics Centre, Creative Commons License).

2.1.2.4.1.2. Translocation trisomy 21

Trisomy 21 or Down syndrome may also result from translocation, where a portion of chromosome 21 may be attached to chromosome 13, 14, 15, or 22 (see Figure 2.61). In some cases, two chromosomes 21 may be fused with each other. This accounts for 4% of Down syndrome cases. In this case, the chromosome number is still 46, since there is no extra chromosome. However, there are three copies of genetic material from chromosome 21. This causes the phenotypes associated with Down syndrome (refer to Section 2.1.1.1.1. Trisomy 21 for a detailed discussion). Additionally, 33% of the translocation Down syndrome cases have been inherited from a normal parent who is a translocation carrier (Stanford Children's Health, 2020; National Down Syndrome Society, 2020).

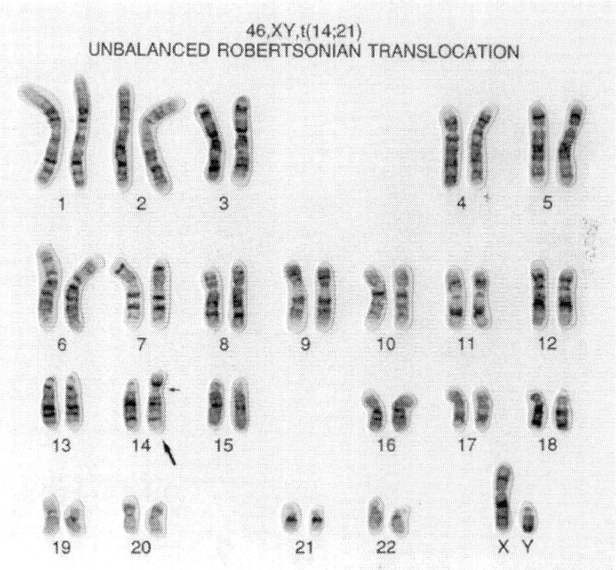

Figure 2.61. Down syndrome may be a result of unbalanced translocation where a portion of chromosome 21 is fused to another chromosome, such as chromosome 14. Since there is already a pair of chromosome 21, the fused segment is an extra copy of genetic material causing the syndrome. (Source: Wessex Reg. Genetics Centre, Creative Commons License).

2.1.2.4.1.3. Chronic myeloid leukemia

Chronic myeloid leukemia (CML) can also be called chronic myelogenous leukemia or chronic granulocytic leukemia. It is a rare cancer type in which the bone marrow produces an abnormal quantity of white blood cells. In more than 90% of CML cases, the **Philadelphia chromosome** is observed. This was named after the city where it was discovered. It is a product of the reciprocal translocation of small segments at the ends of the q arms of chromosome 9 and chromosome 22, (i.e., 9q34 and 22q11, respectively) (see Figure 2.62). The shortened chromosome 22 is called the Philadelphia chromosome. The translocation fuses the coding sequence of the **BCR** (breakpoint cluster region) gene on chromosome 22 with the coding sequence of the *ABL* protooncogene (a gene that codes for tyrosine kinase) on the translocated region from chromosome 9. The resulting fusion gene *BCR-ABL* is an oncogene that codes for a tyrosine kinase with increased enzymatic activity. Consequently, it constitutively activates signaling pathways involved in cell growth and proliferation. This leads to the uncontrolled growth of white blood cells (O'Connor, 2008d; Mayo Clinic, 2019c; Brody, 2016).

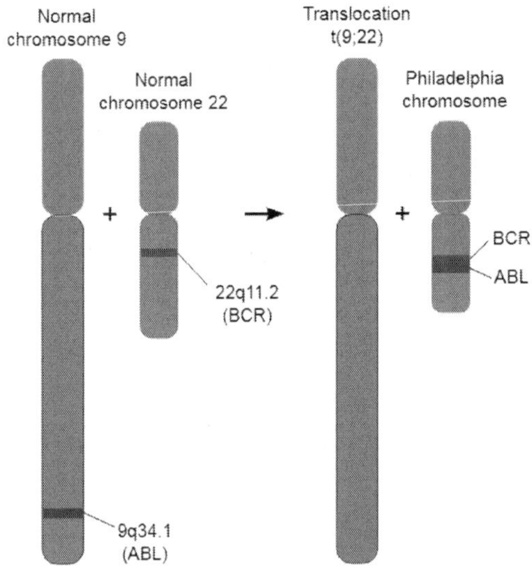

Figure 2.62. A reciprocal translocation may occur between the q34 region of chromosome 9, which contains the ABL gene, and the q11 region of chromosome 22. This results in a shorter chromosome 22 derivative called the Philadelphia chromosome. This chromosome contains the hybrid *BCR-ABL* gene. (Source: Aryn89, Creative Commons License).

The Philadelphia chromosome is also seen in about 5% of children and 20% of adults with acute lymphoblastic leukemia (ALL). It also accounts for approximately 1% of acute myeloid leukemia (AML) cases. The presence of the Philadelphia chromosome in acute leukemia is correlated with a poor outcome. This chromosome rearrangement is not inherited but arises after birth (Harrison, 2001; Aplenc *et al.*, 2011).

2.1.2.4.1.4. Burkitt lymphoma

Burkitt lymphoma is a rare but highly aggressive cancer involving mature B lymphocytes. Most cases are the endemic type that mostly affects young children in Central Africa (see Figure 2.63). Affected individuals usually have a mass in the jaw or facial bones that spreads to other extranodal sites, like bone marrow, peripheral blood, ovary, kidney, breast, and meninges. The sporadic type is seen all over the world, mainly affecting children and young adults. The mass observed among patients is in the abdominal area. Another type of Burkitt lymphoma is immunodeficiency-associated that affects human immunodeficiency virus (HIV) patients. This usually involves the lymph nodes (O'Connor, 2008d; Naeim *et al.*, 2013).

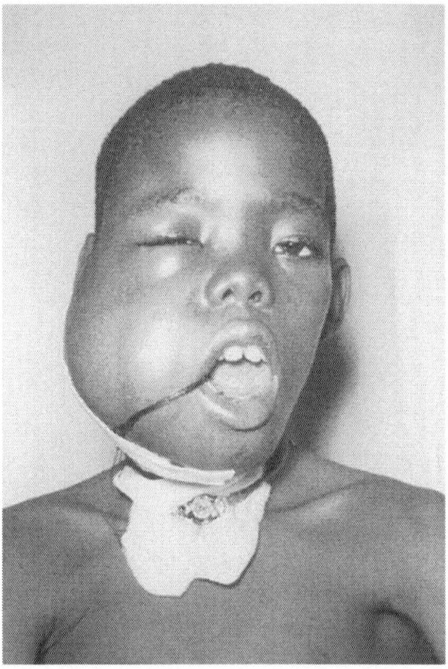

Figure 2.63. A Nigerian boy with a large facial tumor due to Burkitt lymphoma. The tumor is made up predominantly of aberrant B lymphocyte cells. (Source: Robert S. Craig, Center for Disease Control and Prevention, Public Domain).

This results from translocation involving chromosomes 8 and 14 (see Figure 2.64). In particular, the *MYC* proto-oncogene from chromosome 8 is transferred so that it is now under the control of a powerful *IGH* (immunoglobulin heavy chain) gene promoter on chromosome 14. (National Center for Biotechnology Information (US), 1998; O'Connor, 2008d). The region containing the *MYC* gene (i.e. 8q24) may also be translocated to certain regions of chromosome 2 and 22, although these occur less commonly. In these cases, the *MYC* gene transcription comes under the regulation of *IGHK* (immunoglobulin light chain kappa) and *IGHL* (immunoglobulin light chain lambda) gene promoters, respectively (Chaganti, 2001). The *MYC* translocation is also observed in other types of lymphoma (Naeim *et al.*, 2013).

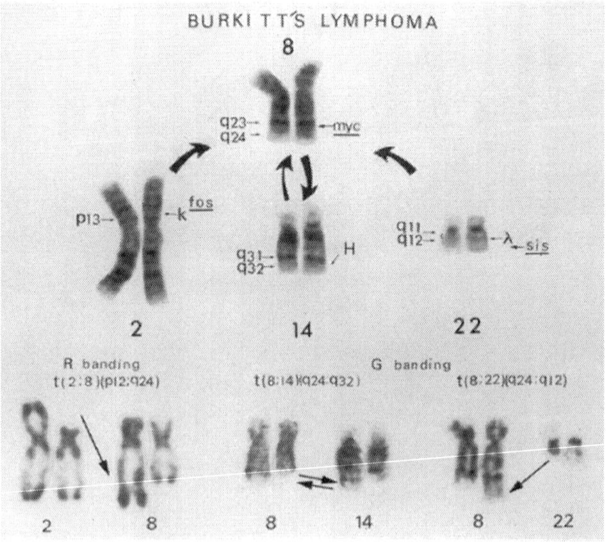

Figure 2.64. Burkitt lymphoma results from the translocation of *MYC* protooncogene at region q24 of chromosome 8 to breakpoints in chromosomes 2, 14, and 22 where the genes for *IGHK*(k), *IGH*(H), and *IGHL*(λ)are located, respectively. This places the MYC gene under the transcriptional control of the strong promoter of these immunoglobulin genes. (Source: National Cancer Institute, Public Domain).

The *MYC* gene codes for a transcription factor so that its expression is precisely regulated. On the other hand, immunoglobulin genes are continuously expressed in B lymphocytes. Therefore, the translocation of the *MYC* gene removes its regulation and is now overexpressed in the B lymphocytes. This, in turn, activates other genes that should not be expressed and eventually leads to malignancy (O'Connor, 2008d; Chaganti, 2001).

2.1.2.5. Diagnosis of structural abnormalities in chromosomes

Section 2.1.1.3. describes some techniques for determining aneuploidy. Some of these can also be used to diagnose changes in chromosome structure. Therefore, diagnosis may also be done prenatally or postnatally. This section focuses on the genetic testing methods employed for detecting chromosomal aberrations.

2.1.2.5.1. Chromosome banding

Chromosome banding techniques produce reproducible banding patterns that enable the identification of the location of breaks and alterations in the chromosome as well as the location of certain genes. These make use of specialized stains applied to metaphase chromosomes to highlight certain regions in the chromosome such as heterochromatin or repeats. Basic banding techniques include Q-banding, G-banding, R-banding, and C banding. These produce similar patterns along the length of the chromosome but may highlight different bands or polymorphic regions (Schreck and Distèche, 1994).

2.1.2.5.1.1. Q-banding technique

Q-banding is the first to be performed and described. It was the first method used to identify the 46 human chromosomes, assigning the p and q arms that are subdivided by numbers. It is named such because it uses the fluorescent dye quinacrine (e.g. quinacrine mustard), which alkylates DNA. This means that it adds methyl or other alkyl groups to the DNA molecule (O'Connor, 2008a; Nabil & Sarra, 2017). The bands fluoresce on exposure to UV light. These bands can be correlated with those produced by G-banding, i.e. intense Q-bands correspond with intense G-bands. A major disadvantage of this technique is that the fluorescence intensity fades rapidly and so a karyograph must be made within a few minutes after staining. Also, although the technique itself is simple to perform, it requires fluorescence microscopy equipment and photomicroscope to capture the short-lived fluorescence pattern on film (Carr, 2008; Moore & Best, 2001).

The occurrence of banding is suggested to be due to the reaction of the alkylating group with the nitrogen atom of guanine and by the intercalation of quinacrine within the DNA double helix. However, the alkylation hypothesis was dispelled upon the discovery that other fluorochromes without alkylating groups, such as quinacrine dihydrochloride and ethidium bromide, can induce similar bands as quinacrine mustard. Another possible reason is the different strengths of binding by quinacrine to different chromosome locations. One factor that influences quinacrine binding is the steric relations between DNA and chromosomal proteins. Quinacrine can bind about 98% of the total length of a human chromosome (Nabil &Sarra, 2017).

Q-banding is most useful in identifying translocations involving various chromosomes (O'Connor, 2008a). In one study, this method was used to perform cytogenetic analysis on peripheral blood lymphocytes from women with premature ovarian failure (POF), wherein ovarian function stops before the age of 40. There were aneuploidies observed, but a greater proportion are structural abnormalities, and nearly half of these involve the X chromosome. In particular, Xq deletions and translocations between the X chromosome and either an autosome or the Y chromosome were observed (see Figure 2.65) (Baronchelli *et al.*, 2011).

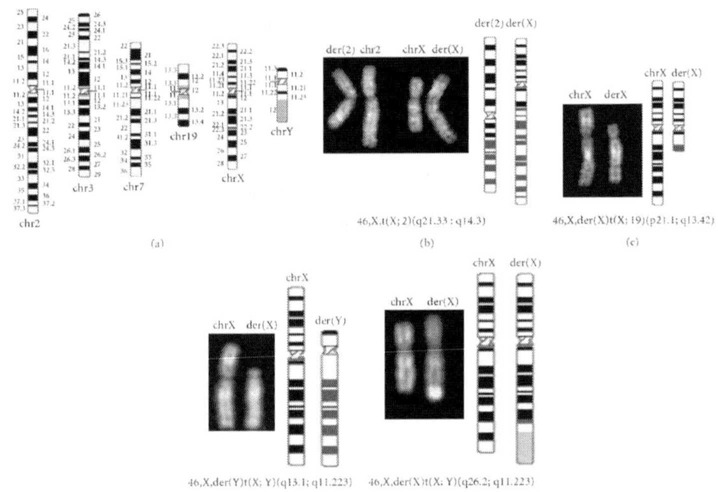

Figure 2.65. (a) Chromosomes have their distinct banding patterns as shown by these illustrations of normal chromosomes. (b-e) Five patients with premature ovarian failure (POV) exhibit translocations involving chromosome X as shown in the partial Q-banding karyotype on the left and the illustrations of the derivative chromosomes on the right. Specifically, the X chromosome was observed to have undergone translocation w ith (b) chromosome 2, (c) chromosome 19, and (d and e) Y chromosome. (Source: Baronchelli, S.; Conconi, D.; Panzeri, E.; Bentivegna, A.; Redaelli, S.; Lissoni, S.; Saccheri, F.; Villa, N.; Crosti, F.; Sala, E.; Martinoli, E.; Volontè, M.; Mrozzi, A.; Dalprà, L. Journal of Biomedicine and Biotechnology 370195, 2011, Creative Commons License).

2.1.2.5.1.2. G-banding technique

G-banding uses Giemsa stain to produce banding patterns on the chromosome. Metaphase chromosomes are treated with the enzyme trypsin to partially degrade the chromosomal proteins. This relaxes the chromatin structure so that Giemsa can access the DNA when the chromosomes are stained. Generally, heterochromatic regions, which tend to be rich in A-T, contain few genes and are late replicating, are stained darker by Giemsa. In contrast, euchromatin, which is less condensed, relatively G-C rich, earlier replicating, and transcriptionally active, incorporates less Giemsa stain. As a result, euchromatic regions appear as light bands, which is sometimes referred to as the interband. A G-band represents up to 10 million base pairs of DNA so that it can contain hundreds of genes (O'Connor, 2008a; Shen, 2019). This technique provides the advantage of producing permanent slides of samples that can be viewed under a standard light microscope. Other stains such as Wright stain and Leishman stain can be used and still produce banding patterns identical to those obtained with Giemsa (Moore and Best, 2001).

After producing the karyotype, band by band analysis can be performed to identify structural changes or rearrangements in the chromosome (see Figure 2.66).

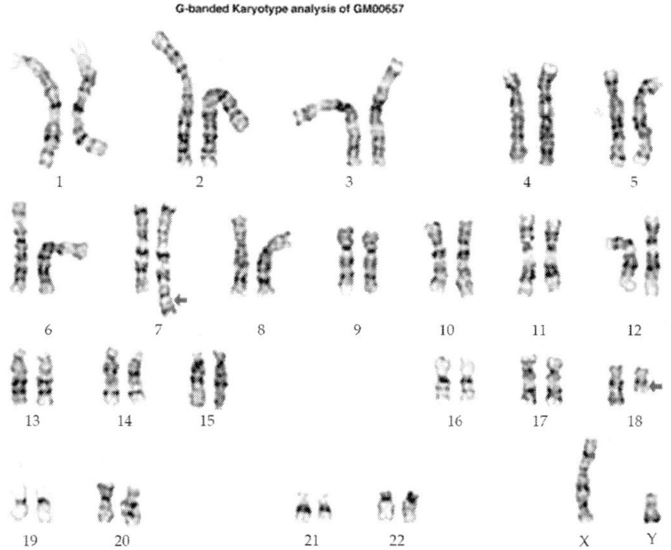

Figure 2.66. A karyotype produced after performing the G-banding technique showing translocation between chromosomes 7 and 18 (red arrows) with po-

tential microdeletion on chromosome 7 or 18 or both. (Source: Tang, Z.; Berlin, D.S.; Toji, L.; Toruner, G.A.; Beiswanger, C.; Kulkarni, S.; Martin, C.L.; Emanuel, B.S.; Christman, M.; Gerry, N.P. Genes, Genomes, Genetics 3 (7), 1143-1149, 2013, Creative Commons License).

2.1.2.5.1.3. R-banding technique

R-banding uses Giemsa stain but it generates a banding pattern that is the reverse that of G-banding (see Figure 2.67). The chromosomes are heated before the Giemsa stain or acridine orange is applied. The heat treatment is hypothesized to denature the A-T rich heterochromatic regions leaving the G-C rich euchromatic region to take up the stain. Therefore, the G-C rich region that produces light-colored bands in the G-banding method generates dark-colored bands in the R-banding method. Since the euchromatin is also rich in genes, R-banding is often used to study gene-rich regions that are located near the telomeres (O'Connor, 2008a). Apart from heat and Giemsa treatment, fluorescent R-banding patterns can also be generated by using dyes with high affinity to the G-C base pair like chromomycin A3, olivomycin, and mithramycin (Moore and Best, 2001).

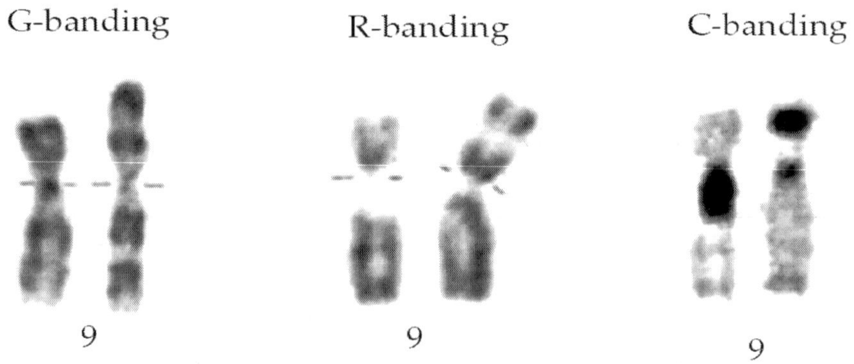

Figure 2.67. Partial karyotype from the same individual was constructed using G-banding, R-banding, and C-banding methods. All techniques reveal that one copy of chromosome 9 has additional bands on its short arm compared to its homologue. The C-banding indicates that the added chromosome material is heterochromatic. It also produces dark centromeric bands. (Source: Kasakyan, S.; Lohmann, L.; Aboura, A.; Quimsiyeh, M.; Menezo, Y.; Tachdjian, G.; Benkhalifa, M. Molecular Cytogenetics 1, 27, 2008, Creative Commons License).

2.1.2.5.1.4. C-banding technique

C-banding can be used to specifically stain noncoding constitutive heterochromatin such as the centromeres (see Figure 2.67 and Figure 2.68) and short arms of acrocentric chromosomes, both of which have A-T rich satellite DNA, as well as telomeres, pericentric regions, and most of the long arm of the Y chromosome. The highly condensed heterochromatin stains darkly while all other chromatin remains lighter (see Figure 2.68). C-banding is generated by treating the chromosomes with acidic and then basic solutions followed by Giemsa staining. It is rarely used for making a clinical diagnosis. Instead, it is primarily used to identify the gene coding potential of chromosomal segments and for studying chromosomal polymorphisms in the population (O'Connor, 2008a; Moore and Best, 2001).

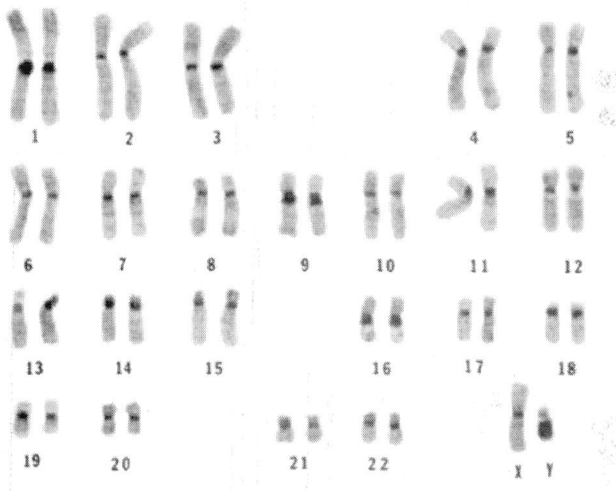

Figure 2.68. C-banding pattern exhibits dark bands for heterochromatic regions such as centromeres, pericentric regions (e.g., chromosomes 1, 9, and 16), and the long arm of the Y chromosome. (Source: Di Tomaso, M.V.; Liddle, P.; Lafon-Hughes, L.; Reyes-Ábalos, A.L.; Folle, G. IntechOpen, 2013, Creative Commons License).

2.1.2.5.1.5. High-resolution banding

Aneuploidy can be detected at relatively low band resolution, i.e. with 350-550 discernible bands per haploid metaphase chromosome spread. However, microdeletions, microduplications, and other subtle structural rearrangements, such as determining whether a particular translocation is balanced or unbalanced, can be diagnosed at higher band resolution levels (i.e., 650 or more bands per haploid genome). The number of identifiable bands in a chromosome spread depends on the degree of chromosome condensation before harvest, the cell type, and the banding method used (Moore and Best, 2001). Lymphocyte cultures and established cell lines are proven to be suitable for high-resolution banding techniques as opposed to samples from leukemic bone marrow and solid tumors (Rønne, 1988).

There are several methods of producing high-resolution banding. One method is by fixing cells in late prophase or early metaphase stage of cell division instead of the usual mid-metaphase stage. At these stages, the chromosomes show minimal chromatin condensation and maximal band resolution. This is produced when cell cultures are synchronized or brought to the same phase of the cell cycle and then exposed for a short time to a low concentration of colcemid. The same results can be obtained by using agents that intercalate into or bind to the DNA molecule so that condensation of the chromosomes is prevented. Some examples of reagents used for this purpose include ethidium bromide, acridine orange, actinomycin D, and Hoechst 33258. Another method used is called **replication banding**, which can produce as high as 1400 bands per haploid chromosome spread. This employs reagents that inhibit chromosome condensation as well as the DNA base analog, 5-bromo-2'deoxyuridine (BrdU), that is taken up at different rates by early-replicating euchromatin and late-replicating heterochromatin segments in the chromosome. Specifically, there is BrdU substitution of thymine in late replicated DNA. This can produce both G- and R-banded patterns after Giemsa staining depending on the timing of the pulse addition of BrdU into the growing, synchronized cell cultures. In the case of the R-banding pattern, the BrdU-substituted segments appear light and the unsubstituted segments appear dark. The decreased condensation of the BrdU-substituted regions enhances band resolution and band contrast (Moore and Best, 2001; Rønne, 1988).

Synchronization of cell cultures and achieving chromosome spreads with minimal condensation are time-consuming and labor- intensive so high-resolution banding techniques are normally reserved for suspected subtle chromosome abnormalities that are not detected by conventional cytogenetic analysis (Moore and Best, 2001).

2.1.2.5.1.6. Sister chromatid exchange

Sister chromatid exchange, where sister chromatids exchange small sections with each other, occasionally occurs during the early stages of cell division. It is thought that this happens in response to DNA damage during replication. However, in certain individuals with genetic disorders like Bloom syndrome or in cases of exposure to genotoxic agents, which causes sister chromatid exchange to occur much more frequently, the chromosomes are more prone to breakage. As a result, there are gaps and breaks in the genetic material (Genetics Home Reference, 2015b; Moore and Best, 2001).

In order to monitor the frequency of sister chromatid exchange, BrdU is added during cell culture. The cells are made to undergo two rounds of DNA replication in the presence of BrdU. After the first round of replication, all the daughter DNA molecules have one unsubstituted parental strand and one strand with BrdU-substituted DNA. As a result, all of the chromatids exhibit the same degree of staining. After the next round of cell division, the sister chromatids exhibit different staining intensity. The original parental DNA strand remains in one chromatid, paired with a newly-synthesized BrdU-substituted strand. The other chromatid is made up of two strands with BrdU incorporated in them. Exposure of the chromosomes to the fluorescent stain Hoechst 33258 and UV light causes loss of chromatin in the chromatid with two BrdU-substituted strands. Consequently, the chromatid stains lightly with Giemsa. On the other hand, the chromatid with just one BrdU-substituted strand is more stable and loses less chromatin, and so it stains more darkly with Giemsa. Therefore, the chromosomes appear to be made up of a light-stained chromatid and a dark-stained one (see Figure 2.69A). Since the sister chromatids show differential staining, exchanges between the two are evidenced by discontinuous light and dark staining regions along the length of the chromatids with one showing the opposite staining pattern as its sister (see Figure 2.69B) (Moore and Best, 2001).

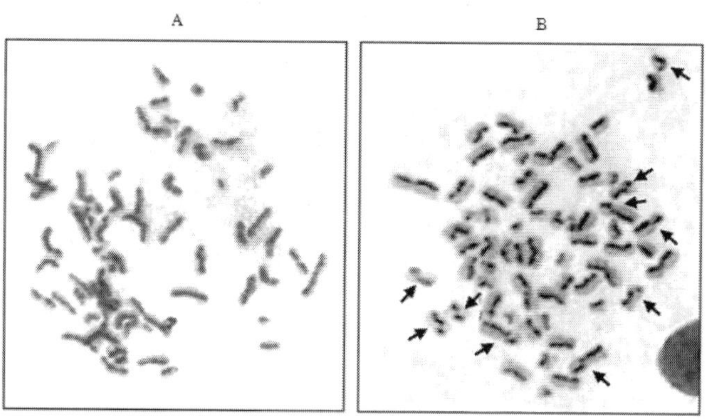

Figure 2.69. (A) After two rounds of cell division with 5 bromo-2' deoxyuri-

dine (BrdU) and then treating the chromosomes with Giemsa stain, the sister chromatids are seen as a pair of light and dark-staining structures. (B) Sister chromatid exchange occurred if the chromatids exhibit a mixture of light and dark segments along their length (arrows). (Source: Sharma, S.; Brosh, R.M. Jr.PloS ONE 2(12), e1297, 2007, Public Domain).

2.1.2.5.2. Multicolor chromosome painting

Chromosome painting can be accomplished by using FISH (refer to section 2.1.1.3.3. Genetic testing: Fluorescence *in situ* hybridization for a detailed discussion). Apart from whole chromosome painting probes, methods for generating additional probes were developed. One way to produce high quality probes is by flow sorting fluorescence-activated metaphase chromosomes from cell samples. A sorting experiment can generate about 500 copies of each chromosome in the human genome. These can be subsequently amplified by degenerate oligonucleotide-primed-PCR (DOP-PCR). This uses degenerate primers, which are a mix of similar oligonucleotides with several possible bases at certain positions. Using this type of primers ensures that a sufficient number of fluorochromes are bound to the region of interest. The flow-sorted probes have the advantage of improving signal-to-noise ratio and staining homogeneity as well as expanding the target number of subsequent DNA amplification so that the high complexity of the probes is ensured. Probes can also be produced by microdissection of metaphase chromosomes. This involves isolating whole chromosomes or certain chromosome parts followed by using micromanipulators or laser-equipped devices. Only 25-50 copies of each chromosome can be isolated. These are then amplified by DOP-PCR and subsequently labeled fluorescently. These can be designed to target a specific chromosomal arm or bands. Aberrant but unidentified chromosomes may also be isolated via flow sorting or microdissection, amplified, and then labeled to be used as the probe on samples from a normal individual. This is called **reverse chromosome painting** (Ried *et al.*, 1998; Kjeldsen and Kølvraa, 2002).

Constructing a multicolored karyotype of the human chromosomes, with each homologue uniquely labeled, had been a long-standing goal of cytogeneticists. Realizing this goal started with using three fluorochromes, fluorochrome combinations, and ratio labeling. Using ratio labeling and visualization with triple bandpass optical filters enabled 12 chromosomes to be visualized simultaneously. Then sensitive imaging devices enabled the addition of suitable dyes towards the near- infrared. This can distinguish up to seven different targets. Finally, in 1996 the 24 human chromosomes were visualized via multicolor FISH (see Figure 2.70). Five fluorochromes that can be combined in various ways were used to produce 24 artificial colors to label chromosome-specific probes. Then the probes were simultaneously hybridized to metaphase chromosome spreads in a single hybridization experiment. There are two possible approaches in achieving this, multiplex FISH and spectral karyotyping (Ried *et al.*, 1998).

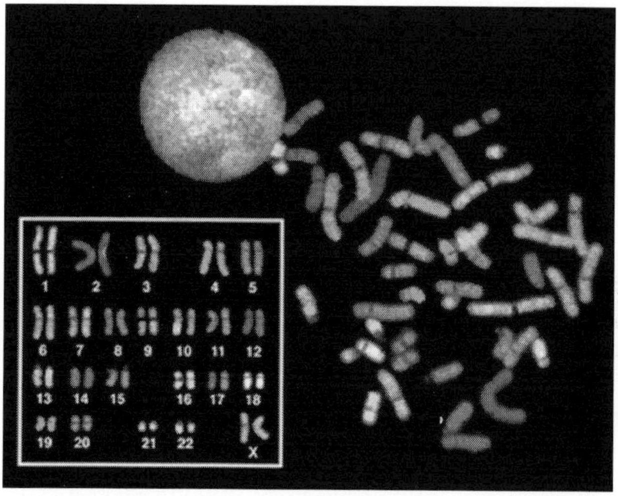

Figure 2.70. It is now possible to combine fluorescent dyes in order to produce 23 colors to label the 22 autosomes and the X chromosomes in a human female karyotype. In the case of a male karyotype, an additional label can be used for the Y chromosome. (Source: National Human Genome Research Institute, Public Domain).

Multicolor karyotyping has certain limitations. It cannot detect intrachromosomal duplications and deletions. Also, it does not provide landmarks or bands to precisely determine chromosome breakpoints and translocated regions. There might be misinterpretations, too when overlapping chromosomes distort the fluorescent color of an adjacent chromosome, making it appear that three chromosomes are involved in translocation when, in reality, there are only two (Shen, 2019).

2.1.2.5.2.1. Multiplex-FISH

Multiplex FISH(m-FISH) uses five different fluorochromes and their specific optical filters, which acquire images sequentially. Each of these filters is a narrow or single bandpass filter designed to allow maximum differentiation of the fluorochromes. Computer software is then used for image shift correction. Another software for image analysis calculates a chromosome segmentation mask based on the 4', 6-diamidino-2-phenylindole (DAPI) image. DAPI is a blue-fluorescent stain that is one of the fluorochrome dyes used. **Chromosome segmentation** aims to separate the chromosomes from noise or the background. The intensities of these pre-defined areas in all subsequently acquired images will be recorded and measured. Based on the labeling schemes, the program then decides whether a single pixel contained signals from one fluorochrome or from two or more. This forms the basis for chromosome identification and color assignment (Ried *et al.*, 1998; Kjeldsen and Kølvraa, 2002). Contrast enhancement, pseudocoloring, and overlaying may be performed using a photoediting software. The automation of the image-capturing procedure, the use of a motorized epifluorescence microscope, and the development of multichannel image-analysis methods facilitate the performance of m-FISH so that results can be available within 24 hours (Uhrig *et al.*, 1999).

This method is usually used to study complex interchromosomal rearrangements, i.e. several aberrations in a single karyotype (see Figure 2.71), as well as to detect small chromosomal changes that are not detected by conventional cytogenetics (Kearney, 2006). In cases of intrachromosomal deletion, duplication or inversion, chromosome-specific multicolor bar codes are used, or m-FISH is supplemented by CGH (Uhrig *et al.*, 1999). Sometimes a karyotype shows an extra chromosome that is structurally abnormal, called the **marker chromosome** or the smaller (relative to chromosome 20) **supernumerary marker chromosomes (sSMCs)**. Mostly, the banding pattern of this abnormal chromosome is not sufficient to determine its chromosomal origin. Therefore, banding methods can be supplemented with m-FISH (see Figure 2.72) and aCGH to identify it (Kalkan, 2017).

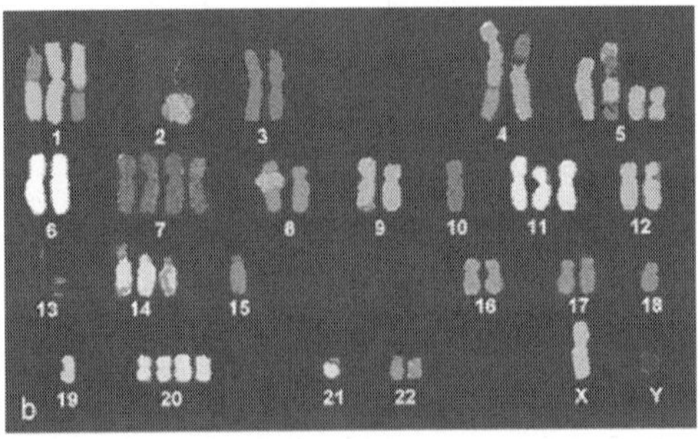

Figure 2.71. A karyotype of a human neck and squamous cell carcinoma cell line

that is labeled via multiplex fluorescence *in situ* hybridization (m-FISH) shows several numerical and structural aberrations of the chromosomes. Aneuploidies are observed for chromosomes 1, 5,7, 11, 19, 20 and 21. Translocation occurred between chromosomes 1 and 15 as well as chromosomes 1 and 18. Deletion is also observed in chromosome 13. (Source: Singchat, W.; Hitakomate, E.; Rerkarmnuaychoke, B.; Suntronpong, A.; Fu, B.; Bodhisuwan, W.; Peyachoknagul, S.; Yang, F.; Koontongkaew, S.; Kornsorn, S.PLoS ONE 11(8), 2016, Creative Commons License).

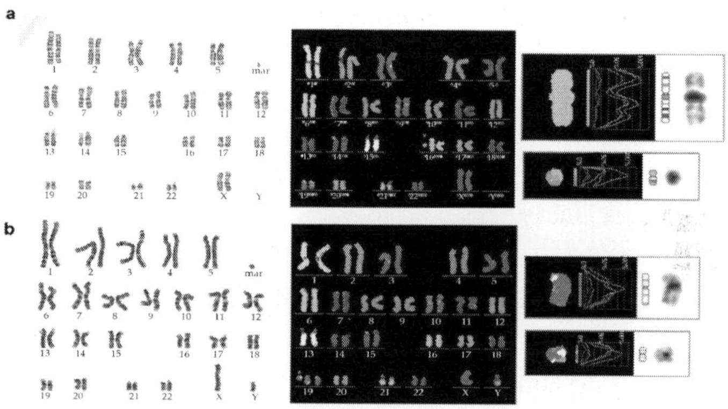

Figure 2.72. Multiplex FISH (m-FISH) was used to determine the chromosomal origin of supernumerary marker chromosomes (sSMCs)in two cases where the sSMC is identified to originate from (a) chromosome 16 and (b) chromosome 19. (Source: Jang, W.; Chae, H.; Jiyeon, K.; Son, J.-O.; Kim, S.K.; Koo, B.K.; Kim, M.; Kim, Y.; Park, I.Y.; Sung, I.K. Molecular Cytogenetics 9(1), 2016, Creative Commons License).

2.1.2.5.2.2. Spectral karyotyping

Spectral karyotyping (SKY) is similar to m-FISH in the sense that it uses five fluorochromes but the methods for detecting and discriminating the different combinations of fluorescence signals are different. SKY is based on spectral imaging, which combines spectroscopy, CCD imaging, and conventional microscopy. It uses a Sagnac interferometer to produce a fluorochrome-specific optical path difference. This is then Fourier transformed to provide spectral information that is distinct for each chromosome. The CCD camera simultaneously recovers the fluorescence emission spectrum at all image points. This produces a differential color display of all chromosomes even by just one exposure of the fluorescently labeled sample. A certain software classifies the image formed by identifying pixels with identical spectra. Such pixels are assigned the same artificial color, which becomes the basis for generating a karyotype (Ried *et al.*, 1998; Kjeldsen and Kølvraa, 2002).

Just like m-FISH, SKY is useful in characterizing complex chromosomal rearrangements and marker chromosomes that are usual in the genome of cancer cells as well as in detecting ambiguous or subtle changes that are not detected by traditional banding analysis such as unbalanced translocation (see Figure 2.73).

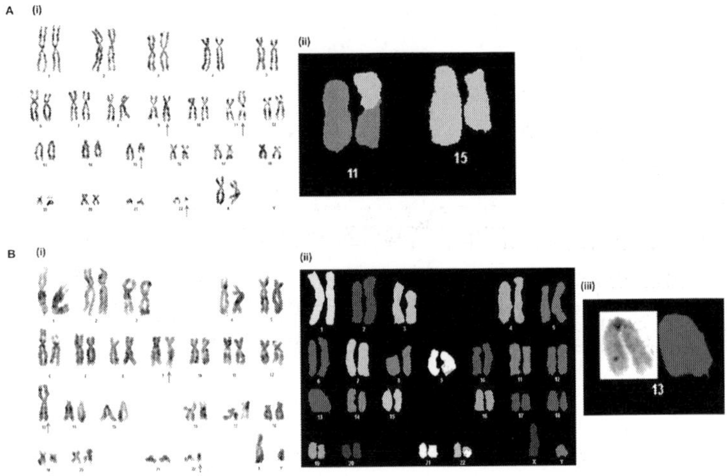

Figure 2.73. Novel chromosomal aberrations in *de novo* chronic myelogenous leukemia (CML) patients in their chronic phase (CP) are detected by G-banding and then confirmed by spectral karyotype. (A) (i) G-banded karyotype shows translocation between chromosomes 9 and 22 and between chromosomes 11 and 15. (ii) Partial spectral karyotype confirms the translocation between chromosomes 11 and 15. (B)(i) G-banded karyotype shows translocation between chromosomes 9 and 22 as well as Robertsonian translocation between the two chromosomes 13. This is confirmed by the (ii)spectral karyotype and (iii) partial spectral karyotype. (Source: Chandran, R.K.; Narayanan, G.; Sakthivel, K.M.; Kumar, R.S.; Krishna, K.M.N.J.; Sreedharan, H. Frontiers in Oncology 9,88, 2019, Creative Commons License).

2.1.2.5.3. PCR-based methods

Apart from detecting aneuploidies, PCR can also be utilized to detect chromosomal rearrangements. Two types of PCR that are used for this purpose are discussed in this section.

2.1.2.5.3.1. Long-distance inverse PCR

Long-distance inverse PCR (LDI-PCR) can be used to identify chromosomal breakpoints and screen *de novo* DNA fusions in rearrangement-prone genomic locations. These detected regions can then be sequenced so that they are identified at the base-pair level. This merely requires approximate sequence information of one chromosome and does not require prior knowledge of the fusion partner chromosome. This proved to be a sensitive and cost-effective method for detecting DNA rearrangements in solid tumors (Pradhan *et al.*, 2016; Schneider *et al.*, 2011).

LDI-PCR can amplify large DNA fragments of interest that are produced after restriction enzyme digestion. This method starts by selecting the sequence covering the genomic region of interest from a genome browser (e.g., ENSEMBL, UCSC, NCBI). The selected sequence is then pasted into the query box of a restriction map generator (e.g. BioEdit, SMS, Restriction Mapper). The chosen restriction enzymes should produce 2-5 kb fragments that are made up of known and unknown sequences. If using a double-digest strategy with enzymes producing sticky ends, the ends must be complementary to each other. After genomic DNA digestion, the restriction fragments are then re-ligated to form circular templates for PCR (see Figure 2.74). A pair of primers are designed based on the known sequence. These should direct amplification opposite each other, i.e. one is directed toward the restriction site while the other one is directed in the opposite direction.

After the PCR, the amplicons comprise of the unknown sequence flanked on both sides by known sequences. When the chromosomal rearrangement results in size difference, called restriction fragment length polymorphism, between the wild type and derivative alleles, their corresponding amplicons can be distinguished by gel electrophoresis. Their gel bands can be excised, purified, and subsequently sequenced. Sequencing with one of the PCR primers directed towards the restriction site allows the immediate identification of the partner gene. Sequencing in the other direction, meanwhile, enables precise mapping of the breakpoint. There might also be discrepant bands on the gel that do not correspond with the calculated amplicon size that is expected. These may be amplicons of translocated fragments, which can also be excised for sequencing to reveal their translocation partner and the rearrangement breakpoint (Schneider *et al.*, 2011).

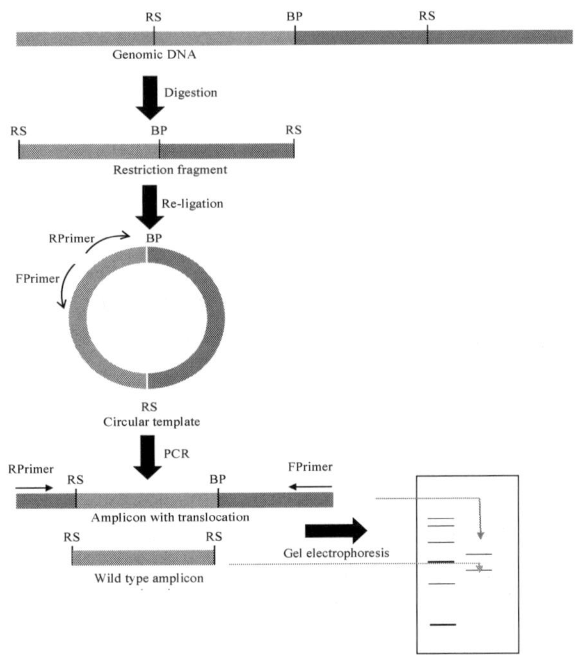

Figure 2.74. Restriction enzyme digestion, at certain restriction sites (RS), is performed on the genomic region prone to rearrangement to produce fragments that cover the breakpoint (BP) and that are composed of known (orange) and unknown (blue) sequences. The generated fragments are re-ligated to form circular templates for PCR with a reverse (RPrimer) and forward primer (Fprimer) relative to the BP. The amplicons are separated via gel electrophoresis and those containing a translocated segment form a higher molecular weight band than those without translocation. (Source: Adapted from Schneider, B.; Drexler, H.G.; MacLeod, R.A.F. Atlas of Genetics and Cytogenetics in Oncology and Haematology 15(1), 106-114, 2011.).

2.1.2.5.3.2. Multiplex delta-PCR

Chromosomal translocations leading to malignancies can be classified into two categories, either a break within genes lead to the formation of a fusion transcript that produces an oncogenic chimeric protein (refer to Section 2.1.2.4.1.3. Chronic myeloid leukemia) or an oncogene becomes overly expressed due to being translocated near an actively transcribed gene (refer to Section 2.1.2.4.1.4. Burkitt lymphoma). The fusion transcript produced via the former mechanism can be detected with high specificity via reverse transcription PCR (RT-PCR). The second mechanism, on the other hand, does not produce a fusion transcript and so it must be detected by PCR at the DNA level (Lin *et al.*, 2010).

Since malignancies usually have several translocations as well as recurrent insertions and deletions, using only a single pair of primers may not detect all breakpoints or mutation loci. Conventional multiplex PCR has the disadvantage of being unclear on agarose gel electrophoresis with weak band intensity or elevated background noise and on capillary electrophoresis with one or several small peaks. One strategy that is developed to circumvent these is **multiplex delta-PCR**, which combines primer sets but ensures specificity allowing the identification of a specific breakpoint from a single multiplexed PCR (Lin *et al.*, 2011).

For each breakpoint, insertion or deletion locus, three primers are designed and used simultaneously (see Figure 2.75). Two of the primers, called the external and internal primers, are oriented in the same direction and are on one side of the target locus. The third primer is oriented in the opposite direction. The internal primer is created at a defined distance Δ (delta) from the external primer. As such, it serves as a confirmatory probe. The amplicons are then separated by capillary gel electrophoresis. A pair of amplicons differing in size by the Δ value indicates a true translocation or mutation (Lin *et al.*, 2011).

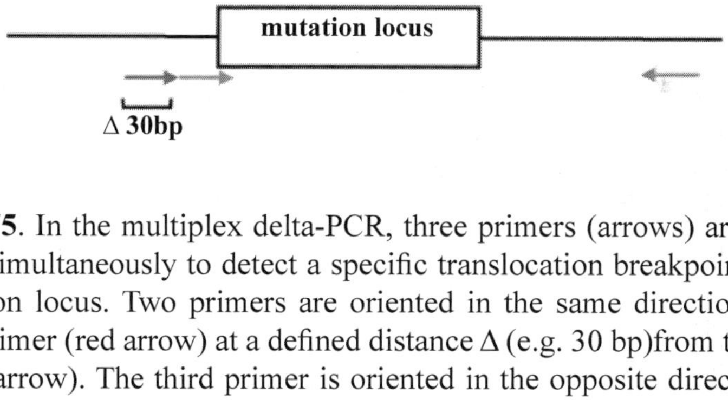

Figure 2.75. In the multiplex delta-PCR, three primers (arrows) are designed and used simultaneously to detect a specific translocation breakpoint or insertion/deletion locus. Two primers are oriented in the same direction with the external primer (red arrow) at a defined distance Δ (e.g. 30 bp)from the internal one (blue arrow). The third primer is oriented in the opposite direction of the other two. (Source: Adapted from Lin, M.-T.; Tseng, L.-H.; Rich, R.G.; Hafez, M.J.; Harada, S.; Murphy, K.M.; Eshleman, J.R.; Gocke, C.D. The Journal of Molecular Diagnostics 13(1), 85-92, 2011.).

2.1.2.5.4. Chromosomal microarray analysis

Chromosomal microarray analysis (CMA) can detect both aneuploidy and unbalanced chromosomal rearrangements. Furthermore, CMA can reveal cryptic chromosomal abnormalities, such as microdeletions, **microduplicationsorCNVs** that are not detected by conventional banding techniques. Not all CNVs result in adverse clinical phenotypes especially if these do not affect any coding region. On the other hand, certain CNVs may involve critical gene/s or an important regulatory region leading to a pathogenic phenotype. In this case, CMA serves as an important diagnostic tool for individuals with unexplained developmental or congenital anomalies. Another advantage of CMA is that it can analyze DNA extracted from uncultured cells of all different types. This shortens the experiment time relative to chromosome analysis and so results can be obtained faster. On the other hand, CMA may not detect mosaicism as well as the delineation of chromosomal rearrangements. For this reason, chromosome analysis by banding methods is still necessary(Levy and Wapner, 2018; Bi *et al.*, 2013).

2.1.2.5.4.1. Array-based CGH

Array-based CGH (aCGH) compares a patient's genome to a normal control genome to identify regions that are either over- or underrepresented in the patient's sample (refer to section 2.1.1.3.3. Genetic testing: Comparative genome hybridization for a detailed discussion). A fluorescence intensity ratio of one means that there are equal contributions of both the patient and the control sample. This means that there is a normal copy number at that region or locus. A ratio that is significantly greater than one indicates that there is duplication in the patient's genetic material so that more of his DNA hybridized with the probe. On the other hand, a ratio that is significantly lesser than one denotes a loss or deletion of genetic material in the patient's genome (Levy and Wapner, 2018). Results obtained in a-CGH can be confirmed by FISH (see Figure 2.77b).

This method proved to be useful in identifying the chromosomal origin of sSMCs (see Figure 2.76) as well as in studying rearrangements at the subtelomeric (see Figure 2.77a) and pericentromeric regions. Subtelomeric regions are present in most chromosomes except for the short arms of chromosomes 13, 14, 15, 21, and 22. These are relatively rich in coding regions (see Figure 2.77c) and are prone to rearrangement. Specifically, interstitial deletions were found to be more common in these regions compared to terminal ones. It has been suggested that structural aberrations in these regions are associated with a high percentage of abnormalities in persons with idiopathic mental retardation (Theise, 2008).

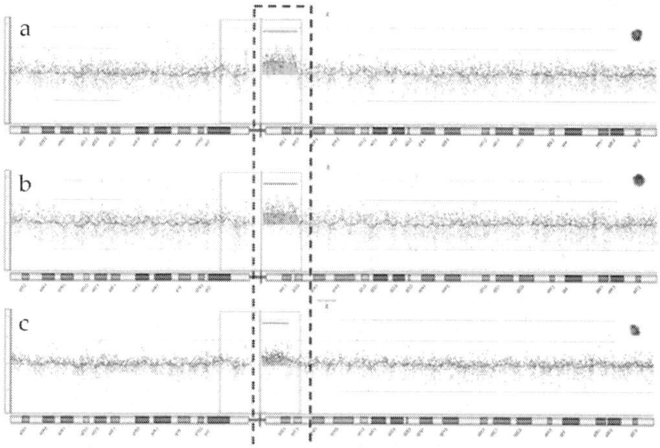

Figure 2.76. In samples with supernumerary marker chromosomes (sSMCs), array CGH detected a (a) gain of 12.6 Mb derived from chromosome 2q11.1q12.3. (b) An identical gain was observed from the mother of the sample source, indicating that the sSMC was maternally inherited. (c) Another sample showed a gain in chromosome 2, but this time at a different region, i.e. 2q11.1q12.1. (Source: Jang, W.; Chae, H.; Jiyeon, K.; Son, J.-O.; Kim, S.K.; Koo, B.K.; Kim, M.; Kim, Y.; Park, I.Y.; Sung, I.K. Molecular Cytogenetics 9(1), 2016, Creative Commons License).

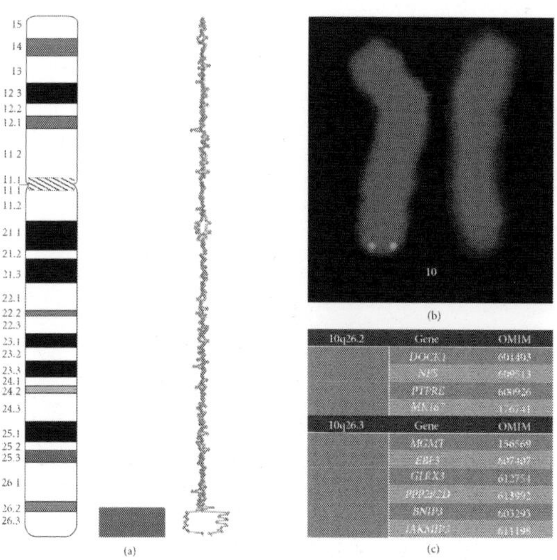

Figure 2.77. Molecular analysis was performed on genetic samples from a male infant with moderate developmental delay and minor dysmorphic features. (a) Array CGH using two alternative arrays, i.e. Cy3/Cy5 (red line) and Cy5/

Cy3 (blue line), showed an interstitial deletion at the subtelomeric region of chromosome 10. (b) The deletion was confirmed by fluorescence *in situ* hybridization where one copy of chromosome 10 does not show a hybridization signal with the subtelomeric probe used. (c) The deleted region is 10q26.2q26.3 which contains a number of genes. (Source: Iourov, I.Y.; Vorsanova, S.G.; Kurinnaia, O.S.; Yurov, Y.B. Case Reports in Genetics 505832, 2014, Creative Commons License).

Rearrangements in the pericentromeric regions can also be detected using microarrays designed to target these regions. These regions are adjacent to the repetitive centromeric regions. These are prone to microdeletions causing a number of genetic disorders such as DiGeorge and Prader-Willi syndromes as well as other syndromes that have not yet been fully characterized. Duplications have also been observed in these regions. The conventional G-banding may not be able to detect rearrangements in these regions because they consist of highly repetitive sequences that show variable banding patterns (Theise, 2008).

2.1.2.5.4.2. Single nucleotide polymorphism microarray

Single nucleotide polymorphism (SNP) microarray analysis (SMA) utilizes arrays with oligonucleotide probes chosen from DNA locations across the genome that exhibit SNPs, i.e. these vary between individuals by a single base pair. SNPs are observed in a substantial portion of the human population and constitute a major portion of human genetic variation, with an estimated 10 million SNPs in the human genome. Each SNP in the array is represented by two oligonucleotide probes, one matching each of the two variant SNP alleles. Each spot on the slide represents either of the two SNP alleles, usually denoted as A and B. Thus, an individual's genotype at a given SNP locus is either AA, AB or BB. These arrays have very high density containing as high as 2.7 million probes with both alleles of each SNP represented multiple times. Therefore, they enable more accurate delineation of breakpoints and even the introns or exons of the gene involved. SNP arrays can be designed with a certain density of SNP probes across different parts of the chromosomes in order to have higher resolution in the regions of interest.

In this method, only the patient's DNA is fluorescently labeled and hybridized to the SNP array (see Figure 2.78). A scanner is used to measure the fluorescence intensity of hybridized probes for each SNP on the array generating the raw intensities of the A and B alleles (RA and RB). A strong A fluorescence intensity and low B fluorescence intensity indicate AA genotype. Conversely, a strong B fluorescence intensity with low A fluorescence intensity indicates BB genotype. If both intensities are similar and are at an intermediate level, then the genotype is AB. The major advantage of SNP arrays over other microarray platforms is it provides additional data called the **B allele frequency (BAF)**. It enables SMA to detect CNVs of even small chromosome segments (see Figure 2.79).

The BAF of each SNP indicates the relative abundance of B allele intensity, which in turn reflects the specific genotype of each SNP variant present in the sample. For a normal diploid individual, it may have three normalized or adjusted values at 0, 0.5, and 1 for AA, AB, and BB, respectively. On the other hand, a BAF close to 0 and 1 but not 0.5 is observed for SNPs in single deletions (copy number =1). For SNPs in single duplications (copy number =3), BAFs are expected to be around 0, 0.33, 0.67, or 1.

Another data derived from raw fluorescence intensities that are correlated with copy number status is the log R ratio (LRR), which is also computed when performing aCGH. This value is obtained by first normalizing the intensity of each SNP (i.e., RA + RB/Rexpected, with Rexpected interpolated by the analysis software) which is then log2-transformed. LRR indicates the relative abundance of the genomic DNA around the SNP (Levy and Wapner, 2018; Sun *et al.*, 2015; Karampetsou *et al.*, 2014; Lin *et al.*, 2014).

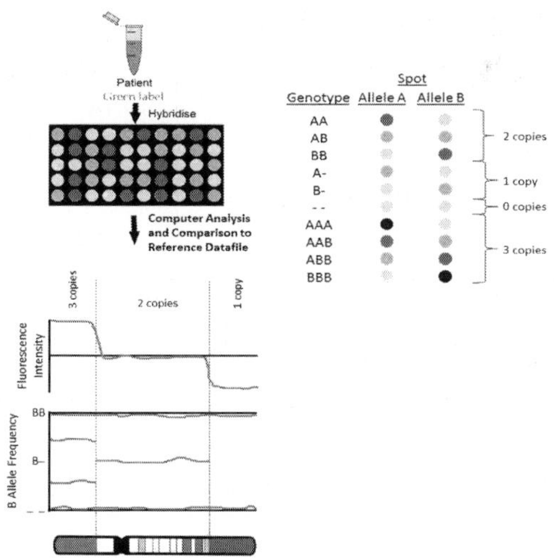

Figure 2.78. When performing single nucleotide polymorphism (SNP) microarray analysis, the patient's sample DNA is fluorescently labeled and then al-

lowed to hybridize with the probes on the array slide. After the DNA hybridizes with probes having its complementary sequence, the fluorescence is measured and the intensity for each SNP allele is calculated. A heterozygous locus (AB) shows equal intensities for both alleles while a homozygous locus (AA or BB) shows higher intensity for the allele present (A or B). Copy number variation can also be extrapolated by comparing the data with a reference data file. Then each probe is plotted along the length of the chromosome depending on its location, fluorescence intensity, and B allele frequency. (Source: Karampetsou, E.; Morrogh, D.; Chitty, L. Journal of Clinical Medicine 3(2), 663-678, 2014, Creative Commons License).

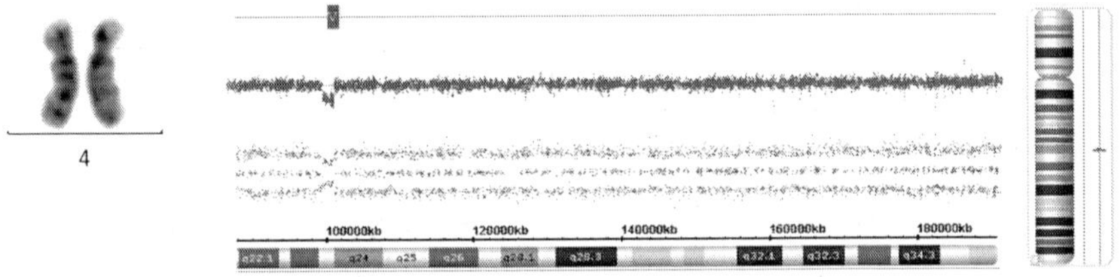

Figure 2.79. Although the banding technique shows normal chromosome 4 in this sample, SNP microarray reveals a loss of a small segment spanning 1.48 kb (red bar) in region 4q24. (Source: Song, Q.; Chu, Y.; Yao, Y.; Peng, M.; Yang, W.; Li, X.; Huang, S. Scientific Reports 7,10354, 2017, Creative Commons License).

Specific BAF patterns can also be used to determine maternal cell contamination, mosaicism (see Figure 2.80), and loss of heterozygosity due to consanguinity (offspring resulting from the mating of close relatives) or uniparental disomy (see Figure 2.81). One advantage of SMA over aCGH is it does not just detect CNVs but can allow a more accurate copy number determination. This information can be extrapolated from the BAF plot, which allows homozygous and heterozygous deletions to be distinguished. Because of this, false positives can be minimized. On the other hand, the SNP array has the disadvantage of having limited genome coverage because of the non-random distribution of SNPs in the genome. Segmental duplications in particular are usually poorly or not covered. For this reason, most SNP arrays used for clinical purposes are actually hybrid arrays that contain both SNP probes and non-polymorphic, SNP-independent CNV probes. These arrays provide more even and consistent genome coverage. (Levy and Wapner, 2018; Sun *et al.*, 2015; Karampetsou *et al.*, 2014). CNV probes do not have distinct alleles and so it provides LRR but not BAF (Lin *et al.*, 2014). Another disadvantage of SMA is its inability to detect balanced rearrangements since these do not alter copy numbers.

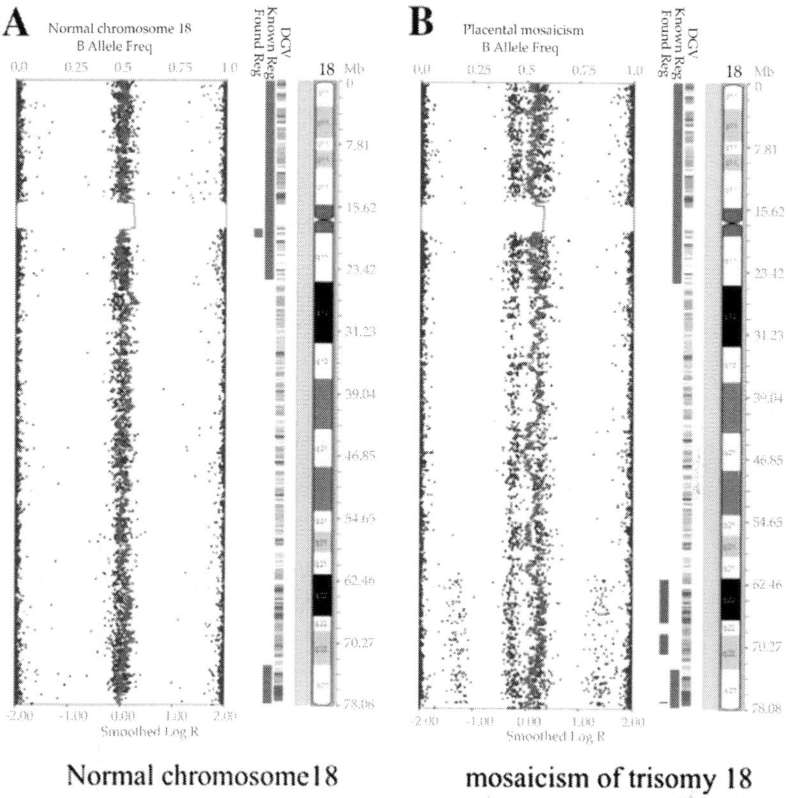

Figure 2.80. SNP array may show the presence of mosaicism, in this case using placental tissue DNA. (A) The upper axis indicates the value of the B allele frequency (BAF, blue dots). For diploid individuals, a normal chromosome (chromosome 18 used as a control in this example) may have three BAF values, i.e. 1 for BB, 0.5 for AB, and 0 for AA. The lower axis indicates the log R ratio (LRR, red line), where R is the ratio of each SNP compared to chromosome 18. The LRR of 0, in this case, means an actual ratio of 1. (B) The SNP array of a placental tissue with mosaic trisomy 18 shows altered BAF values, i.e. 1 for BBB and BB, 0.6 for ABB and AB, 0.4 for AAB and AB, and 0 for AAA and AA. The increased LRR also indicates mosaicism. (Source: Pan, Q.; Sun, B.; Huang, X.; Jing, X.; Liu, H.; Jiang, F.; Zhou, J.; Lin, M.; Yue, H.; Hu, P.; Ning, Y. Molecular Cytogenetics 7(1), 48, 2014, Creative Commons License).

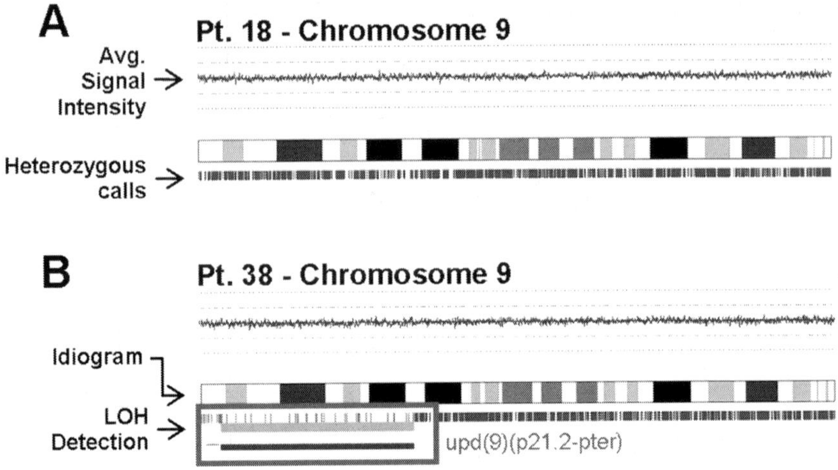

Figure 2.81. (A) The blue line in the SNP microarray karyogram represents the average fluorescent intensity of each SNP while the green tics represent heterozygous calls for each SNP. (B) The absence of heterozygous calls represents areas of uniparental disomy or loss of heterozygosity (LOH) in the chromosome (blue and pink bars). (Source: Gondek, L.P.; Szpurka, H.; McDevitt, M.A.; Maciejewski, J.P. PLoS ONE 2(11), e1225, 2007, Creative Commons License).

2.2. DISEASES DUE TO GENE MUTATION

The previous section discussed in detail chromosomal mutations, which involve alterations in relatively long stretches of DNA. This section discusses gene mutation that refers to a permanent change in a short segment of DNA that codes for an RNA or polypeptide. Some gene mutations may cause disorders, but these are uncommon in the general population since most of these mutations are repaired by enzymes before the gene is expressed. If left unrepaired, the corresponding protein product may not function properly or may not be synthesized altogether. If the affected protein plays a critical role in the body, then the mutation can disrupt normal development or cause a medical condition. In some instances, a gene mutation may be so severe that the embryo cannot survive. This means that the mutation involved affected a gene that is critical for development in the earliest stages of life. In very rare instances, a mutation may actually provide a beneficial effect that helps the individual adapt better to the environment. For example, the novel protein produced from the mutation can make the person more immune to a certain strain of bacteria (Genetics Home Reference, 2020c; 2020d; 2020e).

Other genetic alterations that occur in more than 1% of the population are called **polymorphisms**. These are considered normal variations in the DNA that do not change the function of the protein made by the affected gene. Examples of these include those responsible for normal differences in eye color, hair color, and blood type. In some cases, polymorphisms may affect the risk of developing certain disorders (Genetics Home Reference, 2020c and 2020e).

Mutations can either be hereditary or acquired. **Hereditary mutations** are inherited from a parent. These are also known as **germline mutations** since these are present in the parent's germ cells (i.e., cells that eventually become sperm or egg cells). When an egg or sperm cell carrying a mutation is involved in a fertilization event, the offspring produced will have the mutation throughout his or her life in virtually every cell in his body. Most of the genetic disorders are inherited in a recessive manner, i.e. the two alleles inherited from both parents must bear the mutation, but some are inherited in a dominant manner, i.e. the disease is expressed even if only one allele has the mutation.

Meanwhile **acquired or somatic mutations**, sometimes referred to also as *de novo* mutations, occur at some time during a person's life and are present only in certain cells of the body. These can be caused by mutagens such as ultraviolet radiation, ionizing radiation, or certain chemicals. These may also occur spontaneously if an error is made during DNA replication and such error is not repaired. This type of mutation is not inherited by the offspring (Genetics Home Reference, 2020c). Somatic mutations that affect cell division are the basis of many types of cancer (Scitable, 2014a).

2.2.1. Disorders Due to Point Mutations

A **point mutation** is called such because only one nucleotide is changed in the genetic sequence. A disorder may result due to a point mutation in a single gene or multiple genes. One type of point mutation is **base substitution** where one nucleotide is swapped with another (see Figure 2.82). For instance, a thymine might be inserted during DNA replication instead of cytosine. Since a pyrimidine is substituted with another pyrimidine, this type of substitution is known as **transition**. A purine → purine substitution also falls under the transition type. On the other hand, if it is a purine → pyrimidine substitution (or vice versa), like an adenine replaced with a cytosine, then it is called **transversion**.

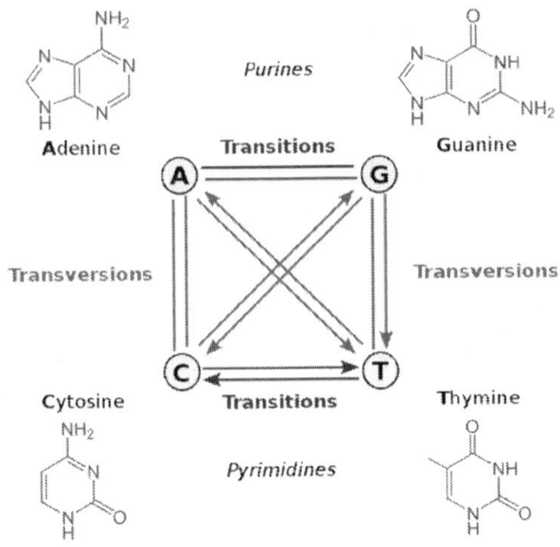

Figure 2.82. Mutation may be a result of base substitution. If a purine is substituted with another purine during replication or if a pyrimidine is replaced with another pyrimidine, then such substitution is called transition. If the substitution is between a purine and a pyrimidine, or vice versa, it is called transversion. (Source: Adapted by Emmanuel J.P. Douzeryfrom Petulda, Creative Commons License).

If the base substitution occurs within exons, it changes the corresponding codon and thus, may change the amino acid sequence. If this occurs, the mutation is called **nonsynonymous**. A nonsynonymous mutation may either be of the missense or nonsense type. If the base substitution changes an amino acid in the protein coded for by the gene, then it is a **missense mutation** (see Figure 2.83). This may or may not affect the function of the resulting protein depending on whether the new amino acid has the same or different chemical properties as the replaced one. Meanwhile, if the base substitution turns a codon for an amino acid into a stop codon, it is called **nonsense mutation** (see Figure 2.83). This results in premature termination of protein synthesis producing a shortened protein that may function improperly or not at all (Genetics Home Reference, 2020f; Loewe, 2008).

In some cases, a base substitution may change the codon, but it won't change the corresponding amino acid (see Figure 2.83). This is called **synonymous mutation** or **same sense mutation**. This is possible because an amino acid may be specified by more than one codon. This is also known as a **silent mutation**. The silent mutation may also include alteration in the intron or intergenic region and polymorphic variation (Loewe, 2008).

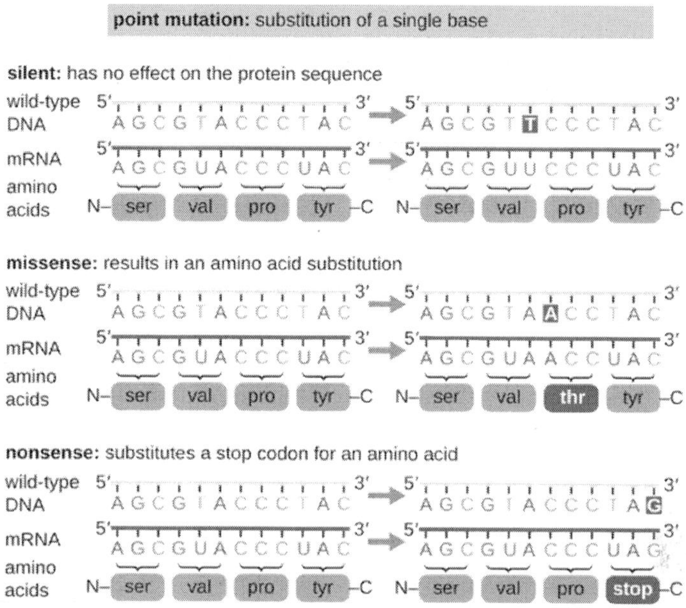

Figure 2.83. Base substitution may change the codon in the mRNA. In turn, this may or may not change the amino acid in the protein product. If the changed codon specifies the same amino acid, then it is a silent mutation. If the amino acid is altered, it is a missense mutation. On the other hand, if the codon becomes a stop codon, it is a nonsense mutation. (Source: OpenStax: Microbiology, Creative Commons License).

Therefore, base substitution may either have a deleterious effect or it may just result in normal variation or polymorphism. In other cases, it may produce a variant of unknown significance (VUS) when it is not known whether it modifies the phenotype or not (Zhou *et al.*, 2005). In some literature, the usage of the term polymorphism is limited to variants produced from base substitution that commonly occur in the human population (i.e., SNPs that occur in more than 1% of the population) while the term mutation is for rare variants (i.e., occurs in less than 0.5-1% of the population) (Stefl *et al.*, 2014).

2.2.1.1. Disorders due to missense mutation

As previously mentioned, replacing a nucleotide in a gene may lead to a different amino acid in the resulting protein. If the replacement has a different chemical property as the original amino acid, it can significantly alter the chemical properties of the protein such as stability, hydrogen bond network, conformation, and function of the protein. A single mutation may affect multiple chemical features since these are usually interrelated. This, in turn, may lead to a genetic disorder and may influence susceptibility to diseases such as cancer and response to drug treatment. Several diseases may also be associated not just with one but with multiple mutations (Stefl *et al.*, 2014).

2.2.1.1.1. Alteration of protein stability

Protein activity is influenced by its three-dimensional conformation and its interaction with its partners, which depends on the stability of monomers and complexes. Thus, a missense mutation that disrupts the stability or binding affinity of a protein complex may alter or completely abolish its function. (Stefl *et al.*, 2014).

An example of a mutation that destabilizes protein molecule is that of the *LMNA* gene. This gene codes for lamins A and C, which are intermediate filaments that form a supportive network underneath the inner nuclear membrane. Certain missense mutations result in lamins that failed to interact with each other properly to form laterally, organized fibers. Instead, they formed small aggregates that are unable to maintain nuclear stability in tissues like the muscles. As a result, these mutations cause a certain type of laminopathies (or lamin-related disorders) that are characterized by muscular dystrophy and dilated cardiomyopathy (Zwerger *et al.*, 2013).

One of these mutations result in *LMNA-associated congenital muscular dystrophy* (L-CMD). This disorder is characterized by decreased fetal or newborn movements and delayed motor developments with weakened axial and proximal muscles. Some affected individuals have neck muscle weakness so that they have a drooping head or poor head control. Cardiac muscles and joints may also be affected in some patients. L-CMD is due to *de novo* mutations in exon 1 and exon 6 of LMNA, which specifically codes for coil 1A and coil 2 of the central rod domain, respectively. The missense mutation in exon 1 involves substituting the 115th nucleotide from adenine to another nucleotide (see Figure 2.84). One possible substitution is c.115A>C (i.e., the adenine which is the 115th nucleotide in the coding or exonic DNA sequence is replaced with cytosine). This results in a missense mutation where the 39th amino acid residue in the protein is changed from asparagine to histidine(i.e., p.Asn39His or N39H).This amino acid is part of the coil1A segment that can form a rigid coiled structure at its hydrophobic

C-terminus and an open structure at its hydrophilic N-terminus. Thus, the coil1A region is thought to regulate lamin structure since it can unwind the lamin molecules and separate these into α-helical monomers or reassemble them into coiled dimers under appropriate conditions (Ishiyama *et al.*, 2018).

The asparagine at the 39th residue is thought to provide hydrophilic properties to the N-terminal region. It may also function in potential head-to-tail association interactions in nuclear lamins. It is part of the segment of coil1A made up mostly of acidic amino acids with negative electrostatic potential. A similar segment is also present in coil2. These acidic patches in coil1A and coil2 may be electrostatically attracted by the positively-charged arginine clusters in the tail and head domains, respectively. Changing asparagine at position 39 with histidine gives a positive charge to the N-terminal acidic patch, which then weakens the head-to-tail association of lamin A dimers and leads to an unstable lamin network (Ishiyama *et al.*, 2018).

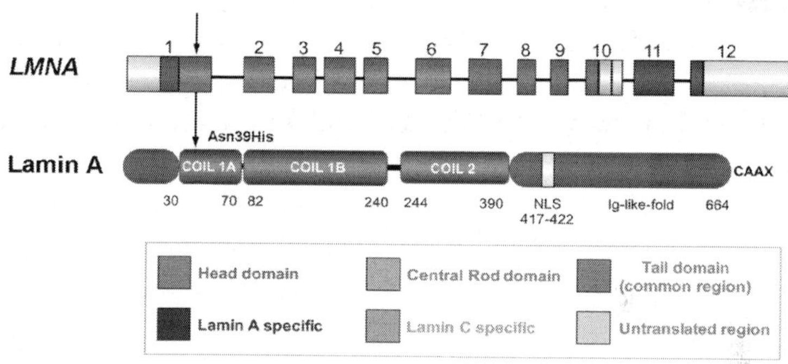

Figure 2.84. Missense mutation in exon 1 of the *LMNA* gene leads to the substitution of the 39th amino acid from asparagine to histidine (black arrow). This affects the electrostatic charge of the coil1A segment and weakens its electrostatic interaction with the tail domain. (Source: Ishiyama, A.; Iida, A.; Hayashi, S.; Komaki, H.; Sasaki, M.; Nonaka, I.; Noguchi, S.; Nishino, I. Human Genome Variation 5, 19, 2018, Creative Commons License).

In some cases, a deleterious mutation may enhance protein stability. This seems to be the case for the *CLIC2* gene that causes *X-linked intellectual disability (XLID)*. XLID accounts for 10-15% of males with intellectual limitations. More than 90 genes located in the X chromosome have been associated with XLID. One of these is *CLIC2* that codes for the chloride intracellular channel 2 (CLIC2). Although the function of this protein is not yet well-established, it is hypothesized to act as a chloride channel. It is also known to inhibit skeletal ryanodine receptor 1 (RyR1) and cardiac RyR2 channels, which both function as major Ca^{2+} release channels in skeletal, cardiac, and in other tissues. It is a soluble membrane with no distinct transmembrane motif. Hence, it is expected that it undergoes a dramatic structural change from soluble to membrane form in order to function as a channel protein. SNPs have been observed in this gene, which do not affect its functionality. However, a disease-causing mutation was identified via sequencing. It is a c.303C>G (i.e., cytosine at position 303 is replaced with guanine) base substitution that leads to a p.H101Q (i.e., histidine at position 101 is replaced with glutamine) change in amino acid. *In silico* simulation demonstrated that one effect of this mutation is to make the three-dimensional structure of CLIC2 more stable so that it decreases the likelihood for the protein to change in conformation as it transitions from soluble to membrane form (Witham *et al.*, 2011).

The enhanced stability may also lead to stronger binding of the mutant CLIC2 to the RyR, which keeps the RyR channel open. As a result, more Ca^{2+} are released. Thus, the mutant CLIC2 functions in reverse compared to the normal CLIC2, which negatively regulates Ca^{2+} release. Since the RyRs are stimulated by the mutant protein, an excess action potential is generated in the cardiac cells of the heart. This may explain why patients with this mutation suffered from cardiomegaly, congestive heart disease, or atrial fibrillation. The nervous system is also affected since RyR2s are also found in the hippocampal neurons, Purkinje cells, and the cerebral cortex. Overactivity of these channels might trigger an excessive release of neurotransmitters and affect other pathways involved in memory and learning. This may account for the seizures and cognitive disability observed in affected individuals (Takano *et al.*, 2012).

2.2.1.1.2. Disruption of hydrogen bond network

Hydrogen bonds are essential to the structure, function, and other properties of macromolecules like proteins. This is formed between a polar (i.e., with a partial positive charge) hydrogen and a negatively-charged hydrogen acceptor such as an oxygen atom. Groups of hydrogen bonds usually form a web of interactions called the *hydrogen bond network*. A missense mutation resulting in the removal, addition or change in position of a hydrogen donor or acceptor may significantly alter the protein's structural integrity, folding, flexibility, pH-dependence, activity, and even cellular localization (Stefl *et al.*, 2014).

For instance, a missense mutation in the gene for lipoxygenase enzyme that is implicated in coronary artery disease disturbed the hydrogen bond network causing decreased protein activity. Lipoxygenase is a lipid metabolizing enzyme that catalyzes the peroxidation of polyunsaturated fatty acids to hydroperoxyl compounds. Specifically, the gene involved is *ALOX15* and the mutation is p.T560M. (i.e., threonine at position 560 is replaced with methionine). Threonine in position 560 of the wild type protein is located in a flexible loop outside the active site of the enzyme. It forms a hydrogen bridge with glutamine at position 294, which is located in a neighboring loop. The replacement of threonine abolished this hydrogen bridge and resulted in a mutant enzyme with significantly reduced catalytic activity (Schurmann *et al.*, 2011).

Among the several missense mutations associated with **familial Alzheimer's disease**, one of them affects the network of hydrogen bonding in a certain fragment (i.e., fragment 21-30) of the amyloid β-protein. Aggregation of oligomers of this protein forms amyloid plaques that are found in the brain tissue of Alzheimer's patients. These oligomers may exist as either a 40 or 42-amino acid peptide (see Figure 2.85) and these are neurotoxic, with the latter form being more neurotoxic. A small number of familial Alzheimer's disease are characterized by early onset of the symptoms. These are associated with missense mutations that change amino acid 22 or 23 of the amyloid β-peptide. The 23rd amino acid is aspartic acid which forms a network of hydrogen bonds that maintains the peptide structure. The p.D23N mutation (i.e., aspartic acid at position 23 is replaced with asparagine) replaces the negatively charged aspartic acid with the neutral asparagine. Consequently, the system of hydrogen bonds between aspartic acid and the surrounding amino acids is disrupted. This changes the peptide structure and increases its susceptibility to enzyme digestion. This may explain the accelerated aggregation of amyloid β-proteins that is associated with familial Alzheimer's disease (Murray *et al.*, 2009).

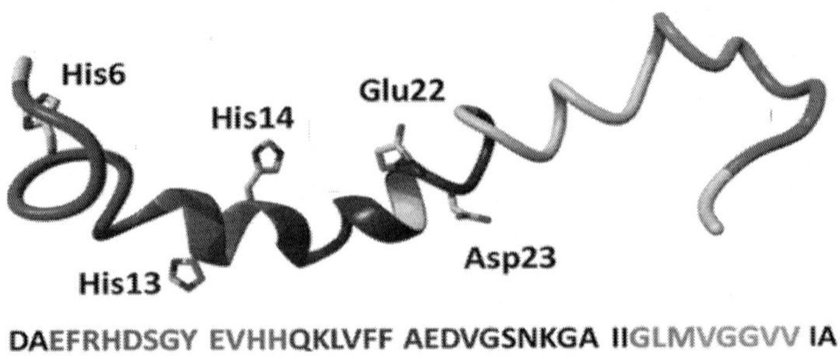

Figure 2.85. The 42-amino acid peptide of amyloid β-protein is neurotoxic and aggregates to form amyloid plaques in the brain of Alzheimer's disease patients. The 23rd amino acid is aspartic acid that forms a network of hydrogen bonds with neighboring residues. (Source: Lermyte, F.; Everett, J.; Lam, Y.P.Y.; Wootton, C.A.; Brooks, J.; Barrow, M.P.; Talling, N.D.; Sadler, P.J.; O'Connor, P.B.; Collingwood, J.F. Journal of The American Society for Mass Spectrometry 30, 2123-2134, 2019, Creative Commons License).

2.2.1.1.3. Disturbance of conformational dynamics

Proteins may change in conformation as they perform their function. Also, several proteins are either intrinsically disordered or have intrinsically disordered regions at certain stages of a biochemical reaction. Missense mutations can affect the flexibility of the protein so that there is a shift in conformational dynamics (Stefl *et al.*, 2014).

This is the case for a missense mutation in *PTEN* gene that is correlated with different disorders like diabetes and cancer. This gene codes for phosphatase and tensin homolog (PTEN) that has significant roles in certain processes like cell survival, proliferation, architecture, and metabolism. When it undergoes p.H61D (i.e., histidine at position 61 is replaced with aspartic acid), PTEN showed decreased number of hydrogen bonds, increased flexibility, and significantly different structure relative to the wild type (Doss and Rajith, 2013). The p.E403K (i.e., glutamic acid at postion 403 is replaced with lysine) missense mutation in the *MCAK* gene has a similar effect on the protein product, which is mitotic centromere-associated kinesin protein. This mutation is associated with colorectal cancer. Using molecular dynamics simulation, the mutant protein showed a significant loss in conformation and stability (Kumar *et al.*, 2013).

Many diseases are characterized by mutant proteins that form polymers, which are accumulated. These include Alzheimer's disease, Lewy body dementia, prion encephalopathies, systemic amyloidosis, and serpinopathies. **Serpinopathies** are caused by the polymerization of a mutant serine protease inhibitor or serine superfamily of proteins. An example of this is α1-antitrypsin deficiency. The α1-antitrypsin is synthesized by hepatocytes and is released into the circulation, where it protects the lung from the actions of neutrophil elastase. Severe deficiency of this protein results from having two copies of the Z allele, which is characterized by the p.E342K (i.e., glutamic acid at position 342 is replaced with lysine) missense mutation (see Figure 2.86). The mutation causes the α1-antitrypsin to assume an intermediate conformation and then to rapidly polymerize. It is hypothesized that the missense mutation disrupted a network of interactions acting as a clasp that regulates change in conformation. The resulting polymers accumulate in the endoplasmic reticulum of hepatocytes. This is associated with neonatal hepatitis, cirrhosis, and hepatocellular carcinoma (Nyon *et al.*, 2012).

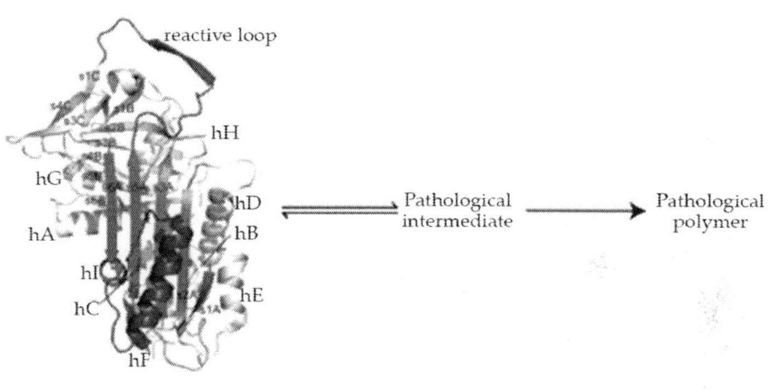

Figure 2.86. A missense mutation changing the 342nd amino acid of α1-antitrypsin from glutamic acid to lysine (encircled area) disrupts a "clasp" that regulates conformational change. As a result, the mutant protein adopts an intermediate conformation that rapidly polymerizes. (Source: Nyon, M.P.; Segu, L.; Cabrita, L.D.; Lévy, G.R.; Kirkpatrick, J.; Roussel, B.D.; Patschull, A.O.M.; Barrett, T.E.; Ekeowa, U.I.; Kerr, R.; Waudby, C.A.; Kalsheker, N.; Hill, M.; Thalassinos, K.; Lomas, D.A.; Christodoulou, J.; Gooptu, B. Structure 20-135(3-3), 504-512, 2012, Creative Commons License).

Another genetic disorder that is caused by a change in protein conformation is **sickle cell anemia** or **sickle cell disease**. This is one of the most common inherited blood disorders with higher prevalence among

African populations. This is caused by a substitution of adenine with thymine leading to p.E6V (i.e., glutamic acid at position 6 is replaced with valine) missense mutation in the *Hgβ* (hemoglobin-beta) gene that encodes the β-globin chain of hemoglobin. Hemoglobin is made up of two α-globin and two β-globin subunits. It is found in red blood cells where it binds oxygen in the lungs and then releases it to various tissues in the body. Normal hemoglobin functions as solitary molecules whether it is bound to oxygen or not. It is a different case for the mutant hemoglobin. The replacement of hydrophilic glutamic acid with a hydrophobic valine allows a weak hydrophobic bond to form between the valine in position 6 and the β-globin chain in an adjacent mutant hemoglobin. This causes the hemoglobin to stack together and polymerize when it releases oxygen (see Figure 2.87).

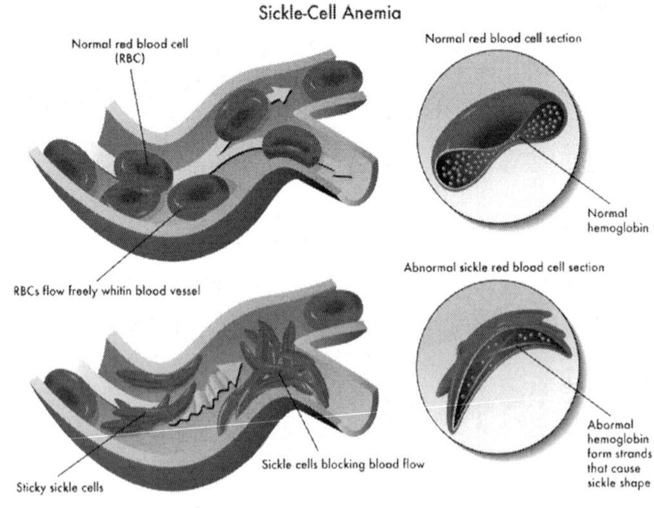

Figure 2.87. Normal hemoglobin molecules occur in red blood cells as solitary units. A missense mutation in the β-globin gene produces a mutant protein product that causes hemoglobin to form long strands when deoxygenated. This distorts the red blood cell so that it may assume a sickle shape that prevents it from flowing freely along small blood vessels. (Source: Diana Grib, Creative Commons License).

Specifically, the hemoglobin molecules aggregate into long bundles made up of 14 strands that twist like a braid. These bundles self-associate into even larger structures. In turn, these stretch and distort the red blood cell so that some of these assume a crescent-like or sickle shape instead of their normal biconcave, disk shape. When it picks up oxygen in the lungs, the mutant hemoglobin depolymerizes and exists as isolated units. These repeated episodes of polymerization and depolymerization damage the hemoglobin and the red blood cell itself, eventually leading to anemia. Moreover, the cyclic polymerization/depolymerization causes the red blood cell membrane to stiffen. These rigid red blood cells cannot flow freely through the small blood vessels so that they pile up and block blood flow to tissues and organs. This leads to painful hypoxic episodes that usually damage organs (National Human Genome Research Institute, 2020; Harvard.edu, 2002).

2.2.1.1.4. Impairment of function

Protein function is largely influenced by its structure and its stability, and vice versa. For instance, some mutations that modify binding specificities of proteins also have destabilizing effects. Also, mutations on functionally important residues, especially polar and charged ones, may often result in more stable structures. This is the case of the *CLIC2* mutation that was previously described (please refer to Section 2.2.2.1.1: Alteration of Protein Stability for a detailed discussion) (Stefl *et al.*, 2014).

Tangier disease is an example of a disorder due to a malfunctioning protein. This is characterized by a very low level of high-density lipoprotein (HDL) circulating in the blood, accumulation of cholesterol-ester in tissue macrophages, and premature atherosclerotic vascular disease. It is caused by a defective ATP-binding cassette transporter 1 protein (ABCA1), which is responsible for transporting cholesterol and phospholipids (i.e., phosphatidylcholine) across the plasma membrane to the cell's exterior (see Figure 2.88). As the name of the protein implies, it also binds and hydrolyses a molecule of ATP in order to translocate lipids. These lipids are then bound by another protein called apolipoprotein A-I (apoA-I) to form HDL. Although apoA-I is located outside the cell, it has a certain portion that binds directly to the extracellular domain of ABCA1 (Nagao *et al.*,2009; Genetics Home Reference, 2012d).

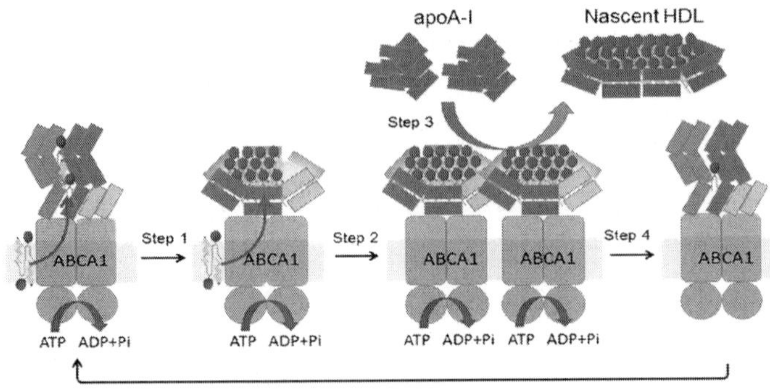

Figure 2.88. (Step 1) ATP-binding cassette transporter 1 (ABCA1) monomers translocate phospholipids and cholesterol across the membrane and then out of the cell. This process is ATP-dependent. (Step 2) The lipids are accumulated in the extracellular domain (ECD) of ABCA1. (Step 3) When enough lipids have been sequestered, ABCA1 forms a dimer and lipid-free apolipoprotein A-I (apoA-I) binds directly to the ECD. (Step 4) The apoA-I picks up the lipids and separates itself from the ECD to form nascent high density lipoprotein (HDL). Then the ABCA1 dimer dissociates into monomers ready to perform step 1 once again. (Source: Ishigami, M.; Ogasawara, F.; Nagao, K.; Hashimoto, H.; Kimura, Y.; Kioka, N.; Ueda, K. Scientific Reports 8, 6160, 2018, Creative Commons License).

Tangier disease is associated with the p.W590S (i.e., tryptophan at position 590 is replaced with serine) missense mutation. This mutation greatly impairs the translocation of cholesterol and phospholipids by ABCA1. This does not affect the binding of ABCA1 with apoA-I, but it slows down the dissociation of these two from each other. It is hypothesized that lipid loading or lipid translocation by ABCA1 changes its conformation so that it will release apoA-I. This means that whenp. W590S mutation impairs lipid loading and translocation, it disrupts a process that is necessary for ABCA1 and apoA-I to be separated from each other (Nagao *et al.*, 2009).

Hypophosphatasia is a rare genetic disorder that is caused by a mutation in the alkaline phosphatase, liver/bone/kidney (*ALPL*) gene. This disrupts the mineralization process during bone and teeth development so that these have low calcium content. Severe forms of the disease may result in death *in utero* while a mild form is characterized by early exfoliation of primary teeth. The most common feature of the disorder is dental deformity (see Figure 2.89). One mutation that leads to hypophosphatasia is p.W29R (i.e., tryptophan at position 29 is replaced with arginine). Molecular dynamics simulation reveals that this causes the N-terminal helix, where amino acid 29 is located, of the resulting protein to be misfolded so that it is separated from the main body and can move freely. This decrease in structural integrity may explain the reduced enzymatic activity observed in the mutant alkaline phosphatase (Chen *et al.*, 2017).

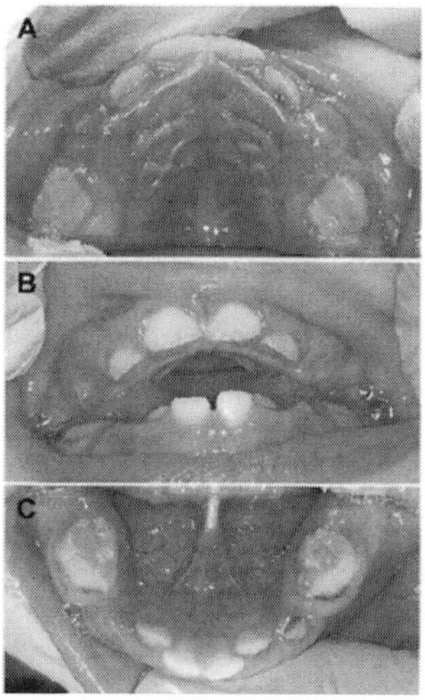

Figure 2.89. An 18-month old boy with hypophosphatasia exhibits (A) high-arched palate, which is an indication of skeletal problems in the maxilla, (B) open bite, and (C) crowding of hypomineralised teeth. (Source: Okawa, R.; Kokomoto, K.; Kitaoka, T.; Kubota, T.; Watanabe, A.; Taketani, T.; Michigami, T.; Ozono, K.; Nakano, K. PLoS ONE 14(10), e0222931, 2019, Creative Commons License).

2.2.1.1.5. Deviation of cellular localization

After being synthesized in the ribosomes, proteins are then transported or sorted into their final location, whether to a specific organelle (e.g., mitochondrion, chloroplast, nucleus, peroxisome, or endoplasmic reticulum) or to be in the cytosol (see Figure 2.90). Upon reaching an organelle, the protein also needs to be sorted properly whether it is to be inserted in the organellar membrane, translocated into a lumen or transported further into a subcompartment of the organelle. Protein transport and sorting is a tightly regulated process involving signal sequences in the protein itself, receptors, energy source, and translocators. A mutation that causes a protein to be in the wrong location can lead to pathological phenotypes.

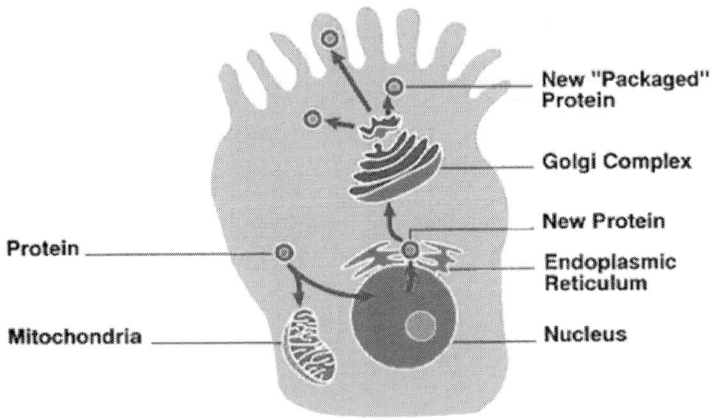

Figure 2.90. After being synthesized, proteins are targeted to their destination, whether to be in the cytosol or to be in a specific part of a certain organelle. (Source: Clip Art, Public Domain).

The previous section discussed a missense mutation in the ABCA1 gene that results in Tangier disease. ABCA1 is synthesized in the endoplasmic reticulum and then transported by vesicles into the plasma membrane (see Figure 2.91). There are a number of missense mutations associated with the Tangier disease that produce mutant ABCA1, which do not migrate into the plasma membrane but accumulate inside the cell. These are p.R587W (i.e., arginine at position 587 is replaced with tryptophan), p.N935S (i.e., asparagine at position 935 is replaced with serine) and p.N1800H, (i.e., asparagine at position 1800 is replaced with histidine) where patients homozygous for these mutant alleles have less than 10% of the plasma HDL levels observed in normal controls; p.A255T, (i.e., alanine at position 255 is replaced with threonine) where homozygotes have approximately10% of the plasma HDL levels in controls; and p.M1091T (i.e., methionine at position 1091 is replaced with threonine), where even heterozygotes have about 30% of the plasma HDL levels observed in the controls. Since the mutant protein is not located in the plasma membrane, there is dramatically reduced cholesterol and phosphatidylcholine efflux (Singaraja *et al.*, 2006).

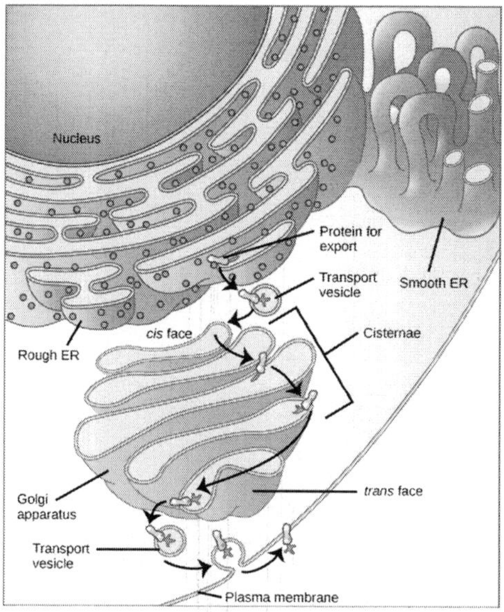

Figure 2.91. Certain proteins like ABCA1, are synthesized by ribosomes on the endoplasmic reticulum and then transported through the Golgi apparatus by vesicles. These vesicles will then bring the proteins to their final location, such as the plasma membrane. (Source: Molnar C.; Gair, J. Concepts of Biology- 1st Canadian Edition, Creative Commons License).

The *AGXT* gene codes for a liver-specific enzyme, alanine-glyoxylate amino transferase (AGT), which metabolizes glyoxylate into glycine. This enzyme is normally targeted into the peroxisome. Certain mutations in the AGXT gene can cause the disorder called **primary hyperoxaluria type 1 (PH1)**. Some of these are missense mutations that cause AGT to be misdirected into the mitochondrion (see Figure 2.92). As a result, there is a deficiency of AGT in the peroxisome so that glyoxylate accumulates and it is converted into oxalate instead of glycine. This leads to progressive deposition of insoluble calcium oxalate initially in the kidney and urinary tract, then eventually all over the body due to chronic kidney failure. The p.P11L (i.e, proline at position 11 is replaced with leucine) missense mutation modifies the N-terminus of the protein so that it now functions as a signal sequence for targeting to the mitochondrion. However, this mitochondrial targeting sequence is normally non-functional since dimerization of the AGT covers the binding site of the mitochondrial import receptor, TOM20 or TOM70, and it keeps the protein in a conformation that does not allow import into the mitochondrion. In some cases, an additional mutation that decreases the efficiency of protein dimerization co-occurs with the mistargeting mutation. These missense mutations include p.G170R (i.e., glycine at position 170 is replaced with arginine), p.I244T (isoleucine at position 244 is replaced with threonine), p.F152I (i.e., phenylalanine at position 152 is replaced with isoleucine), and p.G41R (i.e., glycine at position 41 is replaced with arginine). Since protein dimerization is compromised by any of these mutations, TOM20 can now interact with the mitochondrial targeting sequence in the mutant protein (Schaeffer *et al.*, 2014).

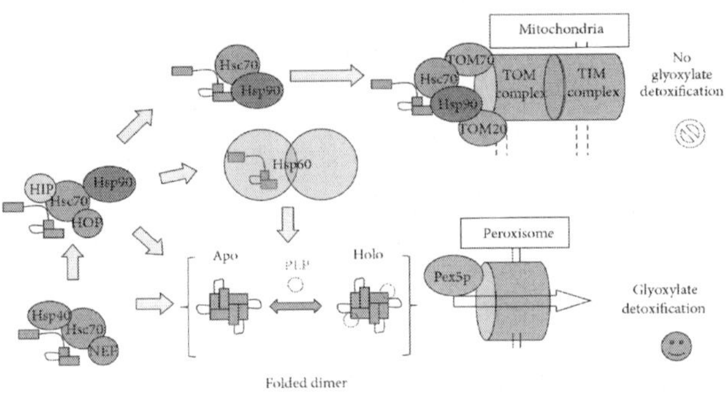

Figure 2.92. After being synthesized by cytosolic ribosomes, the alanine-glyoxylate amino transferase (AGT) (orange) monomer is partially folded when it interacts with Hsp40 and Hsp70 chaperones. Correct folding of the monomer occurs as it is transferred to the Hsp90 and Hsp60 machineries so that it dimerizes. The folded AGT dimer is then imported via the Pex5p import machinery into the peroxisome, where it functions for glyoxylate detoxification. Certain missense mutations generate a targeting sequence that is recognized by the TOM20 or TOM70 receptors leading to the importation of AGT into the mitochondrion. (Source: Pey, A.L.; Albert, A.; Salido, E. BioMed Research International 12, 687658, 2013, Creative Commons License).

2.2.1.2. Disorders due to nonsense mutation

Nonsense mutations may lead to premature termination of proteins synthesis leading to a truncated protein that is nonfunctional. When the nonsense mutation produces a stop codon at least 50 nucleotides upstream of an exon-exon boundary, it may result in **nonsense-mediated RNA decay (NMD)**. This significantly reduces the cytoplasmic level of the affected mRNA and thus, it may result in a complete absence of the protein product. In both scenarios, genetic disorders may result (Antonarakis and Cooper, 2013).

2.2.1.2.1. *Nonsense-mediated RNA decay*

NMD may serve as a mechanism of mRNA quality control so that transcripts containing premature termination codons are degraded. Thus, the cell is protected from the harm brought about by truncated proteins. It may also be beneficial if the potential protein has a deleterious function so that its absence prevents harmful phenotypes from being expressed. Still, NMD may result in a genetic disorder, which is mostly inherited in a recessive manner. This is because, in the heterozygote, the unaffected allele can produce a normal protein whose function is not impeded by mutant ones (Antonarakis and Cooper, 2013). **β-thalassemia** is a disorder due to mutations in the Hgβ gene, which codes for the two β-globin chains that form part of the hemoglobin tetramer along with two α-globin chains (see Figure 2.93). Consequently, hemoglobin production is reduced leading to a lack of oxygen in many parts of the body. Individuals with this disorder may also exhibit anemia (i.e., low levels of red blood cells) leading to pale skin, weakness, and fatigue. They are also at an increased risk of developing abnormal clots (Genetics Home Reference, 2015c). If the mutation is a nonsense type affecting the first or second exon of the Hgβ gene, i.e., shift of codon 17 from AAG to TAG and 39 from CAG to TAG, it leads to NMD. As a result, absent or low levels of mutant mRNAs are observed in the peripheral red blood cells. Correspondingly, mutant proteins are not synthesized or present at very low levels. Heterozygotes for this type of mutation are asymptomatic (Thein, 2013; Frischmeyer and Dietz, 1999).

Figure 2.93. Hemoglobin is a protein component of red blood cells made up of two α-globin chains and two β-globin chains. (Source: OpenStax: Anatomy and Physiology, Creative Commons License).

Another disorder caused by a nonsense mutation that induces NMD is **Marfan syndrome (MFS)**. The affected gene is fibrillin-1 (*FBN1*) coding for a calcium-binding glycoprotein, which is a major component of the 10-12 nm microfibrils of the extracellular matrix. MFS affects the connective tissue, which is found throughout the body. Thus, this disease causes abnormalities in various body parts such as in the heart, blood vessels, eyes, bones, and joints. The two primary features of MFS include vision problems due to ectopia lentis (i.e., displaced lens) (see Figure 2.94), defects in the aorta, and overgrowth of long bones (Genetics Home Reference, 2018f; Frischmeyer and Dietz, 1999). A T → G transversion at nucleotide 26 of exon 51 of the *FBN1* gene creates a premature termination codon TAG at codon 2113. Transcripts containing this stop codon are degraded. The very low levels of mutant mRNAs observed is associated with a mild phenotype that lacked significant vascular or ocular signs and symptoms (Caputi *et al.*, 2002; Frischmeyer and Dietz, 1999).

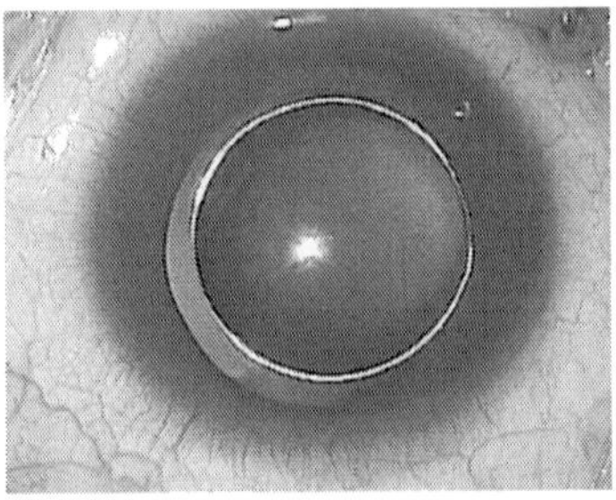

Figure 2.94. One of the common features observed among individuals with Marfan syndrome is having a displaced lens in one or both eyes. (Source: Farat, J.G.; Quieroz, M.F.N.; Lottelli, A.C. Arquivos Brasileiros de Oftalmologia 82(1), 2019, Creative Commons License).

Other examples of genetic disorders that are caused by nonsense mutations include cystic fibrosis, polycystic kidney disease, and muscular dystrophy. Some of these mutations induce NMD of the mutant mRNAs resulting in phenotypes that differ in severity compared to those that produce mutant proteins (Frischmeyer and Dietz, 1999).

2.2.1.2.2. Premature termination of protein synthesis

Some nonsense mutations do not cause NMD so that a protein product is still produced. In this case, the mutant allele may act as a dominant allele since it specifies a protein that lost its function. In a heterozygote, the presence of the normal proteins may not be sufficient to cancel out the pathological phenotype due to the presence of truncated proteins (Antonarakis and Cooper, 2013).

A nonsense mutation of the third or last exon of the Hgβ gene results in a severe form of β-thalassemia in the heterozygotes. Thus, it is considered to be a dominant type of the disorder. The mutation involved is either a c.121G>T (i.e., guanine at position 121 in the exon is replaced with thymine) or c.127C>T (i.e., cytosine at position 127 is replaced with thymine). The resulting mRNAs evade degradation and are translated normally producing truncated proteins, which lack C-terminus and form abnormal dimers with the α-chain. Degradation of these abnormal proteins overwhelms the proteolytic system of the red blood cell precursors. This leads to the formation of inclusions and ineffective erythropoiesis resulting in severe anemia (Hall and Thein, 1994; Frischmeyer and Dietz, 1999).

A severe form of Marfan syndrome also results when a nonsense mutation in the *FBN1* gene gives rise to normal levels of mutant mRNAs that are eventually translated. In the heterozygous condition, the mutant fibrillin-1 monomers disrupt the function of their wild-type counterparts. As a result, the assembly of extracellular microfibrils is negatively affected (Frischmeyer and Dietz, 1999).

2.2.2. Disorders Due to Insertions and Deletions

Mutations due to insertions and deletions are also known as **indels**. As the term indicates, these involve either the addition or loss of one or more nucleotides (see Figure 2.95). Indels are the second most common mutations in the human genome, next to SNPs. Just like SNPs, they can be a source of polymorphism of certain traits, but they can also cause diseases. Indels of one or two nucleotides within the coding sequence of the gene have the greatest effect since these result in frameshift mutation. **Frameshift mutation** refers to a change in the grouping of the nucleotides that make up a codon so that the corresponding amino acid sequence is drastically altered. Consequently, the resulting protein is usually nonfunctional.

On the other hand, the addition or removal of one or more codons will not change the reading frame but will result in a protein with excess or lacking amino acids (Loewe, 2008; Genetics Home Reference, 2020f).

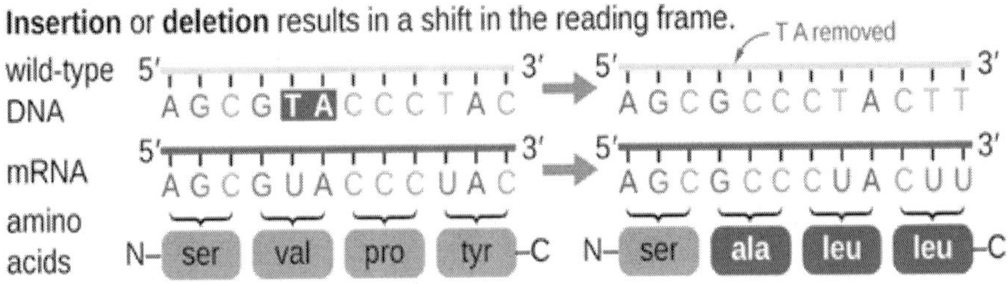

Figure 2.95. Insertion or deletion of nucleotides in the coding sequence of the gene may result in a shift in the reading frame of the gene. This produces a mutant protein with an amino acid sequence that is different from that of the wild type. (Source: OpenStax: Microbiology, Creative Commons License).

2.2.2.1. Disorders caused by frameshift mutation

The addition or removal of one or two nucleotides rearranges the codon with the indel as well as the succeeding codons. This results in a protein with several incorrect amino acids or even a truncated protein if the indel results in the formation of a premature termination codon.

An example of a genetic disorder caused by frameshift mutation is **cystic fibrosis**. This is caused by a mutation in the *CFTR* (cystic fibrosis transmembrane conductance regulator) gene. The CFTR protein functions as a channel in the plasma membrane that allows the transport of chloride in and out of the cell (see Figure 2.96). A dysfunctional or nonfunctional CFTR channel protein disrupts the balance of chloride and fluids at the cell surface causing mucus in various organs to become thick and sticky (see Figure 2.97C). In turn, this leads to lung infections and eventually respiratory failure, poor digestion, and problems in the reproductive system (see Figure 2.97A)(Cystic Fibrosis Foundation, 2020).

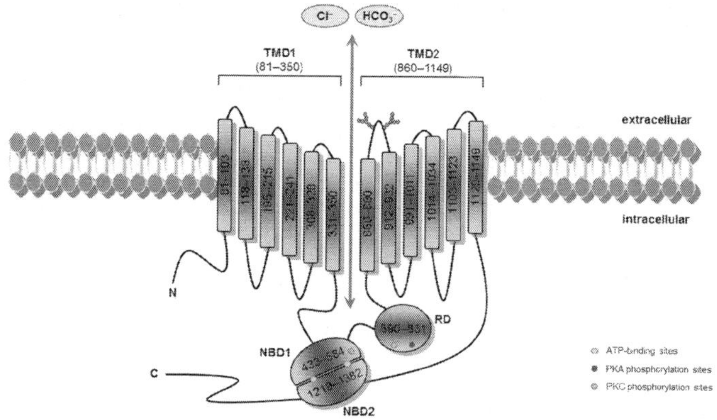

Figure 2.96. The cystic fibrosis transmembrane conductance regulator (CFTR) protein is made up of 1, 480 amino acids with five domains. It has two transmembrane domains (i.e., TMD1 and TMD2) that traverse the plasma membrane. It also has two nucleotide-binding domains (i.e., NBD1 and NBD2) which associate with and hydrolyze ATP. Finally, it has one regulatory domain (RD) which is the site of phosphorylation. CFTR functions to transport chloride (Cl-) across the plasma membrane. (Source: Lopes-Pacheco, M. Frontiers in Pharmacology 7(e98502), 275, 2016, Creative Commons License).

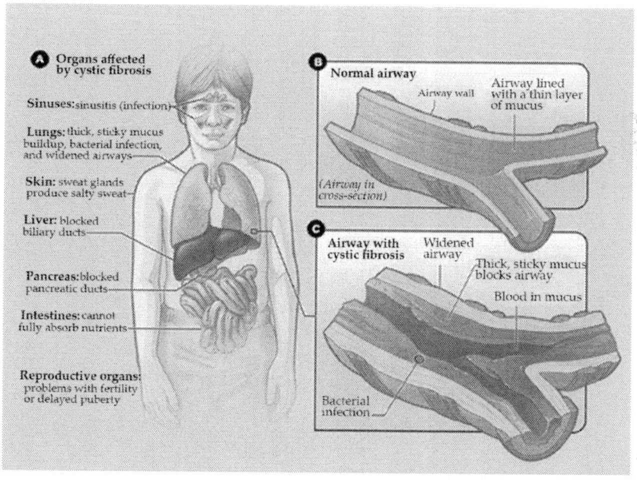

Figure 2.97. (A) Cystic fibrosis affects various organs. (B) A normal airway has a thin layer of mucus, but (C) those with cystic fibrosis have airways that are blocked by thick, sticky mucus containing blood and bacteria. (Source: U.S. National Heart, Lung, and Blood Institute, Public Domain).

There are more than 1,700 mutations in the *CFTR* gene that can cause cystic fibrosis. These include point mutations that lead to missense or nonsense mutations as well as indels inducing frameshift mutations (Cystic Fibrosis Foundation, 2020). Two frameshift mutations that can lead to cystic fibrosis involve nucleotides in exon 7 of the gene. This exon codes for 82 amino acids that form the transmembrane domain of the channel protein. The insertion of TC (i.e., thymine and cytosine) in nucleotide position 1154 shifts the reading frame and introduces a UAA stop codon at codon 369. On the other hand, a deletion of T (i.e., thymine) at position 1213 also shifts the reading frame and generates UAA at codon 368 (Iannuzzi *et al.*, 1991).

Another example of a genetic disorder that may be caused by frameshift mutation is **Tay-Sachs disease**. This is caused by mutations in the *HEXA* gene that specifies the β-hexosaminidase A enzyme. After being synthesized, this enzyme is transported from the ER into the lysosomes where it breaks down the GM2 ganglioside, a lipid found in high amounts in the neurons (see Figure 2.98). Disruptions in the function of β-hexosaminidase A leads to the accumulation of GM2 ganglioside to toxic levels. The buildup of this substance causes the destruction of the neurons in the central nervous system. The most common form of Tay-Sachs disease becomes apparent at infancy. Affected infants have delayed development and weakened muscles at around 3-6 months. As the disease progresses, they experience seizures, vision and hearing loss, and paralysis. They also show intellectual disability. Other forms have milder symptoms(i.e., muscle weakness, ataxia or loss of muscle coordination, speech problems, mental illness)and that appear in childhood, adolescence, or adulthood (Genetics Home Reference, 2012e).

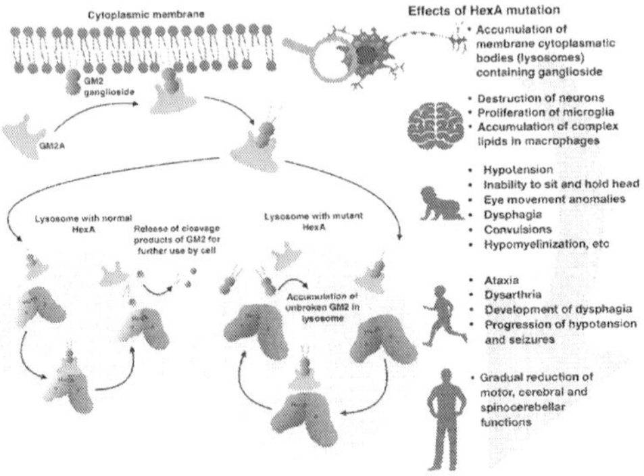

Figure 2.98. β-hexosaminidase A is an enzyme in the lysosome that hydrolyzes GM2 gangliosides. Therefore, a mutation in this enzyme leads to the accumulation of GM2 gangliosides in the lysosome. This leads to progressive destruction

of neurons causing reduced motor and mental functions. (Source: Solovyeva, V.V.; Shaimardanova, A.A.; Chulpanova, D.S.; Kitaeva, K.V.; Chakrabarti, L.; Rizvanov, A.A. Frontiers in Physiology 9, 1663, 2018, Creative Commons License).

The most common mutation in the *HEXA* gene that causes the disease is an insertion of four nucleotides in exon 11. This introduces a premature stop codon after nine nucleotides from the site of insertion. This leads to mRNA instability so that there are severely reduced levels of mutant mRNAs (Boles and Proia, 1995). Another frameshift mutation that leads to Tay-Sachs disease is due to a deletion of cytosine at nucleotide position 1510 that introduces a premature termination codon at codon 508. This results in an α-subunit of β-hexosaminidase with missing 22 amino acids. This abnormal protein is not transported to the lysosome but is retained in the ER and eventually degraded (Lau and Neufeld, 1989).

2.2.2.2. Disorders caused by non-frameshifting indels

In some instances, indels may not change the way that the mRNA is translated into amino acids. This happens when one or more codons are deleted or added (see Figure 2.99). In this case, a protein is usually produced but it has extra or missing amino acids. While some non-frameshifting indels have a neutral effect on the phenotype, several of these result in human disorders. A pathogenic non-frameshifting indel commonly occurs at conserved sites, functional sites, or located in buried regions of the protein. It is usually longer than a neutral indel and results in a more hydrophobic and more apolar protein (Bermejo-Das-Neves *et al.*, 2014).

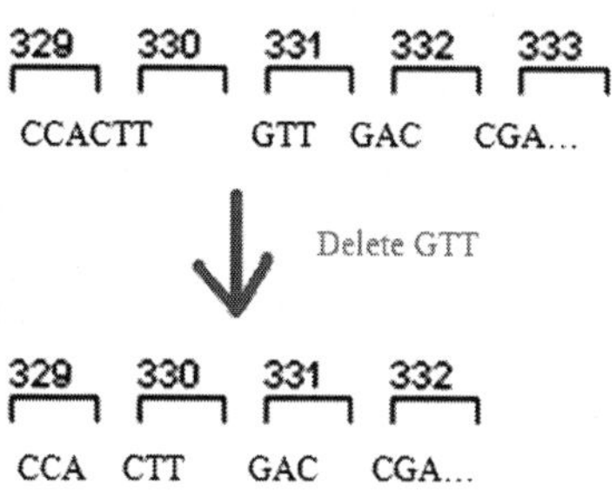

Figure 2.99. The deletion of a 3-nucleotide codon (e.g. codon GTT) will not change the way that the rest of the mRNA is translated but it will produce a protein with a missing amino acid. (Source: Adapted from Sumukal, Creative Commons License).

The most common mutation that causes cystic fibrosis is a deletion of a codon that specifies phenylalanine in position 508 (F508del) of the protein. This mutation is present in about 85% of cystic fibrosis patients. The resulting mutant protein fails to stay in the correct three-dimensional form. As a result, it is degraded before it reaches the plasma membrane (Cystic Fibrosis Foundation, 2020).

2.2.3. Disorders Due to Variation in Gene Copy Number

Humans are diploid, which means that they have two copies of each gene in their genome. One allele comes from the mother while the other one is from the father. However, a person may have more than two copies or less commonly, less than two copies of certain genes. This is also considered as CNV but at the genetic level. This may be a result of insertion, deletion, or duplication of segments, which are large enough to include whole genes (Genetics Home Reference, 2020g). Some of these variations are categorized as **copy number polymorphisms (CNPs)**, which are mostly less than 10 kb in length and occur in the general human population at greater than 1% frequency (Eichler, 2008).

More than 10% of the human genome contain these differences in gene copy number. This variation can influence the activity of genes and may affect body processes. However, much of these do not affect health or development. For instance, CNPs are usually observed in genes that code for proteins that are important in drug detoxification and immunity. However, some differences may influence a person's risk of diseases and response to certain drugs. CNV of certain genes has been associated with complex neurodegenerative disorders as well as various types of cancer. In addition, CNPs associated with immune response genes were correlated with susceptibility to certain genetic disorders such as psoriasis and Crohn's disease (Genetics Home Reference, 2020g; Eichler, 2008).

Duplication and triplication of the locus containing the *SNCA* gene were observed in a greater proportion of patients with **Parkinson's disease (PD)** that have a family history of the disease. The gene multiplication is thought to be a result of intra-allelic duplication and inter-allelic recombination with unequal crossing-over. The SNCA gene codes for α-synuclein. **Lewy bodies**, the pathological hallmark of PD, is composed mainly of α-synuclein (see Figure 2.100). The severity of the clinical phenotype of the patient is correlated with the number of copies of *SNCA*. This may indicate that abnormal expression of the gene, resulting in abnormally high levels of α-synuclein, is the key event in PD rather than a qualitative mutation of the wild-type protein. Therefore, SNCA dosage is correlated with age of onset, progression, and severity of symptoms which include parkinsonism (i.e., tremor, slow movement, impaired speech, or muscle stiffness), autonomic dysfunction, and dementia observed in the affected individuals (Singleton *et al.*, 2003; Ross *et al.*, 2008).

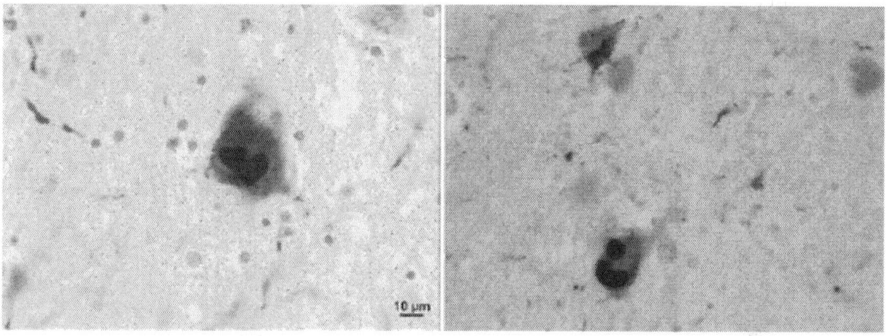

Figure 2.100. Regions of substantia nigra in a patient in Parkinson's disease that were magnified 60 times. It shows the α-synuclein inclusions in the neurons that aggregated to form Lewy bodies. (Source: Suraj Rajan, Creative Commons License).

Certain breast cancers are associated with overexpression of the *ERBB2* gene. This codes for human epidermal growth factor receptor 2 protein. High copies of *ERBB2* can cause overexpression and are associated with aggressive forms of breast cancer. Thus, determining the copy number of this gene can be used to diagnose breast cancer (Lobo, 2008).

The copy number of the *HBD-2* gene is also associated with certain disorders. This codes for human β-defensin 2, which is a member of a family of antimicrobial peptides that protect the intestinal lining against bacterial invasion (Fellermann *et al.*, 2006). A person may have 2-12 copies of the β-defensin genes in his genome (Hollox *et al.*, 2008). It was determined that healthy individuals have an average of four copies of the *HBD-2* gene. On the other hand, those with colonic **Crohn's disease (CD)** have three or lesser copies of the gene. This relatively low gene copy number is associated with reduced *HBD-2* mRNA levels in the intestinal mucosa. This, in turn, predisposes the individual to colonic CD. CD is a severe chronic inflammatory bowel disease that is characterized by ulcers in the intestine, mostly in the ileum and colon. The inflammation may lead to abdominal pain, severe diarrhea, fatigue, weight loss, and malnutrition (Fellermann *et al.*, 2006).

On the other hand, an increased copy number of the *HBD* genes is significantly associated with the risk of developing **psoriasis**. Production of human β-defensin 2 is induced as part of the inflammatory response in the skin. Psoriasis is a common inflammatory skin disease, which is characterized by red scaling elevated plaques on the elbows, knees, and trunk. Examination of the skin tissue with psoriasis shows inflammation and deviant differentiation of the epidermis. A study among Dutch and German patients with psoriasis indicates that each additional copy of the *HBD* genes above two copies increases the risk of having psoriasis and that each copy adds 34% to the relative risk (Hollox *et al.*, 2008).

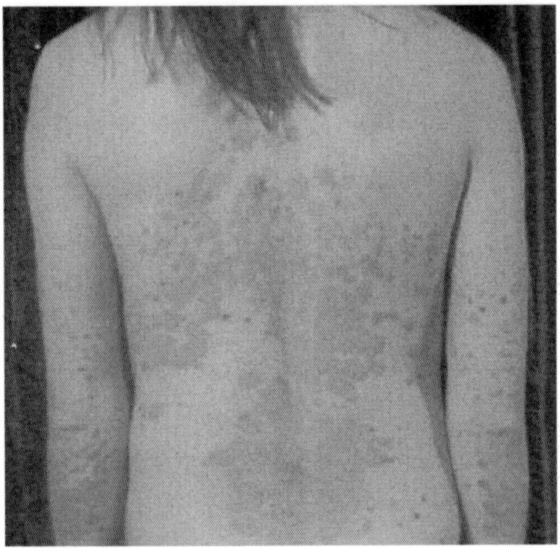

Figure 2.101. Psoriasis is a skin inflammatory disorder that is characterized by red elevated plaques on the arms and torso. (Source: James Heilman and The Wednesday Island, Creative Commons License).

2.2.4. Diagnosis of Gene Mutation

Determining variation in gene copy numbers and the presence of a mutant allele can be accomplished via techniques that are described in the previous sections such as the various types of PCR and electrophoresis, FISH, and microarrays. On the other hand, determining the identity of the nucleotides that were changed in a mutated gene can be done primarily by sequencing methods.

2.2.4.1. DNA sequencing as a diagnostic tool

The development of sequencing tools revolutionized the field of genetics, including its medical applications. Although in the early periods, sequencing was tedious and expensive. Thus, it was not commonly used as a laboratory or clinical tool for diagnosis. **Next- generation sequencing technologies** that enabled the parallel sequencing of millions of DNA or RNA nucleotides at a time were eventually developed. This made it plausible to employ the equencing method as a widely used tool for genetic testing.

Prior to determining its sequence, genomic DNA must be extracted. Then it is usually fragmented by restriction enzyme digestion (see Figure 2.102). The DNA fragments are then inserted into vectors, usually a plasmid, which are used to transform bacterial cells or virions. This collection of bacterial cells or virions containing the fragments of genomic DNA is the **DNA library**. In the case of **next-generation sequencing (NGS)** technology, preparing DNA for sequencing depends on the platform to be used. In general, it involves cutting the extracted DNA into fragments of the desired length. Fragmentation may be carried out by the usual enzymatic digestion or by physical methods, such as acoustic shearing and sonication (Head *et al.*, 2014).

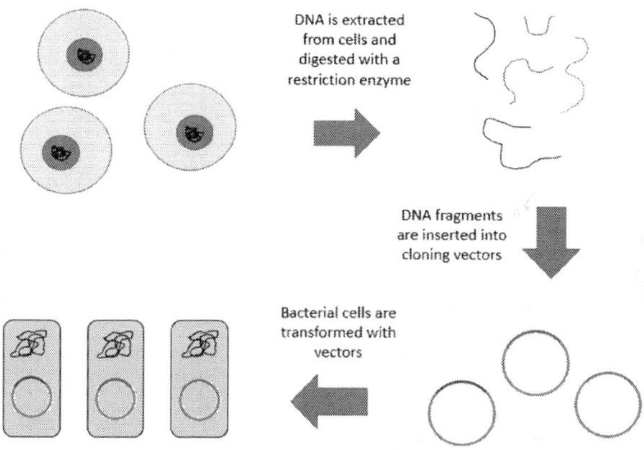

Figure 2.102. A DNA library is prepared from the extracted genomic DNA that is eventually digested by restriction enzymes. The resulting fragments are inserted into vectors which are eventually taken up by bacterial cells. (Source: Aluquette, Creative Commons License).

2.2.4.1.1. Sanger sequencing

Sanger sequencing is one of the first-generation sequencing technologies. This was named after Frederick Sanger who developed it in 1975. This essentially determines the sequence of nucleotides as it is replicated. The DNA to be sequenced is used as a template for DNA synthesis, which is catalyzed by DNA polymerase. A DNA primer is also required to initiate the synthesis. Four DNA synthesis reactions are prepared, which contain normal A, G, C, and T dNTPs. These are the building blocks that are joined together by DNA polymerase to form a new DNA strand as it replicates the template. Each reaction also contains a low amount of one of the four dideoxynucleotide triphosphates (ddNTPs), i.e. ddATP, ddGTP, ddCTP, or ddTTP. These are similar to dNTPs but lack a 3' hydroxyl (-OH) group (see Figure 2.103). Each of these is labeled with a different fluorescent dye. When a ddNTP is incorporated into the strand during the reaction, synthesis is terminated because the 3' OH group which is needed to form the link with the next nucleotide is missing (Scitable, 2014b). For this reason, Sanger sequencing is also called the chain termination method. At the end of each reaction, DNA fragments of various lengths are generated, all of them ending with a fluorescently labeled ddNTP (Merck, 2019).

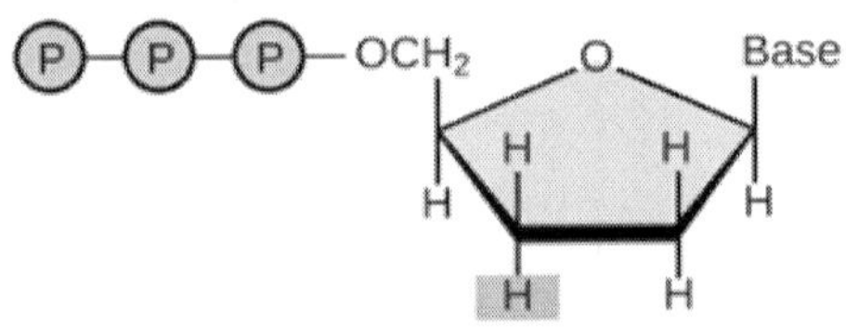

Dideoxynucleotide (ddNTP)

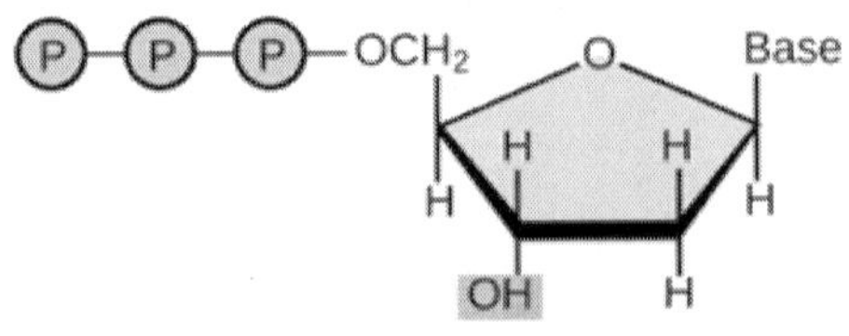

Deoxynucleotide (dNTP)

Figure 2.103. A dideoxynucleotide (ddNTP) is similar in structure to a deoxynucleotide (dNTP) except for its lack of 3' hydroxyl (OH) group (shaded light blue). The 3' OH group is needed in forming a bond with the next nucleotide during DNA replication. For this reason, elongation of DNA stops when a ddNTP is incorporated in the DNA strand being synthesized. (Source: OpenStax CNX, Creative Commons License).

The DNA fragments produced from the four reactions are then separated by electrophoresis. Early sequencing efforts used polyacrylamide gels for this process. These were eventually replaced by capillary electrophoresis, which automated the procedure and thus, increased the output. In this step, the reaction products are injected into a capillary, which is a long glass tube filled with gel polymer (see Figure 2.104). The fragments are separated based on their size or molecular weight. Just before reaching the positive electrode at the end of the capillary, the fluorescently labeled DNA fragments cross the path of a laser beam, which activates and causes the dye on the fragment ends to fluoresce. Expectedly, the smallest fragment will move across the laser beam first, followed by the next smallest, and so on (Schoales, 2015; Khan Academy, 2016).

Since the four types of ddNTP are each labeled with a different fluorescent dye, they emit fluorescence at different wavelengths. These are detected by an optical device and recorded in a chromatogram or electrophoretogram which shows the fluorescence peak of each nucleotide (see Figure 2.103). Certain software converts the fluorescence signals to digital data, translating a certain signal to a corresponding nucleotide so that the DNA sequence is eventually determined (Schoales, 2015; Applied Biosystems, 2009; Khan Academy, 2016).

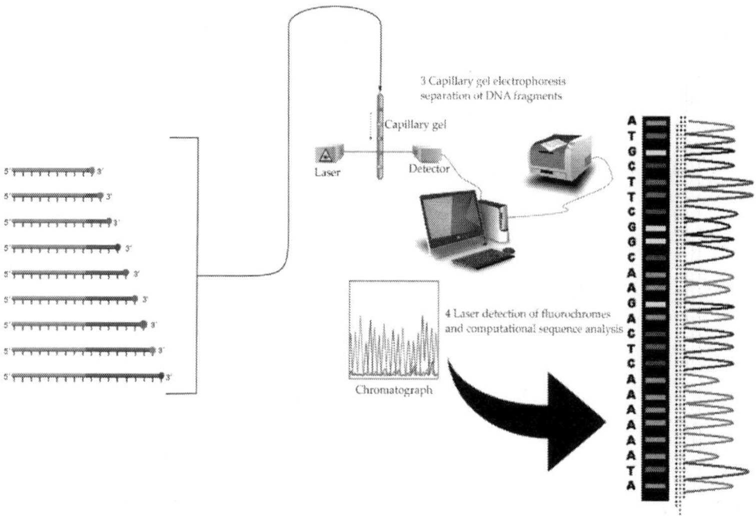

Figure 2.104. Capillary electrophoresis separates the DNA fragments from the Sanger reactions based on molecular weight. As each fragment moves past a laser, the fluorescently labeled dideoxynucleotide (ddNTP) at the end of the fragment gets activated. Each type of ddNTP emits fluorescence at a different

wavelength. The resulting chromatogram is translated by computer software into a DNA sequence. (Sources: (Left image) Estevesj, Creative Commons License; (Right image) Modified from Abizar, Creative Commons License; (Arrow) SimpleIcon, Creative Commons License).

In the case of a heterozygous locus with both the normal and mutant allele, two fluorescent peaks of the same intensity will be observed. On the other hand, if a locus has two copies of the mutant allele, then the expected fluorescent color will be replaced completely by the color of the new nucleotides. Sanger sequencing can detect base substitutions and indels (Gomes and Korf, 2018). As such, it can be used for targeted sequencing of exons or areas within exons to determine the presence of a certain mutation that is known to be associated with a disease. For example, DNA from a tissue biopsy sample from melanoma may be sequenced to determine if it harbors a mutation in the *BRAF* gene, which is found in more than 50% of melanomas. Patients diagnosed to have such a mutation may be treated with drugs that target this mutation instead of the standard chemotherapy (Lab Tests Online, 2020).

Sanger sequencing is also used to sequence entire genes, encompassing all the exons as well as certain non-coding areas such as the promoter and introns. Some genetic disorders are caused by several mutations in various regions of a gene. Thus, it is important to assess the entire sequence of the concerned gene (Lab Tests Online, 2020).

This sequencing technology has also been utilized for several decades in identifying the complex mutations present in certain tumors. This is done by designing primers that cover several regions of interest or several genes in the sample DNA (Gomes and Korf, 2018). For example, Sanger sequencing is used to routinely assess DNA samples from certain brain tumor specimens (e.g., gliomas, astrocytomas, or glioblastomas). Specifically, it is used to detect the presence of point mutations in genes relevant to such brain cancers like *IDH1, IDH2, H3F3A, HIST1H3B*, and *BRAF* as well as *TERT* promoter region (Solomon, 2018).In other cases, Sanger sequencing is used to determine mutations in genes that increase the risk of having cancer. This is the case for gene mutations that cause **Lynch syndrome**, which is an inherited condition. Mutations in the *MLH1, MSH2, MSH6, PMS2* and *EPCAM* genes cause the syndrome, which predisposes the affected person to many types of cancer such as colon cancer and endometrial cancer (Lab Tests Online, 2020).

A distinct disadvantage of this method is that it is more expensive compared to other multiplex testing systems. Therefore, its use is usually limited to sequencing a specific gene or a small subset of genes, and not the whole genome (Gomes and Korf, 2018). It is also usually limited to assessing known mutation hot spots. In the case of disease due to novel or rare gene mutations or a complex disorder with undetermined causative genes, Sanger sequencing is not an ideal diagnostic tool (Aleem, 2019). It can also be used to determine mosaic mutations. However, Sanger sequencing cannot provide a precise quantity of mosaic mutation, e.g. whether a mutation is present in 30% or 50% of cells. Additional testing methods should be used for quantification (Gomes and Korf, 2018). Another limitation of Sanger sequencing is its sensitivity of detection, which is approximately 15%-20% mutant allele frequency. This means that 15%-20% of the DNA molecules being sequenced need to contain the mutation in order to be reliably detected. Considering that human cells are diploid, the tissue sample used for DNA extraction must contain at least 30%-40% tumor cells to detect mutations in oncogenes that are heterozygous (i.e., present only in one copy) and fully clonal (i.e., present in all tumor cells). For certain tumors, it is challenging to obtain a tissue sample with sufficient tumor cells since the sample may contain abundant non-tumor cells or inflammatory infiltrate. In such cases, Sanger sequencing may provide a false negative result for the mutation being tested (Solomon, 2018).

2.2.4.2.2. Next-generation sequencing

As mentioned previously, Sanger sequencing is a laborious method that it took more than a decade to complete the sequencing of the human genome. In addition, it is very costly at a price of US$500,000-1,000,000/gigabytes (Gbp). On the other hand, post-Sanger sequencing technologies, which are collectively called next-generation sequencing (NGS), enable sequencing of a DNA sample within days at a cost that is 1,7000 times lower than Sanger sequencing (Saunders and Kingsmore, 2011). NGS methods are much faster than Sanger sequencing because they sequence millions of small DNA fragments from different parts of the genome at the same time instead of reading a single DNA fragment from one part of the genome. For this reason, NGS is also called **massively parallel sequencing**. NGS is also more sensitive in detecting mutations than Sanger sequencing. It can detect as low as 2-5% mutant allele frequency (Lab Tests Online, 2020).

NGS include second- and third-generation sequencing. **Second-generation sequencing techniques** convert the starting DNA fragments into double-stranded molecules that are flanked by adapters, which are oligonucleotides that enable the binding of target DNA to a substrate. The substrate is either a bead or a flowcell (see Figure 2.105). Then amplification by PCR is done to generate thousands of DNA on the substrate, and the

DNA sequences are then determined by detecting light, fluorescence, or pH (Knief, 2014). This amplification step increases signal intensity which may reduce sequencing error. Multiple sequencing reactions can occur and be monitored at once. However, not all of the reactions are 100% efficient. For instance, some of the DNA fragments may have missed a base incorporation or ligation causing these to be out of synch with those who have. This asynchronization becomes amplified with each additional cycle creating noise during sequencing. This is the major factor why second-generation sequencers initially generated short sequence reads, about 300-500 bp, but these were increased by improvements in the technology (Peterson and Arick II, 2018).

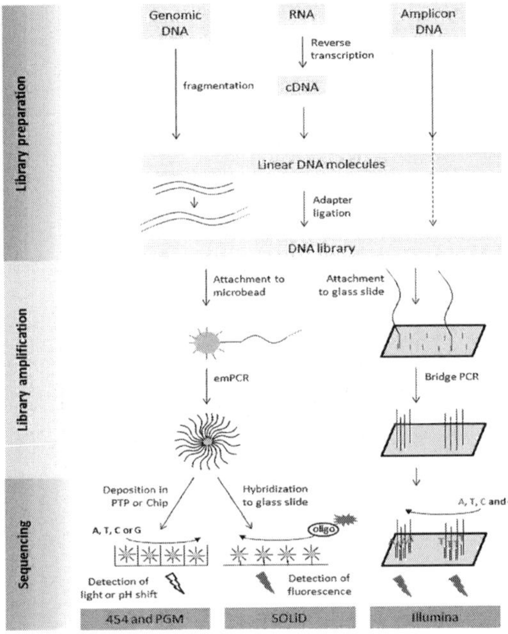

Figure 2.105. The commonly used second-generation sequencing platforms share certain similarities. They prepare their DNA library by making their starting DNA material double-stranded and ligating it to adapters. These are then amplified into thousands of copies by PCR prior to sequencing. The sequencing method varies among the platforms in terms of the data that is collected, whether it is fluorescence, light, or pH. (Source: Knief, C., Frontiers in Plant Science 5, 216, 2014, Creative Commons License).

The short reads generated by second-generation sequencing cannot fully span over repetitive regions creating gaps in the genome sequence. **Third-generation sequencing technologies** provide much longer reads of more than 10 kbp which can close these gaps. However, these produce lower quality reads and lower throughputs compared to second-generation techniques. These are also more expensive (Peterson and Arick II, 2018; Li and Harkess, 2018; Rhoads and Au, 2015). One major difference between third-generation sequencing and the second-generation techniques is that the former no longer amplifies the DNA fragments but determines the sequence directly from the template DNA making it useful for *de novo* sequencing (Park and Kim, 2016).

NGS can also be used to sequence the transcriptome (the total RNA sequence). PacBio sequencing, a third-generation sequencing platform, has been used to analyze multiple mRNA isoforms derived from a single gene (Smith, 2017).

2.2.4.2.2.1. Whole exome sequencing

Because NGS is both high speed and high throughput, it can be used for **whole exome sequencing(WES)**. **Exome** refers to all the exons or protein -coding regions of all the genes in the genome. The human genome (i.e., both the nuclear and mitochondrial DNA) is made up of an estimated 180, 000 exons, approximately 1% of the entire genome. These exons make up about 22, 000 genes. Most of the pathogenic mutations (approximately 85%) occur in the exome. The WES approach can cover 95% of the exons. Therefore, it is an efficient and thorough method of determining the mutations that cause rare, monogenic genetic disorders as well as in identifying multiple candidate genes that predispose individuals to common diseases and cancers (Oristaglio *et al.*, 2015; Mahdieh and Rabbani, 2013).

In WES, the determined exon sequence is compared with reference sequences representing what is normal in the population. Variations in the patient's DNA sequence relative to the standards can then be identified and evaluated whether it is common (or polymorphic) or rare. These variants can be helpful to confirm a clinical diagnosis or to assess the risk of having a disease (Oristaglio *et al.*, 2015; Aleem, 2019).

Using WES to find the specific DNA mutation has proven to be useful in determining the correct medical management to be recommended to a patient. As an example, a 15-month old male child presented features associated with Crohn's disease, i.e. diarrhea, peri-anal abscesses and proctitis (i.e., inflammation of the inner lining of the inner rectum), transmural pancolitis (i.e., universal colitis or inflammation of the five tissue layers of the intestine) with colocutaneous fistulae (i.e., abnormal openings or tracts between the colon and the skin). He was given therapy for Crohn's disease namely, nasoenteric feeding and the drug infliximab. However, the perineal fistulae (i.e., misplaced and narrowed anal opening) persisted, new fistulae developed, and the entire colon was still inflamed. Various treatments and surgeries to address his symptoms were done until he was 4 years old. Finally, exome sequencing revealed that he has a G→A substitution at a highly conserved position in the *XIAP* (X-linked inhibitor of apoptosis protein) gene. This base substitution resulted in a p.C203Y (i.e., cysteine at position 203 is replaced with tyrosine) missense mutation. *XIAP* was not associated with inflammatory bowel disease like Crohn's disease. However, *XLP2*, a disorder, caused by *XIAP* mutation, has colitis as one of its symptoms. Based on this information, allogeneic hematopoietic stem cell transplantation was performed on the patient to address the *XIA* Pprotein deficiency. Within 42 days after the transplant, the child was able to eat and drink and there has been no recurrence of gastrointestinal disease (Worthey *et al.*, 2011).

Just like other diagnostic tools, WES has some limitations. For one, the exons in the genome may not be adequately covered. Depending on the sequencing platform used, repeat sequences are not sufficiently represented and deletion of a certain exon may be missed. Also, WES detects indels less reliably and it cannot readily detect small CNVs. Targeted WES can also be performed focusing on a subset of genes that are associated with a certain disorder. However, not all the genes relevant to diseases are currently known (Smith, 2017).

2.2.4.2.2.2. Whole genome sequencing

NGS can also be used for whole genome sequencing (WGS), where the sequence of the individual's entire genome is determined. This aims to determine the sequence even of non-coding DNA, such as introns and intergenic regions. Even if these regions do not code for proteins, they may have a regulatory function on gene expression. Therefore, analyzing their sequence relative to a reference genome may still provide helpful information on the genetic basis of diseases (Oristaglio *et al.*, 2015; Aleem, 2019).

WGS is helpful in identifying a specific mutation carried by a patient whose disease may be caused by more than one type of mutation. An example is Charcot-Marie-Tooth (CMT) disease, which may be caused by mutations in 37 genes with different inheritance patterns (Kingsmore and Saunders, 2011).

WGS approach has also been key in arriving at a correct diagnosis despite puzzling clinical phenotypes or those that have characteristics of more than one type of genetic disorder. This is the case for two siblings that displayed a phenotype resembling that of a genetically heterogeneous group of diseases called **hereditary sensory and autonomic neuropathies (HSANs)**. HSAN is characterized by pain sensitivity, hypohydrosis (i.e., decreased sweating), and mild muscle weakness in a subset of patients. The two siblings exhibited severe pain insensitivity and hypohydrosis since birth. They had been examined by several local and international experts and the initial clinical diagnosis given was an immune-inflammatory disease. This diagnosis was based on the severe destruction of large and small joints as well as the vertebrae of the two patients. However, the severe pain insensitivity remained unexplained. WGS was performed and it revealed a homozygous mutation in the leukemia inhibitory factor receptor (*LIFR*) gene. Mutation in this gene is associated, not with HSAN, but with **Stüve-Wiedemann syndrome (SWS)** (Aleem, 2019). This is an autosomal recessive disorder characterized by severe bone abnormalities, dysfunction of the autonomic nervous system, and reduced ability to feel pain. Many of those born with this condition do not survive past infancy. However, some survive into adolescence or later just like the two siblings who were aged 10 and 14 years (Genetics Home Reference, 2016).

REFERENCES

1. Ahmad, M., Silvera-Redondo, C. & Rodriguez, M.H., (2010). Nondisjunction and chromosomal anomalies. SaludUninorte, 26(1), 117-133
2. Ahn, J.Y. & Lee, J.T., (2008). X Chromosome: X Inactivation. Nature Education, 1(1), 24.
3. Aleem, A.A., (2019). DNA Sequencing Resolves Misdiagnosed and Rare Genetic Disorders. In Biochemical Analysis Tools-Methods for Bio-Molecules Studies; Boldura O.-M.; Baltă, C.; Awwad, N.S. (ed.). IntechOpen. DOI: 10.5772/intechopen.86556 (Accessed on 5 July 2020).
4. Antonarakis, S.E. & Cooper, D.N., (2013). Chapter 7-Human Gene Mutation in Inherited Disease: Molecular Mechanisms and Clinical Consequences. In Emery and Rimoin's Principles and Practice of Medical Genetics (6thedn.); Rimoin, D.; Pyeritz, R.; Korf, B. (ed.). Academic Press: Massachusetts. pp.1-48. https://doi.org/10.1016/B978-0-12-383834-6.00007-0 (Accessed on 22 June 2020).
5. Aplan, P.D., (2006). Causes of ontogenic chromosomal translocation. Trends in Genetics, 22(1), 46-55. doi: 10.1016/j.tig.2005.10.002 (Accessed on 15 May 2020).
6. Aplenc, R., Lipsitz, E.G. & Adamson, P.C., (2011). Chapter 15-Oncology. In Pediatric Secrets (5thedn.); Polin, R.A.; Ditmar, M.F. (ed.). Mosby Elsevier: Philadelphia. pp. 584-613. https://doi.org/10.1016/B978-0-323-06561-0.00015-X (Accessed on 20 May 2020).
7. Applied Biosystems, (2009). DNA Sequencing by Capillary Electrophoresis (2ndedn.). https://tools.thermofisher.com/content/sfs/manuals/cms_041003.pdf?icid=lsg_gsd_CP1236_PJT8083_blog_seqitout-howsangerworks (accessed August 7, 2019).
8. Baronchelli, S., Conconi, D., Panzeri, E., Bentivegna, A., Redaelli, S., Lissoni, S., Saccheri, F., Villa, N., Crosti, F., Sala, E., Martinoli, E., Volontè, M., Mrozzi, A. & Dalprà, L., (2011). Cytogenetics of Premature Ovarian Failure: An Investigation on 269 Affected Women. Journal of Biomedicine and Biotechnology, 370195. doi: 10.1155/2011/370195 (Accessed on 26 May 2020).
9. Benítez-Burraco, A., Barcos-Martínez, M., Espejo-Portero, I., Fernández-Urquiza, M., Torres-Ruiz, R., Rodríguez-Perales, S. &

Jiménez-Romero, M.S., (2018). Narrowing the Genetic Causes of Language Dysfunction in the 1q21.1 Microduplication Syndrome. Frontiers in Pediatrics, 6, 163. doi: 10.3389/fped.2018.00163 (Accessed on 28 April 2020).

10. Bermejo-Das-Neves, C., Nguyen, H.-N., Poch, O. & Thompson, J.D., (2014). A comprehensive study of small non-frameshift insertions/deletions in proteins and prediction of their phenotypic effects by a machine learning method (KD4i). BMC Bioinformatics, 15,111. https://doi.org/10.1186/1471-2105-15-111 (Accessed on 29 June 2020).
11. Bi, W., Borgan, C., Pursley, A.N., Hixson, P., Shaw, C.A., Bacino, C.A., Lalani, S.R., Patel, A., Stankiewicz, P., Lupski, J.R., Beaudet, A.L. & Cheung, S.W., (2013). Comparison of chromosome analysis and chromosomal microarray analysis: what is the value of chromosome analysis in today's genomic array era? Genetics in Medicine, 15(6), 450-457.
12. Bishop, R., (2010). Applications of fluorescence in situ hybridization (FISH) in detecting genetic aberrations of medical significance. Bioscience Horizons, 3(1). doi: 10.1093/biohorizons/hzq009 (Accessed on 3 April 2020).
13. Blair, I.P., Nash, J., Gordon, M.J. & Nicholson, G.A., (1996). Prevalence and origin of de novo duplications in Charcot-Marie-Tooth disease type 1A: first report of a de novo duplication with maternal origin. American Journal of Human Genetics, 58(3), 472-476.
14. Boles, D.J. & Proia, R.L., 1995. The Molecular Basis of HEXA mRNA Deficiency Caused by the Most Common Tay-Sachs Disease Mutation. The American Journal of Human Genetics, 56, 716-724.
15. Brody, T., (2016). Chapter 18- Hematological Cancers. In Clinical Trials: Study Design, Endpoints and Biomarkers, Drug Safety, and FDA and ICH Guidelines (2nd edn.). Academic Press: London. Pp.331-376. https://doi.org/10.1016/B978-0-12-804217-5.00018-7 (Accessed on 20 May 2020).
16. Caputi, M., Kendzior, R.J., Jr. & Beemon, K.L., (2002). A nonsense mutation in the fibrillin-1 gene of a Marfan syndrome patient induces NMD and disrupts an exonic splicing enhancer. Genes and Development, 16(14), 1754-1759. doi: 10.1101/gad.997502 (Accessed on 24 June 2020).
17. Carr, S.M., (2008). Q-Banding: Fluorescent staining of chromosomes. https://www.mun.ca/biology/scarr/Fig17_05a.html (Accessed on 25

May 2020).

18. Cassidy, S.B., Schwartz, S., Miller, J.L. & Driscoll, D.J., (2011). Prader-Willi syndrome. Genetics in Medicine, 14, 10-26. https://doi.org/10.1038/gim.0b013e31822bead0 (Accessed on 24 April 2020).
19. Centers for Disease Control and Prevention, (2019a). Birth Defects: Down Syndrome. https://www.cdc.gov/ncbddd/birthdefects/downsyndrome.html (Accessed on 17 March 2020).
20. Centers for Disease Control and Prevention, (2019b). Birth Defects: Data and Statistics. https://www.cdc.gov/ncbddd/birthdefects/downsyndrome/data.html (Accessed on 17 March 2020).
21. Cereda, A. & Carey, J.C., (2012). The trisomy 18 syndrome. Orphanet Journal of Rare Diseases, 7 (81). https://doi.org/10.1186/1750-1172-7-81 (Accessed on 18 March 2020).
22. Chaganti, R., Burkitt's Lymphoma (2001). In Encyclopedia of Genetics; Brenner, S.; Miller, J.H. (ed.). Academic Press: Massachusetts. pp.246-248. https://doi.org/10.1006/rwgn.2001.0151 (Accessed on 24 May 2020).
23. Chen, B., Li, L., Ren, W., Yi, L., Wang, Y. & Yan, F., (2017). A novel missense mutation in the ALPL gene causes dysfunction of the protein. Molecular Medicine Reports, 16(1), 710-718. DOI: 10.3892/mmr.2017.6668 (Accessed on 17 June 2020).
24. Clancy, S. & Shaw, K.M., (2008). DNA Deletion and Duplication and the Associated Genetic Disorders. Nature Education, 1(1), 23.
25. Cystic Fibrosis Foundation, 2020. Types of CFTR Mutations. https://www.cff.org/What-is-CF/Genetics/Types-of-CFTR-Mutations/ (Accessed on 29, June 2020).
26. Elucigene, (2014). What is QF-PCR? https://www.elucigene.com/what-is-qf-pcr/ (Accessed on 31 March 2020).
27. De Jong, S., Cjepelev, I., Janson, E., Strengman, E., van den Berg, L.H., Veldink, J.H. & Ophoff, R.A., (2012). Common inversion polymorphism at 17q21.31 affects expression of multiple genes in tissue-specific manner. MBC Genomics, 13, 458. https://doi.org/10.1186/1471-2164-13-458 (Accessed on 12 May 2020).
28. De Rycke, M. & Sermon, K., (2010). Preimplantation Genetic Diagnosis. In Molecular Diagnostics (2nd edn.); Patrinos, G.P.; Ansorge, W.J. (ed.). Academic Press: Massachusetts. pp. 485-500. https://doi.org/10.1016/B978-0-12-374537-8.00033-X (Accessed on 4 April 2020).

29. Dolskiy, A.A., Lemskaya, N.A., Maksimova, Y.V., Shorina, A.R., Kolesnikova, I.S. & Yudkin, D.V., (2018). Robertsonian translocation 13/14 associated with rRNA genes overexpression and intellectual disability. Egyptian Journal of Medical Human Genetics, 19(2), 141-145. https://doi.org/10.1016/j.ejmhg.2017.11.002 (Accessed on 19 May 2020).

30. Doss, C.G.P. & Rajith, B., (2013). A New Insight Into Structural and Functional Impact of Single-Nucleotide Polymorphisms in PTEN Gene. Cell Biochemistry and Biophysics, 66, 249-263. DOI: 10.1007/s12013-012-9472-9 (Accessed on 17 June 2020).

31. Downie, S.E., Flaherty, S.P. & Matthews, C.D., (1997). Detection of chromosomes and estimation of aneuploidy in spermatozoa using fluorescence in-situ hybridization. Molecular Human Reproduction, 3(7), 585-598.

32. Dutta, P., Bhosale, S., Singh, R., Gubrellay, P., Patil, J., Sehdev, B., Bhagat, S. & Bansal, T., (2017). Amelogenin Gene-The Pioneer in Gender Determination from Forensic Dental Samples. Journal of Clinical & Diagnostic Research, 11(2): ZC56-ZC59. doi: 10.7860/JCDR/2017/22183.9407 (Accessed on 31 March 2020).

33. Eichler, E.E., (2008). Copy Number Variation and Human Diseases. Nature Education, 1(3),1.

34. Fellermann, K., Stange, D.E., Schaeffeler, E., Schmalzl, H., Wehkamp, J., Bevins, C.L., Reinisch, W., Teml, A., Schwab, M., Lichter, P., Radlwimmer, B. & Stange, E.F., (2006). A Chromosome 8 Gene-Cluster Polymorphism with Low Human Beta-Defensin 2 Gene Copy Number Predisposes to Crohn Disease of the Colon. The American Journal of Human Genetics, 79(3), 438-448. doi: 10.1086/505915 (Accessed on 1 July 2020).

35. Ferguson-Smith, M.A., (2001a). Monosomy. In Encyclopedia of Genetics; Brenner, S.; Miller, J.H. (ed.). Academic Press: Massachusetts. p. 1239. https://doi.org/10.1006/rwgn.2001.0848 (Accessed on 23 March 2020).

36. Ferguson-Smith, M.A., (2001b). Chromosome Aberrations. In Encyclopedia of Genetics; Brenner, S.; Miller, J.H. (ed.). Academic Press: Massachusetts. pp. 345-348. https://doi.org/10.1006/rwgn.2001.0203 (Accessed on 13 April 2020).

37. Frischmeyer, P.A. & Dietz, H.C., (1999). Nonsense-Mediated mRNA Decay in Health and Disease. Human Molecular Genetics, 8(10),

1893-1900. https://doi.org/10.1093/hmg/8.10.1893 (Accessed on 24 June 2020).

38. Gao, Y., Ma, Y.-C., Ju, Y.-J. & Li, Y.-N., (2018). Mosaicism trisomy 10 in a 14-month old child with additional neurological abnormalities and literature review. BMC Pediatrics, 18 (266). https://doi.org/10.1186/s12887-018-1237-1 (Accessed on 18 March 2020).
39. Gatt, M.E. &Izraeli, S., (2019). 78-Lymphoid Leukemias. In Clinical Immunology: Principles and Practice (5thedn.); Rich, R.R.; Fleisher, T.A.; Shearer, W.T.; Schroeder, H.W., Jr.; Frew, A.J.; Weyand, C.M. (ed.). Elsevier Limited: Philadelphia. pp/1049-1063.el. https://doi.org/10.1016/B978-0-7020-6896-6.00078-8 (Accessed on 20 May 2020).
40. Genetic Alliance, (2010). Diagnosis of a Genetic Disease. In Understanding Genetics: A District of Columbia Guide for Patients and Health Professionals. District of Columbia Department of Health: Washington D.C. https://www.ncbi.nlm.nih.gov/books/NBK132142/ (Accessed on 25 March 2020).
41. Genetics Home Reference, (2008). GTF21 gene. https://ghr.nlm.nih.gov/gene/GTF2I (Accessed on 29 April 2020).
42. Genetics Home Reference, (2012a). Trisomy 18. https://ghr.nlm.nih.gov/condition/trisomy-18#inheritance (Accessed on 18 March 2020).
43. Genetics Home Reference, (2012b). Wolf-Hirschhorn syndrome. https://ghr.nlm.nih.gov/condition/wolf-hirschhorn-syndrome#resources (Accessed on 15 April 2020).
44. Genetics Home Reference, (2012c). ELN gene. https://ghr.nlm.nih.gov/gene/ELN (Accessed on 29 April 2020).
45. Genetics Home Reference, (2012d). ABCA1 gene. https://ghr.nlm.nih.gov/gene/ABCA1 (Accessed on 17 June 2020).
46. Genetics Home Reference, (2013). Trisomy 13. https://ghr.nlm.nih.gov/condition/trisomy-13#genes (Accessed on 18 March 2020).
47. Genetics Home Reference, (2014a). Triple X syndrome. https://ghr.nlm.nih.gov/condition/triple-x-syndrome#genes (Accessed on 19 March 2020).
48. Genetics Home Reference, (2014b). 1q21.1 microduplication. https://ghr.nlm.nih.gov/condition/1q211-microduplication#genes (Accessed on 28 April 2020).
49. Genetics Home Reference, (2014c). Prader-Willi syndrome. https://

ghr.nlm.nih.gov/condition/prader-willi-syndrome#genes (Accessed on 24 April 2020).

50. Genetics Home Reference, (2015a). Jacobsen syndrome. https://ghr.nlm.nih.gov/condition/jacobsen-syndrome#statistics (Accessed on 21 April 2020).
51. Genetics Home Reference, (2015b). Bloom syndrome. https://ghr.nlm.nih.gov/condition/bloom-syndrome#genes (Accessed on 29 May, 2020).
52. Genetics Home Reference, (2015c). Beta thalassemia. https://ghr.nlm.nih.gov/condition/beta-thalassemia#genes (Accessed on 24 June, 2020).
53. Genetics Home Reference, (2016). Stüve-Wiedemann syndrome. https://ghr.nlm.nih.gov/condition/stuve-wiedemann-syndrome#genes (Accessed on 5 July 2020).
54. Genetics Home Reference, (2017a). Turner syndrome. https://ghr.nlm.nih.gov/condition/turner-syndrome#statistics (Accessed on 24 March 2020).
55. Genetics Home Reference, (2017b). 7q11.23 duplication syndrome. https://ghr.nlm.nih.gov/condition/7q1123-duplication-syndrome#genes (Accessed on 29 April 2020).
56. Genetics Home Reference, (2017c). MECP2 duplication syndrome. https://ghr.nlm.nih.gov/condition/mecp2-duplication-syndrome#genes (Accessed on 6 May 2020).
57. Genetics Home Reference, (2018a). X chromosome. https://ghr.nlm.nih.gov/chromosome/X (Accessed on 19 March 2020).
58. Genetics Home Reference, (2018b). 47, XYY syndrome. https://ghr.nlm.nih.gov/condition/47xyy-syndrome#genes (Accessed on 21 March 2020).
59. Genetics Home Reference, (2018c). What is non-invasive prenatal testing (NIPT) and what disorders can it screen for? https://ghr.nlm.nih.gov/primer/testing/nipt (Accessed on 1 April 2020).
60. Genetics Home Reference, (2018d). Distal 18q deletion syndrome. https://ghr.nlm.nih.gov/condition/distal-18q-deletion-syndrome#inheritance (Accessed on 16 April 2020).
61. Genetics Home Reference, (2018e). Proximal 18q deletion syndrome. https://ghr.nlm.nih.gov/condition/proximal-18q-deletion-syndrome#inheritance (Accessed on 16 April 2020).

62. Genetics Home Reference, (2018f). Marfan syndrome. https://ghr.nlm.nih.gov/condition/marfan-syndrome (Accessed on 24 June 2020).

63. Genetics Home Reference, (2019). 22q11.2 deletion syndrome. https://ghr.nlm.nih.gov/condition/22q112-deletion-syndrome#inheritance (Accessed on 15 April 2020).

64. Genetics Home Reference, (2020a). How are genetic conditions diagnosed? https://ghr.nlm.nih.gov/primer/consult/diagnosis (Accessed on 24 March 2020).

65. Genetics Home Reference, (2020b). Cri-du-chat syndrome. https://ghr.nlm.nih.gov/condition/cri-du-chat-syndrome#inheritance (Accessed on 13 April 2020).

66. Genetic Home Reference, (2020c). What is a gene mutation and how do mutations occur? https://ghr.nlm.nih.gov/primer/mutationsanddisorders/genemutation#:~:text=A%20gene%20mutation%20is%20a,chromosome%20that%20includes%20multiple%20genes (Accessed on 11 June 2020).

67. Genetics Home Reference, (2020d). How can gene mutations affect health and development? https://ghr.nlm.nih.gov/primer/mutationsanddisorders/mutationscausedisease (Accessed on 11 June 2020).

68. Genetics Home Reference, (2020e). Do all gene mutations affect health and development? https://ghr.nlm.nih.gov/primer/mutationsanddisorders/neutralmutations (Accessed on 11 June 2020).

69. Genetics Home Reference, (2020f). What kinds of gene mutations are possible? https://ghr.nlm.nih.gov/primer/mutationsanddisorders/possiblemutations (Accessed on 15 June 2020).

70. Genetics Home Reference, (2020g). Can a change in the number of genes affect health and development? https://ghr.nlm.nih.gov/primer/mutationsanddisorders/copynumbervariation (Accessed on 1 July 2020).

71. Gomes, A. & Kprf, B., (2018). Genetic Testing Techniques. In Pediatric Cancer Genetics (1st edn.); Robin, N.; Farmer, M. (ed.). Elsevier, Inc.: Philadelphia. pp.47-64. https://doi.org/10.1016/B978-0-323-48555-5.00005-3 (Accessed on 2 July 2020).

72. Goyal, M., Jain, M., Singhal, S. & Nandimath, K., (2017). 18p Deletion Syndrome: Case Report with Clinical Consideration and Management. Contemporary Clinical Dentistry, 8(4), 632-636. doi: 10.4103/ccd.

ccd_129_17 (Accessed on 17 April 2020).

73. Gravholt, C.H., (2013). Sex-Chromosome Abnormalities. In Emery and Rimoin's Principles and Practice of Medical Genetics (6th edn.); Rimoin, D.; Pyeritz, R.; Korf, B. (ed.). Academic Press: Massachusetts. pp. 1-32. https://doi.org/10.1016/B978-0-12-383834-6.00050-1 (Accessed on 21 March 2020).
74. Griffiths, A.J.F., Miller, J.H., Suzuki, D.T., Lewontin, R.C. & Gelbart, W.M., (2000). An Introduction to Genetic Analysis (7th edn.). W.H. Freeman: New York.
75. Hall, G.W. & Thein, S., (1994). Nonsense Codon Mutations in the Terminal Exon of the Beta-Globin Gene Are Not Associated With a Reduction in beta-mRNA Accumulation: A Mechanism for the Phenotype of Dominant Beta-Thalassemia. Blood, 83(8), 2031-2037.
76. Harrison, C.J., (2001). Philadelphia Chromosome. In Encyclopedia of Genetics; Brenner, S.; Miller, J.H. (ed.). Academic Press: Massachusetts.pp.1449-1450. https://doi.org/10.1006/rwgn.2001.0991 (Accessed on 20 May 2020).
77. Harvard.edu. (2002). How Does Sickle Cell Cause Disease? https://sickle.bwh.harvard.edu/scd_background.html (Accessed on 17 June 2020).
78. Head, S.R., Komori, H.K., LaMere, S.A., Whisenant, T,; Van Nieuwerburgh, F., Salomon, D.R. & Ordoukhanian, P., (2014). Library construction for next-generation sequencing: Overviews and challenges. BioTechniques, 56(2), 61-77. doi 10.2144/000114133 (Accessed on 7 July 2020).
79. Healthline, (2018). Robertsonian Translocation Explained in Plain Language. https://www.healthline.com/health/robertsonian-translocation(Accessed on 19 May 2020).
80. Herman, R.K., (2001). Aneuploid. In Encyclopedia of Genetics; Brenner, S.; Miller, J.H. (ed.). Academic Press: Massachusetts, U.S.A. pp.65-66.
81. Hirota, T., Ohta, H., Powell, B.E., Mahadevaiah, S.K., Ojarikre, O.A., Saitou, M. & Turner, J.M.A., (2017). Fertile offspring from sterile sex chromosome trisomic mice. Science, 357 (6354), 932-935. doi:10.1126/science.aam9046 (Accessed on 18 March 2020).
82. Hollox, E.J., Huffmeier, U., Zeeuwen, P.L.J.M., Palla, R., Lascorz, J., Rodjik-Olthuis, D., van de Kerkhof, P.C.M., Traupe, H., de Jongh,

G., den Heijer, M., Reis, A., Armour, J.A.L. & Schalkwijk, J., (2008). Psoriasis is associated with increased beta-defensin genomic copy number. Nature Genetics, 40(1), 23-25. doi: 10.1038/ng.2007.48 (Accessed on 1 July 2020).

83. Horii, M., Boyd, T.K. & Parast, M.M., (2018). Placental Development and Complications of Previable Pregnancy. In Diagnostic Gynecologic and Obstetric Pathology (3rd edn.); Crum, C.P.; Nucci, M.R.; Howitt, B.E.; Granter, S.R.; Parast, M.M.; Boyd, T.K. (ed.). Elsevier: Philadelphia. pp.1070-1102. https://doi.org/10.1016/B978-0-323-44732-4.00029-7 (Accessed on 30 March 2020).

84. Hu, J., Ou, Z., Infante, E., Kochmar, S.J., Madan-Khetarpal, S., Hoffner, L., Parsazad, S. & Surti, U., (2017). Chromosome 12q13.13q13.13 microduplication and microdeletion: a case report and literature review. Molecular Cytogenetics, 10, 24. https://doi.org/10.1186/s13039-017-0326-4 (Accessed on 1 May 2020).

85. Huber, D., Voith von Voithenberg, L. & Kaigala, G.V., (2018). Fluorescence in situ hybridization (FISH): History, limitations and what to expect from micro-scale FISH? Micro and Nano Engineering, 1, 15-24. https://doi.org/10.1016/j.mne.2018.10.006 (Accessed on 3 April 2020).

86. Huo, P., Luo, Q., Li, J., Jiao, B., Rong, L., Zhang, J. & Wu, X., (2019). High accuracy of quantitative fluorescence polymerase chain reaction combined with non-invasive pre-natal testing for mid-pregnancy diagnosis of common fetal aneuploidies: A single-center experience in China. Experimental and Therapeutic Medicine, 18(1), 711-721. https://doi.org/10.3892/etm.2019.7625 (Accessed on 31 March 2020).

87. Hutaff-Lee, C., Cordeito, L. & Tartaglia, N., (2013). Cognitive and medical features of chromosomal aneuploidy. In Handbook of Clinical Neurology Vol. 111; Dulac, O.; Lassonde, M.; Sarnat, H.B. (ed.). Elsevier B.V.: Amsterdam. pp. 273-279. https://doi.org/10.1016/B978-0-444-52891-9.00030-0 (Accessed on 21 March 2020).

88. Hwang, S.H., Kim, H.-J. & Kim, H.S., (2012). Chapter 1 Profiling of Mutations in the F8 and F9. Causative Genes of Hemophilia A and Hemophilia B. In Hemophilia; Batorova, A. (ed). InTech: Croatia. http://www.intechopen.com/books/hemophilia/profiling-of-mutations-in-thef8-and-f9-causative-genes-of-hemophilia-a-and-hemophilia-b (Accessed on 13 May 2020).

89. Iannuzzi, M.C., Stern, R.C., Collins, F.S., Tom Hon, C., Hidaka, N.,

Strong, T., Becker, L., Drumm, M.L., White, M.B., Gerard, B. & Dean, M., (1991). Two Frameshift Mutations in the Cystic Fibrosis Gene. The American Journal of Human Genetics, 48, 227-231.

90. Interchim, (2020). Cot-1 Human DNA. https://www.interchim.fr/ft/C/CE6040.pdf (Accessed on 5 April 2020).
91. Ishiyama, A., Iida, A., Hayashi, S., Komaki, H., Sasaki, M., Nonaka, I., Noguchi, S. & Nishino, I., (2018). A novel LMNA mutation identified in a Japanese patient with LMNA-associated congenital muscular dystrophy. Human Genome Variation, 5, 19. https://doi.org/10.1038/s41439-018-0018-6 (Accessed on 15 June 2020).
92. Jeong, S.-Y., Kim, B.-Y. & Yu, J.E., (2010). De Novo Pericentric Inversion of Chromosome 9 in Congenital Anomaly. Yonsei Medical Journal, 51(5), 775-780. DOI: 10.3349/ymj.2010.51.5.775 (Accessed on 12 May 2020).
93. Kalkan, R., (2017). The Use of Molecular Cytogenetic Techniques for the Identification of Chromosomal Abnormalities. In Chromosomal Abnormalities-A Hallmark Manifestation of Genomic Instability; Larramendy, M.L.; Soloneski, S. (eds.). IntechOpen. DOI: 10.5772/67415 (Accessed on 2 June 2020).
94. Kan, A.S.Y., Lau, E.T., Tang, W.F., Chan, S.S.Y., Ding, S.C.K., Chan, K.Y.K., Lee, C.P., Hui, P.W., Chung, B.H.Y., Leung, K.Y., Ma, T., Leung, W.C. & Tang, M.H.Y., (2014). Whole-Genome Array CGH Evaluation for Replacing Prenatal Karyotyping in Hong Kong. PLoS ONE, 9(2), e87988. https://doi.org/10.1371/journal.pone.0087988 (Accessed on 5 April 2020).
95. Karampetsou, E., Morrogh, D. & Chitty, L., (2014). Microarray Technology for the Diagnosis of Fetal Chromosomal Aberrations: Which Platform Should We Use? Journal of Clinical Medicine 3(2), 663-678. https://doi.org/10.3390/jcm3020663 (Accessed on 9 June 2020).
96. Kearney, L., (2006). Multiplex-FISH (M-FISH): technique, development and applications. Cytogenetic Genome Research, 114, 189-198. https://doi.org/10.1159/000094202 (Accessed on 31 May 2020).
97. Khan Academy, (2016). Biotechnology: DNA sequencing. https://www.khanacademy.org/science/high-school-biology/hs-molecular-genetics/hs-biotechnology/a/dna-sequencing (accessed August 7, 2019).

98. Kingsmore, S.F. & Saunders, C.J., (2011). Deep Sequencing of Patient Genomes for Disease Diagnosis: When Will It Become Routine? Science Translational Medicine, 3(87), 87ps23. DOI: 10.1126/scitranslmed.3002695 (Accessed on 5 July 2020).

99. Kjeldsen, E. & Kølvraa, S., (2002). FISH Techniques, FISH Probes and Their Applications in Medicine and Biology – An Overview. In FISH Technology; Rautenstrauβ, B.; Liehr, T. (eds.). Springer- Verlag Berlin Heidelberg: New York. pp. 3-50.

100. Knief, C, (2014). Analysis of plant microbe interactions in the era of next generation sequencing technologies. Frontiers in Plant Science, 5, 216. https://doi.org/10.3389/fpls.2014.00216 (Accessed on 7 July 2020).

101. Kumar, A., Rajendran, V., Sethumadhavan, R. & Purohit, R., (2013). Evidence of Colorectal Cancer-Associated Mutation in MCAK: A Computational Report. Cell Biochemistry and Biophysics, 67, 837-851. DOI: 10.1007/s12013-013-9572-1 (Accessed on 17 June 2020).

102. Lab Tests Online, (2020). DNA Sequencing. https://labtestsonline.org/dna-sequencing (Accessed on 3 July 2020).

103. Lau, M.M.H. & Neufeld, E.F., (1989). A Frameshift Mutation in a Patient with Tay-Sachs Disease Causes Premature Termination and Defective Intracellular Transport of the α-Subunit of β-Hexosaminidase. The Journal of Biological Chemistry, 264(35), 21376-21380.

104. Lester, S.C., (2010). Manual of Surgical Pathology (3rd edn.). Elsevier Saunders: Philadelphia. pp. 423-472. https://doi.org/10.1016/B978-0-323-06516-0.10022-0 (Accessed on 30 March 2020).

105. Levy, B. & Wapner, R., (2018). Prenatal diagnosis by chromosomal microarray analysis. Fertility and Sterility, 109(2), 201-212. https://doi.org/10.1016/j.fertnstert.2018.01.005 (Accessed on 8 June 2020).

106. Li, F.W. & Harkess, A., (2018). A guide to sequence your favorite plant genomes. Applications in Plant Science, 6(3), e1030. https://doi.org/10.1002/aps3.1030 (Accessed on 7 July 2020).

107. Lin, C.-F., Naj, A. C. & Wang, L.-S., (2014). Analyzing Copy Number Variation using SNP Array Data: Protocols for Calling CNV and Association Tests. Current Protocols in Human Genetics, 79(1), 1.27.1-1.27.15. https://doi.org/10.1002/0471142905.hg0127s79 (Accessed on 10 June 2020).

108. Lin, M.-T., Tseng, L.-H., Rich, R.G., Hafez, M.J., Harada, S., Murphy,

K.M., Eshleman, J.R. & Gocke, C.D., (2011). Δ-PCR, A Simple Method to Detect Translocations and Insertion/Deletion Mutations. The Journal of Molecular Diagnostics, 13(1), 85-92. https://doi.org/10.1016/j.jmoldx.2010.11.004 (Accessed on 4 June 2020).

109. Lobo, I., (2008). Copy Number Variation and Genetic Disease. Nature Education, 1(1), 65.

110. Loewe, L., (2008). Genetic Mutation. Nature Education, 1(1), 113.

111. Maddieh, N. & Rabbani, B., (2013). An Overview of Mutation Detection Methods in Genetic Disorders. Iranian Journal of Pediatrics, 23(4), 375-388.

112. Mainardi, P.C., (2006). Cri du chat syndrome. Orphanet Journal of Rare Diseases, 1, 33. doi:10.1186/1750-1172-1-33 (Accessed on 13 April 2020).

113. Maranda, B., Lemieux, N. & Lemyre, E., (2006). Familial deletion 18p syndrome: case report. BMC Medical Genetics, 7, 60. doi: 10.1186/1471-2350-7-60 (Accessed on 17 April 2020).

114. Mattina, T., Perrotta, C.S. & Grossfeld, P., (2009). Jacobsen syndrome. Orphanet Journal of Rare Diseases, 4,9. https://doi.org/10.1186/1750-1172-4-9 (Accessed on 23 April 2020).

115. Mayo Clinic, (2019a). Amniocentesis. https://www.mayoclinic.org/tests-procedures/amniocentesis/about/pac-20392914 (Accessed on 25 March 2020).

116. Mayo Clinic, (2019b). Chorionic villus sampling. https://www.mayoclinic.org/tests-procedures/chorionic-villus-sampling/about/pac-20393533 (Accessed on 25 March 2020).

117. Mayo Clinic, (2019c). Chronic myelogenous leukemia. https://www.mayoclinic.org/diseases-conditions/chronic-myelogenous-leukemia/symptoms-causes/syc-20352417 (Accessed on 20 May 2020).

118. Mennuti, M.T., (2019). Cytogenetics: Part2, Structural Chromosome Rearrangements and Reproductive Impact. In Perinatal Genetics; Norton, M.E.; Kuller, J.A.; Dugoff, L. ed. Elsevier, Inc.: St. Louis, Missouri. PP.39-48. https://doi.org/10.1016/B978-0-323-53094-1.00005-9(Accessed on 10 May 2020).

119. Merck, (2019). Sanger Sequencing Steps and Methods. https://www.sigmaaldrich.com/technical-documents/articles/biology/sanger-sequencing.html (accessed August 7, 2019).

120. Moore, C.M. & Best, R.G., (2001). Chromosome Preparation and

Banding. In Encyclopedia of Life Sciences. https://onlinelibrary.wiley.com/doi/pdf/10.1038/npg.els.0001444 (Accessed on 26 May 2020).

121. Munné, S., (2012). Preimplantation Genetic Diagnosis for Aneuploidy and Translocations Using Array Comparative Genomic Hybridization. Current Genomics, 13(6), 463-470. doi: 10.2174/138920212802510457 (Accessed on 4 April 2020).
122. Murray, M.M., Krone, M.G., Bernstein, S.L., Baumketner, A., Condron, M.M., Lazo, N.D., Teplow, D.B., Wyttenbach, T., Shea, J.-E. & Bowers, M.T., (2009). The Amyloid β-Protein: Experiment and Theory on the 21-30 Fragment. The Journal of Physical Chemistry, 113(17), 6041-6046. doi: 10.1021/jp808384x (Accessed on 16 June 2020).
123. Myers, K.A., Innes, A.M. & Mah, J.K.-W., (2016). Familial Congenital Facial Synkinesis Due to 12q Duplication: A Case Report and Literature Review. Pediatrics, 138(6), e20161724. DOI: https://doi.org/10.1542/peds.2016-1724 (Accessed on 1 May 2020).
124. Nabil, A. & Sarra, F., (2017). Q-Banding. In Reference Module in Life Sciences. DOI: 10.1016/B978-0-12-809633-8.06986-7 (Accessed on 25 May 2020).
125. Naeim, F., Rao, P.N., Song, S.X. & Grody, W.W., (2013). Atlas of Hemapathology: Morphology, Immunophenotype, Cytogenetics, and Molecular Approaches. Academic Press: London. pp.465-471. https://doi.org/10.1016/B978-0-12-385183-3.00039-5 (Accessed on 24 May 2020).
126. Nagao, K., Zhao, Y., Takahashi, K., Kimura, Y. & Ueda, K., (2009). Sodium taurocholate-dependent lipid efflux by ABCA1: effects of W590S mutation on lipid translocation and apolipoprotein A-I dissociation. Journal of Lipid Research, 50(6), 1165-1172. doi: 10.1194/jlr.M800597-JLR200 (Accessed on 17 June 2020).
127. Nagaoka, S.I., Hassold, T.J. & Hunt, P.A., (2012). Human aneuploidy: mechanisms and new insights into an age-old problem. Nature Review Genetics, 13, 493-504.
128. National Center for Biotechnology Information (US), (1998). Burkitt lymphoma. In Genes and Diseases. National Center for Biotechnology Information (US): Bethesda, Maryland. https://www.ncbi.nlm.nih.gov/books/NBK22257/ (Accessed on 24 May 2020).
129. National Down Syndrome Society, (2020). Down Syndrome. https://

www.ndss.org/about-down-syndrome/down-syndrome/ (Accessed on 21 May 2020).

130. National Health Service, (2018). Overview: Prader-Willi syndrome. https://www.nhs.uk/conditions/prader-willi-syndrome/ (Accessed on 24 April 2020).
131. National Human Genome Research Institute (2020). About Sickle Cell Disease. https://www.genome.gov/Genetic-Disorders/Sickle-Cell-Disease (Accessed on 17 June 2020).
132. National Institutes of Health, (2007). Charcot-Marie-Tooth Disease Fact Sheet. https://www.ninds.nih.gov/Disorders/Patient-Caregiver-Education/Fact-Sheets/Charcot-Marie-Tooth-Disease-Fact-Sheet (Accessed on 8 May 2020).
133. National Institutes of Health, (2013). Chromosome 1q21.1 duplication syndrome. https://rarediseases.info.nih.gov/diseases/10591/chromosome-1q211-duplication-syndrome (Accessed on 28 April 2020).
134. National Institutes of Health, (2015a). Cri du chat syndrome. https://rarediseases.info.nih.gov/diseases/6213/cri-du-chat-syndrome (Accessed on 13 April 2020).
135. National Institutes of Health, (2015b). Chromosome 4p deletion. https://rarediseases.info.nih.gov/diseases/6090/chromosome-4p-deletion (Accessed on 15 April 2020).
136. National Institutes of Health, (2015c). Chromosome 12 q duplication. https://rarediseases.info.nih.gov/diseases/1926/chromosome-12q-duplication#ref_10247 (Accessed on 1 May 2020).
137. National Institutes of Health, (2015d). Chromosome Xq duplication. https://rarediseases.info.nih.gov/diseases/5369/chromosome-xq-duplication (Accessed on 6 May 2020).
138. National Institutes of Health, (2016). Trisomy 13. https://rarediseases.info.nih.gov/diseases/7341/trisomy-13 (Accessed on 18 March 2020).
139. National Institutes of Health, (2017a). What are common symptoms of Down syndrome? https://www.nichd.nih.gov/health/topics/down/conditioninfo/symptoms (Accessed on 17 March 2020).
140. National Institutes of Health, (2017b). FAQs About Chromosome Disorders. https://rarediseases.info.nih.gov/guides/pages/73/faqs-about-chromosome-disorders (Accessed on 13 April 2020).
141. National Institutes of Health, (2017c). 22q11.2 deletion syndrome.

https://rarediseases.info.nih.gov/diseases/10299/22q112-deletion-syndrome#ref_7348 (Accessed on 15 April 2020).

142. National Institutes of Health, (2017d). Chromosome 9 inversion. https://rarediseases.info.nih.gov/diseases/10765/chromosome-9-inversion (Accessed on 12 May 2020).

143. National Institutes of Health, (2018a). Klinefelter syndrome. https://rarediseases.info.nih.gov/diseases/8705/klinefelter-syndrome (Accessed on 22 March 2020).

144. National Institutes of Health, (2018b). Chromosome 18p deletion. https://rarediseases.info.nih.gov/diseases/8631/chromosome-18p-deletion (Accessed on 17 April 2020).

145. National Institutes of Health, (2018c). Charcot-Marie-Tooth disease type 1. https://rarediseases.info.nih.gov/diseases/12433/charcot-marie-tooth-disease-type-1 (Accessed on 8 May 2020).

146. National Institutes of Health, (2018d). Hemophilia A. https://rarediseases.info.nih.gov/diseases/6591/hemophilia-a (Accessed on 13 May 2020).

147. National Organization of Rare Disorders, (2009). Rare Disease Databases: Chromosome 18q- Syndrome. https://rarediseases.org/rare-diseases/chromosome-18q-syndrome/ (Accessed on 16 April 2020).

148. National Organization of Rare Disorders, (2017). Rare Disease Database: MECP2 Duplication Syndrome. https://rarediseases.org/rare-diseases/mecp2-duplication-syndrome/ (Accessed on 6 May 2020).

149. National Organization of Rare Disorders, (2020). Rare Disease Database: Turner Syndrome. https://rarediseases.org/rare-diseases/turner-syndrome/ (Accessed on 24 March 2020).

150. Nicholson, J.M. & Cimini, D., (2011). How Mitotic Errors Contribute to Karyotypic Diversity in Cancer. In Advances in Cancer Research (Vol.112); Gisselson, D. (ed). Academic Press: San Diego, California. Pp. 43-75.

151. Nikiforova, M.N. & Nikiforov, Y.E., (2011). Molecular Anatomic Pathology: Principles, Techniques, and Application to Immunohistologic Diagnosis. In Diagnostic Immunohistochemistry: Theranostic and Genomic Applications (3rd edn.); Dabbs, D.J. (ed.). Saunders: Philadelphia. pp. 42-57. https://doi.org/10.1016/B978-1-4160-5766-6.00006-6 (Accessed on 4 April 2020).

152. Nyon, M.P., Segu, L., Cabrita, L.D., Lévy, G.R., Kirkpatrick, J., Roussel, B.D., Patschull, A.O.M., Barrett, T.E., Ekeowa, U.I., Kerr, R., Waudby, C.A., Kalsheker, N., Hill, M., Thalassinos, K., Lomas, D.A., Christodoulou, J. & Gooptu, B., (2012). Structural Dynamics Associated with Intermediate Formation in an Archetypal Conformational Disease. Structure 20-135(3-3), 504-512. doi: 10.1016/j.str.2012.01.012 (Accessed on 17 June 2020).

153. O'Connor, C., (2008a). Karyotyping for Chromosomal Abnormalities. Nature Education, 1(1), 27.

154. O'Connor, C., (2008b). Chromosomal Abnormalities: Aneuploidies. Nature Education, 1(1), 172.

155. O'Connor, C., (2008c). Fluorescence In Situ Hybridization (FISH). Nature Education, 1(1), 171.

156. O'Connor, C., (2008d). Human Chromosome Translocations and Cancer. Nature Education, 1(1), 56.

157. Oristaglio, S.F., Levy, S.E., Hakonarson, H., Sullivan, N., Fontanarosa, J. & Schoelles, K.M., (2015). Genetic Testing for Developmental Disorders, Intellectual Disability, and Autism Spectrum Disorder. Technical Briefs No. 23. Agency for Healthcare Research and Quality (US): Maryland.

158. Orpha.net, (2015). Wolf-Hirschhorn syndrome. https://www.orpha.net/consor/cgi-bin/Disease_Search.php?lng=EN&data_id=147&Disease_Disease_Search_diseaseType=ORPHA&Disease_Disease_Search_diseaseGroup=280&Disease(s)/group%20of%20diseases=Distal-deletion-4p&title=Distal-deletion-4p&search=Disease_Search_Simple (Accessed on 16 April 2020).

159. Orpha.net, (2016). Charcot-Marie-Tooth disease type 1. https://www.orpha.net/consor/cgi-bin/OC_Exp.php?Lng=EN&Expert=65753 (Accessed on 8 May 2020).

160. Orpha.net, (2020). Mosaic trisomy 10. https://www.orpha.net/consor/cgi-bin/OC_Exp.php?lng=EN&Expert=96063 (Accessed on 18 March 2020).

161. Ozen, F., Saglam, Z.A., Kurtulgan, H.K. &Ozen, A.V., (2014). A Balanced Robertsonian Translocation (13;14) of a Male: A Case Report. Journal of Medical Cases, 5(12), 615-617. doi: http://dx.doi.org/10.14740/jmc1921w (Accessed on 20 May 2020).

162. Park, S.T. & Kim, J., (2016). Trends in Next-Generation Sequencing

and a New Era for Whole Genome Sequencing. International Neurourology Journal, 20 (Suppl 2), S76-83. https://doi.org/10.5213/inj.1632742.371 (accessed on 7 July 2020).

163. Peterson, D.G. & Arick II, M., (2018). Sequencing Plant Genomes. In Progress in Botany Vol. 80; Cánovas, F.; Lüttge, U.; Matyssek, R.; Pretzsch, H., (ed.). Springer: Cham: Switzerland, pp.109-193.
164. Pradhan, B., Sarvilinna, N., Matilainen, J., Aska, E., Sjöberg, J. & Kauppi, L., (2016). Detection and screening of chromosomal rearrangements in uterine leiomyomas by long-distance inverse PCR. Genes, Chromosomes & Cancer, 55(3), 215-226. doi: 10.1002/gcc.22317 (Accessed on 3 June 2020).
165. Puig, M., Casillas, S., Villatoro, S. & Cáceres, M., (2015). Human inversions and their functional consequences. Briefings in Functional Genomics, 14(5), 369-379. https://doi.org/10.1093/bfgp/elv020 (Accessed on 12 May 2020).
166. Rafi, I., Hill, M., Hayward, J. & Chitty, L.S., (2017). Non-invasive prenatal testing: use of cell-free fetal DNA in Down syndrome screening. British Journal of General Practice, 67(660), 298-299. doi: 10.3399/bjgp17X691625 (Accessed on 1 April 2020).
167. Rhoads, A. & Au, K.F., (2015). PacBio Sequencing and Its Applications. Genomics, Proteomics & Bioinformatics, 13(5), 278-289. https://doi.org/10.1016/j.gpb.2015.08.002 (Accessed on 7 July 2020).
168. Ried, T., Schröck, E., Ning, Y. & Wienberg, J., (1998). Chromosome Painting: A Useful Art. Human Molecular Genetics, 7(10), 1619-1626. https://doi.org/10.1093/hmg/7.10.1619 (Accessed on 31 May 2020).
169. Rønne, M., (1988). Chromosome Preparation and High Resolution Banding Techniques: A Review. Journal of Dairy Science, 72(5), 1363-1377. https://doi.org/10.3168/jds.S0022-0302(89)79243-2 (Accessed on 27 May 2020).
170. Ross, O.A., Braithwaite, A.T., Skipper, L.M., Kachergus, J., Huliham, M.M., Middleton, F.A., Nishioka, K., Fuchs, J., Gasser, T., Maraganore, D.M., Adler, C.A., Larvor, L., Chartier-Harlin, M.-C., Nilsson, C., Langston, J.W., Gwinn, K., Hattori, N. & Farrer, M.J., (2008). Genomic Investigation of Alpha-Synuclein Multiplication and Parkinsonism. Annals of Neurology, 63(6). doi: 10.1002/ana.21380 (Accessed on 1 July 2020).
171. Sanlaville, D., Schluth-Bolard, C. & Turleau, C., (2009). Distal Xq

duplication and functional Xq disomy. Orphanet Journal of Rare Diseases 4, 4. https://doi.org/10.1186/1750-1172-4-4 (Accessed on 7 May 2020).

172. Sara, S., (2008). For People With Down Syndrome, Longer Life Has Complications. The New York Times. https://www.nytimes.com/2008/06/01/nyregion/01down.html (Accessed on 17 March 2020).
173. Schaeffer, C., Creatore, A. & Rampoldi, L., (2014). Protein trafficking defects in inherited kidney diseases. Nephrology Dialysis Transplantation, 29(suppl.4), iv33-iv44. https://doi.org/10.1093/ndt/gfu231 (Accessed on 22 June 2020).
174. Schneider, B., Drexler, H.G. & MacLeod, R.A.F., (2011). LDI-PCR in Cancer Translocation Mapping. Atlas of Genetics and Cytogenetics in Oncology and Haematology, 15(1), 106-114. http://AtlasGeneticsOncology.org/Deep/LDI-PCRinCancerID20087.htm (Accessed on 3 June 2020).
175. Schoales, J., (2015). How Does Sanger Sequencing Work? https://www.thermofisher.com/blog/behindthebench/how-does-sanger-sequencing-work/ (accessed August 7, 2019).
176. Schreck, R.R. & Distèche, C.M., (1994). Chromosome Banding Techniques. Current Protocols in Human Genetics, 00(1). https://doi.org/10.1002/0471142905.hg0402s00 (Accessed on 25 May 2020).
177. Schurmann, K., Anton, M., Ivanov, I., Richter, C., Kuhn, H. & Walther, M., (2011). Molecular Basis for the Reduced Catalytic Activity of the Naturally Occurring T560M Mutant of Human12/15-Lipoxygenase That Has Been Implicated in Coronary Artery Disease. Journal of Biological Chemistry, 286(27), 23920-23927. doi: 10.1074/jbc.M110.211821 (Accessed on 16 June 2020).
178. Scitable, (2014a). DNA Is Constantly Changing through the Process of Mutation. https://www.nature.com/scitable/topicpage/dna-is-constantly-changing-through-the-process-6524898/#:~:text=Base%20substitution,for%20another%20during%20DNA%20replication.&text=With%20base%20substitution%20mutations%2C%20only,is%20affected%20(Figure%201) (Accessed on 11 June 2020).
179. Scitable, (2014b). Definition: DNA sequencing. https://www.nature.com/scitable/definition/dna-sequencing-205 (accessed August 7, 2019).

180. Scriven, P.N., Flinter, F.A., Braude, P.R. & Ogilvie, C.M., (2001). Robertsonian translocations – reproductive risks and indications for preimplantation genetic diagnosis. Human Reproduction, 16(11), 2267-2273. https://doi.org/10.1093/humrep/16.11.2267 (Accessed on 20 May 2020).

181. Shen, C.-H., (2019). Diagnostic Molecular Biology. Academic Press: London. pp.331-358. https://doi.org/10.1016/B978-0-12-802823-0.00013-4 (Accessed on 26 May 2020).

182. Singaraja, R.R., Visscher, H., James, E.R., Chroni, A., Coutinho, J.M., Brunham, L.R., Kang, M.H., Zannis, V.I., Chimini, G. & Hayden, M.R., (2006). Specific Mutations in ABCA1 Have Discrete Effects on ABCA1 Function and Lipid Phenotypes Both In Vivo and In Vitro. Circulation Research, 99, 389-397. https://doi.org/10.1161/01.RES.0000237920.70451.ad (Accessed on 18 June 2020).

183. Singleton, A.B., Farrer, M., Johnson, J., Singleton, A., Hague, S., Kachergus, J., Hulihan, M., Peuralinna, T., Dutra, A., Nussbaum, R., Lincoln, S., Crawley, A., Hanson, M., Maraganore, D., Adler, S., Cookson, M.R., Muenter, M., Baptista, M., Miller, D., Blancato, J., Hardy, J. & Gwinn-Hardy, K., (2003). A-Synuclein Locus Triplication Causes Parkinson's Disease. Science, 302(5646), 841. DOI: 10.1126/science.1090278 (Accessed on 1 July 2020).

184. Smith, M., (2017). DNA Sequence Analysis in Clinical Medicine, Proceeding Cautiously. Frontiers in Molecular Biosciences, 4, 24. doi: 10.3389/fmolb.2017.00024 (Accessed on 7 July 2020).

185. Socolov, D., Socolov, R., Gorduza, V.E., Butureanu, T., Stanculescu, R., Carauleanu, A. & Pavaleanu, I., (2017). Increased nuchal translucency in fetuses with a normal karyotype—diagnosis and management. Medicine (Baltimore), 96(29), e7521. doi: 10.1097/MD.0000000000007521 (Accessed on 26 May 2020).

186. Solomon, D.A., (2018). Integrating Molecular diagnostics With Surgical Neuropathology. In Practical Surgical Neuropathology: A Diagnostic Approach (2nd edn.); Perry, A.; Brat, D.J. (ed.). Elsevier, Inc.: Philadelphia. pp. 71-89. https://doi.org/10.1016/B978-0-323-44941-0.00005-9 (Accessed on 2 July 2020).

187. Stanford Children's Health, (2020). Translocation Down Syndrome. https://www.stanfordchildrens.org/en/topic/default?id=translocation-down-syndrome-90-P02153 (Accessed on 21 May 2020).

188. Stefl, S., Nishi, H., Petukh, M., Panchenko, A.R. & Alexov, E., (2014).

Molecular mechanisms of disease-causing missense mutations. Journal of Molecular Biology, 425(21), 3919-3936. https://doi.org/10.1016/j.jmb.2013.07.014 (Accessed on 15 June 2020).

189. Stochholm, K., Bojesen, A., Jensen, A.S., Juul, S. & Gravholt, C.H., (2012). Criminality in men with Klinefelter's syndrome and XYY syndrome: a cohort study. BMJ Open, 2, e000650. doi:10.1136/bmjopen-2011-000650 (Accessed on 23 March 2020).

190. Sun, F., Oristaglio, J., Levy, S.E., Hakonarson, H., Sullivan, N., Fontanarosa, J. & Schoelles, K.M., (2015). Genetic Testing for Developmental Disabilities, Intellectual Disability, and Autism Spectrum Disorder. Technical Brief No. 23.Agency for Healthcare Research and Quality: Rockville, Maryland. www.effectivehealthcare.ahrq.gov/reports/final.cfm (Accessed on 8 June 2020).

191. Takano, K., Liu, D., Tarpey, P., Gallant, E., Lam, A., Witham, S., Alexov, E., Chaubey, A., Stevenson, R.E., Schwartz, C.E., Board, P.G. & Dulhunty, A.F., (2012). An X-linked channelopathy with cardiomegaly due to a CLIC2 mutation enhancing ryanodine receptor channel activity. Human Molecular Genetics, 21(20), 4497-4507. doi: 10.1093/hmg/dds292(Accessed on 16 June 2020).

192. Tartaglia, N.R., Howell, S., Sutherland, S., Wilson, R. & Wilson, L., (2010). A review of trisomy X (47, XXX). Orphanet Journal of Rare Diseases, 5(8). https://doi.org/10.1186/1750-1172-5-8 (Accessed on 19 March 2020).

193. Templado, C., Uroz, L. & Estop, A., (2013). New insights on the origin and relevance of aneuploidy in human spermatozoa. Molecular Human Reproduction, 19(10), 634-643. https://doi.org/10.1093/molehr/gat039 (Accessed on 17 March 2020).

194. Thein, S.L., (2013). The Molecular Basis of β-Thalassemia. Cold Spring Harbor Perspectives in Medicine. 3(5), a011700. doi: 10.1101/cshperspect.a011700(Accessed on 24 June 2020).

195. ThermoFisher Scientific, (2020). Quantitative Fluorescence PCR (qf-pcr). https://www.thermofisher.com/ph/en/home/life-science/sequencing/fragment-analysis/quantitative-fluorescence-pcr.html (Accessed on 31 March 2020).

196. Thiesen, A., (2008). Microarray-based Comparative Genomic Hybridization (aCGH). Nature Education, 1(1):45.

197. Tsai, A.C-H., Pickler, L., Tartaglia, N. & Hagerman, R., (2009).

Chromosomal Disorders and Fragile X Syndrome. In Developmental-Behavioral Pediatrics (4th edn.); Carey, W.B.; Crocker, A.C.; Coleman, W.; Elias, E.; Feldman, H. (ed.). Elsevier, Inc.: Philadelphia. pp.224-234. https://doi.org/10.1016/B978-1-4160-3370-7.00024-9 (Accessed on 23 March 2020).

198. Tulay, P., (2017). Chromosomal Abnormalities in Preimplantation Embryos and Detection Strategies in PGD and PGS. In Chromosomal Abnormalities-A Hallmark Manifestation of Genomic Instability; Larramendy, M.L.; Soloneski, S. (ed.). IntechOpen. DOI: 10.5772/67419 (Accessed on 21 May 2020).

199. Turleau, C., (2008). Monosomy 18p. Orphanet Journal of Rare Diseases, 3(4). https://doi.org/10.1186/1750-1172-3-4 (Accessed on 17 April 2020).

200. UCFS Health, (2020). FAQ: Cell-Free DNA Screening. https://www.ucsfhealth.org/education/faq-cell-free-dna-screening (Accessed on 1 April 2020).

201. Uhrig, S., Schuffenhauer, S., Fauth, C., Wirtz, A., Daumer-Haas, C., Apacik, C., Cohen, M., Müller-Navia, J., Cremer, T., Murken, J. & Speicher, M.R., (1999). Multiplex-FISH for Pre- and Postnatal Diagnostic Applications. The American Journal of Human Genetics, 65(2), 448-462. https://doi.org/10.1086/302508 (Accessed on 31 May 2020).

202. Vallat, J.-M., (2003). Dominantly Inherited Peripheral Neuropathies. Journal of Neuropathology and Experimental Neurology, 62(7), 699-714. DOI: 10.1093/jnen/62.7.699 (Accessed on 10 May 2020).

203. Voullaire, L., Wilton, L., Slater, H. & Williamson, R., (1999). Detection of aneuploidy in single cells using comparative genomic hybridization. Prenatal Diagnosis, 19(9). https://doi.org/10.1002/(SICI)1097-0223(199909)19:9<846::AID-PD657>3.0.CO;2-%23 (Accessed on 5 April 2020).

204. Weiss, M.M., Hermsen, M.A.J.A., Meijer, G.A., van Grieken, N.C.T., Baak, J.P.A., Kuipers, E.J. & van Diest, P.J., (1999). Comparative genomic hybridization. Journal of Clinical Pathology: Molecular Pathology, 52, 243-251.

205. Witham, S., Takano, K., Schwartz, C. & Alexov, E., (2011). A Missense Mutation in CLIC2 Associated with Intellectual Disability is Predicted by In Silico Modeling to Affect Protein Stability and Dynamics. Proteins, 79(8), 2444-2454. doi: 10.1002/prot.23065 (Accessed on 16

June 2020).

206. Worthey, E.A., Mayer, A.N., Syverson, G.D., Helbling, D., Bonacci, B.B., Decker, B., Serpe, J.M., Dasu, T., Tschannen, M.R., Veith, R.L., Basehore, M.J., Broeckel, U., Tomita-Mitchell, A., Arca, M.J., Casper, J.T., Margolis, D.A., Bick, D.P., Hessner, M.J., Routes, J.M., Verbsky, J.W., Jacob, H.J. & Dimmock, D.P., (2011). Making a definitive diagnosis: Successful clinical application of whole exome sequencing in a child with intractable inflammatory bowel disease. Genetics in Medicine, 13, 255-262. https://doi.org/10.1097/GIM.0b013e3182088158 (Accessed on 7 July 2020).
207. Xie, X. & Liang, Q., (2014). Establishment of a 10-Plex Quantitative Fluorescent-PCR Assay for Rapid Diagnosis of Sex Chromosome Aneuploidies. PLoS ONE, 9(9), e106307. doi:10.1371/journal.pone.0106307 (Accessed on 31 March 2020).
208. Zhou, X., Iversen, E.S., Jr. & Parmigiani, G., (2005). Classification of Missense Mutations of Disease Genes. Journal of American Statistics Association, 100, 51-60. doi: 10.1198/016214504000001817 (Accessed on 15 June 2020).
209. Zwerger, M., Jaalouk, D.E., Lombardi, M.L., Isermann, P., Mauermann, M., Dialynas, G., Hermann, H., Wallrath, L.L. & Lammerding, J., (2013). Myopathic lamin mutations impair nuclear stability in cells and tissue and disrupt nucleo-cytoskeletal coupling. Human Molecular Genetics 22(12), 2335-2349. doi: 10.1093/hmg/ddt079 (Accessed on 15 June 2020).

INDEX

E

F

G

H

I

J

K

L

M

S

T

U

V

W

Z